SPORT AND EXERCISE PHYSIOLOGY TESTING GUIDELINES

Sport and Exercise Physiology Testing Guidelines is a comprehensive, practical sourcebook of principles and procedures for physiological testing in sport and exercise.

Volume I: specific guidelines for physiological testing in over 30 sports disciplines.
Volume II: guidelines for exercise testing in key clinical populations.

Each volume also represents a full reference for informed, good practice in physiological assessment, covering:

General Principles of Physiological Testing including health and safety, and blood sampling.
Methodological Issues including reliability, scaling and circadian rhythms.
General Testing Procedures for lung and respiratory muscle function, anthropometry, flexibility, pulmonary gas exchange, lactate testing, RPE, strength testing and upper-body exercise.
Special Populations including children, older people and female participants.

Written and compiled by subject specialists, this authoritative laboratory research resource is for students, academics and those providing scientific support service in sport science and the exercise and health sciences.

Edward M. Winter is Professor of the Physiology of Exercise at Sheffield Hallam University. **Andrew M. Jones** is Professor of Applied Physiology at the University of Exeter. **R.C. Richard Davison** is Principal Lecturer in Exercise Physiology at the University of Portsmouth. **Paul D. Bromley** is Principal Lecturer in Exercise Physiology at Thames Valley University and Consultant Clinical Scientist in the Department of Cardiology at Ealing Hospital, London. **Thomas H. Mercer** is Professor of Clinical Exercise Physiology and Rehabilitation, Department of Physiotherapy, School of Health Sciences, Queen Margaret University College, Edinburgh.

SPORT AND EXERCISE PHYSIOLOGY TESTING GUIDELINES

The British Association of Sport and Exercise Sciences Guide

Volume I: Sport Testing

Edited by Edward M. Winter,
Andrew M. Jones, R.C. Richard Davison,
Paul D. Bromley and Thomas H. Mercer

Routledge
Taylor & Francis Group

LONDON AND NEW YORK

First published 2007
by Routledge
2 Park Square, Milton Park, Abingdon, Oxon OX14 4RN

Simultaneously published in the USA and Canada
by Routledge
270 Madison Ave, New York, NY 10016

Reprinted 2007 (twice)

Routledge is an imprint of the Taylor & Francis Group, an informa business

Typeset in Sabon and Futura by
Newgen Imaging Systems (P) Ltd, Chennai, India
Printed and bound in Great Britain by
TJ International Ltd, Padstow, Cornwall

British Library Cataloguing in Publication Data
A catalogue record for this book is available from the British Library

Library of Congress Cataloging in Publication Data
Sport and exercise physiology testing : guidelines : the British Association of
Sport and Exercise Sciences guide / edited by
Edward M. Winter ... [et al.]. — 1st ed.
p. cm.
Includes bibliographical references and index.
1. Physical fitness—Testing. 2. Exercise—Physiological aspects.
I. Winter, Edward M. II. British Association of Sport and Exercise Sciences.
GV436.S665 2006
613.7'1—dc22   2006011234

ISBN10: 0–415–36140–0 (hbk)
ISBN10: 0–415–36141–9 (pbk)
ISBN10: 0–203–96684–8 (ebk)

ISBN13: 978–0–415–36140–8 (hbk)
ISBN13: 978–0–415–36141–5 (pbk)
ISBN13: 978–0–203–96684–6 (ebk)

CONTENTS

TABLES AND FIGURES

TABLES

FIGURES

NOTES ON CONTRIBUTORS

Les Ansley, Sport and Exercise Science, Kingston University.

Greg Atkinson, Research Institute for Sport and Exercise Sciences, Liverpool John Moores University.

Chris Barnes, Department of Sport Science, Middlesbrough Football Club.

Anthony J. Blazevich, Centre for Sports Medicine and Human Performance, Brunel University.

Simon L. Breivik, PGMOL, FA Premier League.

Paul D. Bromley, Department of Human Sciences, Thames Valley University and Department of Cardiology, Ealing Hospital NHS Trust.

John Buckley, Centre of Exercise and Nutrition Sciences, University of Chester and The Lifestyle Exercise and Physiotherapy Centre, Shrewsbury.

Melonie Burrows, Childrens Health and Exercise Research Centre, University of Exeter.

Chris Byrne, School of Sport and Health Sciences, University of Exeter.

Dale Cannavan, Center for Sports Medicine and Human Performance, Brunel University.

Matt Cosgrove, Sports Science Department, Welsh Institute of Sport.

Polly Davey, Human Performance Centre and Academy of Sport, Physical Activity and Wellbeing, London South Bank University.

R.C. Richard Davison, Department of Sport and Exercise Science, University of Portsmouth.

Jo Doust, Chelsea School, University of Brighton.

Steve Draper, Faculty of Sport, Health and Social Care, University of Gloucestershire.

Roger Eston, School of Sport and Health Sciences, University of Exeter.

Richard J. Godfrey, Centre for Sports Medicine and Human Performance, Brunel University.

Nick Grantham, Lead Strength and Conditioning Coach, English Institute of Sport – West Midlands.

Sarah L. Hardman, Welsh Institute of Sport, Cardiff.

Robert A. Harley, Chelsea School, University of Brighton.

Andy Harrison, English Institute of Sport.

Michael G. Hughes, Cardiff School of Sport, University of Wales Institute, Cardiff.

David V.B. James, Faculty of Sport, Health and Social Care, University of Gloucestershire.

Graham Jarman, Faculty of Organisation and Management, Sheffield Hallam University.

Andrew M. Jones, School of Sport and Health Sciences, University of Exeter.

John B. Leiper, School of Sport and Exercise Sciences, Loughborough University.

Craig A. Mahoney, School of Sport, Performing Arts and Leisure, University of Wolverhampton.

David Macutkiewicz, Olympic Medical Institute, British Olympic Association.

Ron Maughan, School of Sport and Excercise Sciences, Loughborough University.

Alison McConnell, Centre for Sports Medicine and Human Performance, Brunel University.

Thomas H. Mercer, School of Health Sciences, Queen Margaret University College, Edinburgh.

Stuart H. Mills, Chelsea School, University of Brighton.

Jeremy A. Moody, English Institute of Sport (North East Region).

Alan M. Nevill, Research Institute of Healthcare Sciences, University of Wolverhampton.

Steve Olivier, School of Social and Health Sciences, University of Abertay Dundee.

Kate Owen, Olympic Medical Institute, British Olympic Association.

Nicola Phillips, Physiotherapy Department, School of Healthcare Studies, Cardiff University.

Mike J. Price, Department of Biomolecular and Sports Science, Faculty of Health and Life Sciences, Coventry University.

Thomas Reilly, Research Institute for Sport and Exercise Sciences, Liverpool John Moores University.

Leigh E. Sandals, Faculty of Sport, Health and Social Care, University of Gloucestershire.

John M. Saxton, Centre for Sport and Exercise Science, Sheffield Hallam University.

Susan M. Shirreffs, School of Sport and Exercise Sciences, Loughborough University.

Marcus S. Smith, School of Sport, Exercise and Health Sciences, University of Chichester.

Paul M. Smith, Centre for Sport and Exercise Sciences, University of Greenwich.

Richard G. Smith, Sports Science Coordinator, England & Wales Cricket Board National Cricket Centre, Loughborough University.

Neil Spurway, Centre for Exercise Science and Medicine, Institute of Biomedical and Life Sciences, University of Glasgow.

Arthur D. Stewart, School of Health Sciences, The Robert Gordon University, Aberdeen.

Nigel P. Stockill, England Cricket Team, England and Wales Cricket Board.

Gareth Stratton, REACH Group, Research Institute for Sport and Exercise Sciences, Liverpool John Moores University.

Suzan R. Taylor, Sport and Exercise Sciences, North East Wales Institute of Higher Education.

Kevin G. Thompson, English Institute of Sport.

Richard J. Tong, Cardiff School of Sport, University of Wales Institute, Cardiff.

Gregory P. Whyte, Research Institute for Sport and Exercise Sciences, Liverpool John Moores University.

Michael Wilkinson, Division of Sport Sciences, Northumbria University.

Craig A. Williams, Children's Health and Exercise Research Centre (CHERC), School of Sport and Health Sciences, University of Exeter.

Huw D. Wiltshire, National Fitness Director, Welsh Rugby Union.

Edward M. Winter, The Centre for Sport and Exercise Science, Sheffield Hallam University.

Dan M. Wood, Standards and quality analytical team, Department of Health.

Andrea L. Wooles, Great Britain Cycling Team.

ACKNOWLEDGEMENTS

For the last four decades, there is one person in particular who has both guided the development of sport and exercise science in the United Kingdom and been a constant source of inspiration to its practitioners, teachers and researchers. Many of the contributors have benefited directly from his influence.

The publication of this volume has coincided with his retirement and the editors take this opportunity to extend their appreciation and thanks to Professor N.C. Craig Sharp BVMS MRCVS PhD FIBiol FBASES FafPE.

FOREWORD

It is well recognised that scientific support plays a vital part in the preparation both of elite athletes and those for whom physical activity is a goal.

As standards of performance continue their searing rate of progress while simultaneously, concerns are expressed about the activity profile of our children and adolescents, the need for evidence-based approaches to meet these challenges increases in importance. UK Sport and the British Association of Sport and Exercise Sciences (BASES) have always had a close relationship and this fourth edition of Sport and Exercise Physiology Testing Guidelines provides an opportunity to underscore this relationship.

The award of the 2012 Olympic Games to London re-emphasises the link and highlights the dual aim to improve athletic performance and enhance the health of the nation. UK Sport and BASES will be working together to achieve this aim.

Sue Campbell CBE
Chair of UK Sport

FOREWORD

BASES's guidelines on the physiological assessment of all who participate in exercise and sport whether it is at the elite or recreational level of participation represents another significant contribution to the professional development of sport and exercise science in the United Kingdom.

The editorial team has brought together researchers and practitioners in the field of exercise physiology to share with us their knowledge, insight and experience. They have produced a set of guidelines that represent current knowledge and best practice in the field of physiological assessment. The guidelines will be widely regarded as the benchmark publication for all practitioners whether they work in the laboratory or in the field. Therefore I am delighted to commend the guidelines to you because they not only serve to share more widely the available knowledge and so contribute to raising standards but also because they reassure us that the future of sport and exercise science in the United Kingdom is in safe hands.

Clyde Williams
Professor of Sports Science
Loughborough University

INTRODUCTION

*Edward M. Winter, Paul D. Bromley,
R.C. Richard Davison, Andrew M. Jones
and Thomas H. Mercer*

It is 21 years since under the stewardship of Tudor Hale, a working group that also comprised Neil Armstrong, Adrianne Hardman, Philip Jakeman, Craig Sharp and Edward Winter produced the Position Statement on the Physiological Assessment of the Elite Competitor. This was distributed to members of the Sports Physiology Section of the then British Association of Sports Sciences (BASS).

In 1988 the document had metamorphosed into a second edition that was formally published for and on behalf of BASS by White Line Press. BASS's accreditation scheme for physiology laboratories and personnel had been established and the second edition provided a reference frame for the associated criteria. Moreover, undergraduate study of sport and exercise science had continued to gather pace and celebrated its twelfth birthday.

Nine years later in 1997, BASS had evolved into the British Association of Sport and Exercise Sciences (BASES). The addition of 'exercise' into the Association's title clearly reflected acknowledgement that there were exercise scientists whose interests were not restricted to sport. Steve Bird and Richard Davison had assumed the mantle of responsibility to revise the second edition and edited the third edition that had a simplified title: *Physiological Testing Guidelines*. These editors, together with 22 contributing authors, produced 19 chapters that were organised under 4 rubrics: General issues and procedures; Generic testing procedures; Sport-specific testing guidelines; and Specific considerations for the assessment of the young athlete.

This fourth edition shares a common feature with edition three: the nine-year gap from the previous one. These gaps should not be confused with a lack of activity by BASES members. On the contrary, sport and exercise science has undergone astonishing growth since it became degree-standard study in the United Kingdom. The fledgling discipline has developed into a major area and some 10,000 students graduate each year with a sport or exercise-related degree. More than 100 institutions of higher education offer undergraduate sport and exercise-related programmes and approximately a third of these offer masters' courses. In addition, doctoral studies feature prominently and since 1992, sport and exercise-related subjects have had their own panel in the Higher Education Funding Councils' Research Assessment Exercise.

Mirroring this growth has been the increase in vocational applications of sport and exercise science and many enjoy careers in diverse settings. These settings include sport and exercise support work with national governing bodies, professional clubs, the Home Countries' Sport Institutes, and public and private healthcare providers as well as private, governmental and voluntary organisations that are engaged in the provision of exercise for people with, or at high risk of, a myriad of diseases and disabilities. BASES' accreditation scheme for researchers and practitioners in sport and exercise science provides a gold-standard quality assurance mechanism for senior personnel while its supervised experience scheme encourages and nurtures young scientists. Currently, there are some 387 accredited scientists and 376 who are benefiting from supervised experience.

A recent development has been the use of the term physical activity rather than exercise. Physical activity is seen to be less intimidating and encompasses every-day activities like gardening, domestic tasks and walking. These are activities in which most participate at least some time during the day.

The expansion of the evidence base that underpins physiological assessments of athletes is matched by exponential growth in the fields of clinical exercise physiology and medical and health related applications of exercise assessment. Exercise testing and the interpretation of exercise test data, especially those from integrated cardiopulmonary exercise tests, has made important contributions both to research and the clinical management of patients.

In the clinical setting, exercise testing has direct relevance in several applications. It is used in the functional assessment of patients and has implications both for the diagnosis and prognosis of conditions. It helps to determine safe and effective exercise prescription and testing is also used to evaluate the effectiveness of medical, surgical or exercise interventions that are designed to be therapeutic. The use of exercise testing in patients with diseases and disabilities means that associated exercise scientists must appreciate the implications of the disorder being investigated and be able to adapt tests and procedures to take account of existing co-morbidities, for example, obesity and musculoskeletal dysfunction, that might influence performance and results.

The impetus to this fourth edition was provided by Ged Garbutt and the outcome is an ambitious attempt to address several matters in one core text. Specifically, it aims to: address both sport and exercise/physical activity applications; acknowledge psychological aspects of exercise testing; consider technical matters such as health-and-safety issues that impact on the management of laboratories and field-based work; and acknowledge the requirement to ensure that procedures abide by principles for seeking and gaining ethics approval.

The text is intended to be a reference guide for practitioners, researchers and teachers in sport and exercise science. The editorial team are well placed to appreciate the challenges that this triune presents. Moreover, the list of contributors includes many of the United Kingdom's leading sport and exercise scientists who are similarly well placed to appreciate the challenges in applying research to scientific support for sport and exercise.

REFERENCES

Bird, S. and Davison, R. (1997). *Physiological Testing Guidelines*, 3rd edn. Leeds: The British Association of Sport and Exercise Sciences.

Hale, T., Armstrong, N., Hardman, A., Jakeman, P., Sharp, C. and Winter, E. (1988). *Position Statement on the Physiological Assessment of the Elite Competitor*, 2nd edn. Leeds: White Line Press.

PART 1

GENERAL PRINCIPLES

RATIONALE

Edward M. Winter, Paul D. Bromley,
R.C. Richard Davison, Andrew M. Jones
and Thomas H. Mercer

INTRODUCTION

The physiology of exercise can be defined as the study of how the body responds and adapts to exercise and an important part of this study is the identification of physiological characteristics that explain rather than simply describe performance. This identification applies both to competitive athletes and to those whose interests are in the role of physical activity in the promotion and maintenance of health.

The continued rise in performance standards in sport underscores the need to develop knowledge and understanding of related mechanisms to optimise athletes' training. Such optimisation ensures that training maximises adaptations but not at the expense of developing unexplained underperformance syndromes – previously known as overtraining. Similarly, the frequency, intensity and duration of physical activity required to promote and sustain health is important; concerns about the possible inactivity of our children and adolescents give rise to anxiety about possible long-term problems such as diabetes and cardio-vascular disease that might ensue from hypokinesis.

It is also well recognised that exercise can be both prophylactic and therapeutic in clinical populations. Consequently, sport and exercise science graduates are increasingly in demand with such populations because of the expertise they bring to exercise testing and interpretation.

Technological advances and the development and refinement of procedures have been and continue to be a hallmark of sport and exercise science. Furthermore, there are many sports and activities to be considered along with a variety of influential factors such as the age, gender, ability and disability of participants. This presents a major challenge for physiologists: the selection and administration of tests and their subsequent interpretation is intellectually and practically exacting.

Moreover, it is now well established that effective scientific support comprises contributions from several disciplines. Consequently, researchers and practitioners have to be able to recognise the limits of their expertise and know when to seek the advice and guidance of others. Similarly, a detailed understanding of techniques and procedures to assess the validity and reproducibility of measures is an essential competency that sport and exercise scientists must possess. It is important to know the extent to which an apparent change in a measure is meaningful and does not lie within a confidence interval for error. As differences in performance can be measured in small fractions, the identification of what is meaningful can be obscured by random error.

Overarching all this is the increased support provided to athletes by most of the world's leading sporting nations. This emphasises the need for UK-based athletes to have the best possible scientific and medical support if they are to compete effectively in international competition.

WHY ASSESS?

The rationale for assessment remains as it has for some three decades: to develop knowledge and understanding of the exercise capabilities of humans. A practical outcome of this is enhanced performance and exercise tolerance of individuals who are tested.

Assessment should be preceded by a full needs analysis which in turn should be based on a triangulation of the requirements and views of the athlete, coach and scientist. Assessments should be an integral part of an athlete's training and scientific support programmes and should be conducted regularly and frequently.

Moreover, assessments should reflect the movement and other demands of the sport or activity in which the athlete or exerciser participates. Hence, the investigator should have a detailed understanding of the mechanisms of energy release that are challenged.

Specifically, reasons for undertaking physiological tests are (Bird and Davison, 1997) to:

1   Provide an initial evaluation of strengths and weakness of the participant in the context of the sport or activity in which they participate. This information can be used to inform the design and implementation of a training programme.
2   Evaluate the effectiveness of a training programme to see if performance or rehabilitation is improving and intended physiological adaptations are occurring.
3   Evaluate the health status of an athlete or exerciser. This might be part of a joint programme with clinical staff.
4   Provide an ergogenic aid. Often, in the setting of short-term goals for the improvement of fitness for example, the prospect of being tested often acts as a motivational influence.
5   Assist in selection or identify readiness to resume training or competition.

6   Develop knowledge and understanding of a sport or activity for the benefit of coaches, future athletes and scientists.
7   Answer research questions.

In the clinical domain, the utility of exercise testing has expanded from a role that simply categorised the health status of a patient or participant to one that can diagnose functional limitation, that is, whether the origin is cardiac, pulmonary or muscular. This might be in the presence of multiple pathologies where precise diagnosis requires considerable expertise.

TEST CRITERIA

It is recognised that to be effective, assessments should be specific and valid and that resulting measures should be reproducible and sensitive to changes in performance.

Specificity

Assessments should mimic the form of exercise under scrutiny. This is a key challenge for instance in multiple-sprint activities such as field-games and racquet-sports in which changes in speed and direction of movement predominate. Factors that should be considered in the design of test protocols are:

1   Muscle groups, type of activity and range of motion required.
2   Intensity and duration of activity.
3   Energy systems recruited.
4   Resistive forces encountered.

Accordingly, activity-specific ergometers should be used and these might have to be designed to satisfy local requirements. Similarly, field-based as opposed to laboratory-based procedures might provide improved characterisations of patterns of motion. It is worth noting that sacrificing specificity to reduce artefacts is a consideration that should be made test-by-test. For instance, in diagnostic testing, it might be prudent to select a mode of exercise that does not necessarily reflect daily activities. A cycle ergometer could fall into this category but its use would achieve improved signal acquisition in electrocardiographic, pulmonary and metabolic measurements.

Validity

Validity is the extent to which a test measures what it purports to measure. This applies, for instance, to the assessment of mechanisms that might explain endurance and to the appropriate use of mechanical constructs to describe in particular, the outcomes of maximal intensity, that is, all-out exercise.

Reproducibility

An important requirement for data if they are to be considered meaningful is that they must be reproducible. Enthusiastic debate continues about the metric or metrics that most appropriately assess reproducibility (Atkinson and Nevill, 1998 and elsewhere in this text). Consequently, exercise scientists need to have a keen appreciation of these metrics and their respective advantages and disadvantages. In essence, variability in measures can be attributed to technical and biological sources. The former comprise precision and accuracy of instruments coupled with the skill of the operator, hence procedures for calibration are critical. The latter comprise random and cyclic biological variation. Knowledge and understanding of the magnitude of these errors play a key role in the interpretation of measures.

As a result, an indication of error in tests is a requirement if meaningful information is to be provided. This includes comparisons of test results with norms or those of other performers.

Sensitivity

Sensitivity is the extent to which physiological measures reflect improvements in performance. Clearly, reproducibility is implicated but sensitivity is probably at the heart of the matter: it is in itself a key measure of our understanding of mechanisms and the accuracy and precision of our instruments to reflect these mechanisms.

It is highly likely that assessments of the physiological status of athletes and exercisers will continue to be an important part of scientific programmes. Similarly, it is probable that assessments will continue to undergo development and refinement as our knowledge base grows. This will increase the sensitivity with which physiological measures explains changes in performance. As a result, the rationale for assessment will strengthen so the need for knowledgeable, skilled and experienced sport and exercise scientists will increase.

REFERENCES

Atkinson, G. and Nevill, A.M. (1998). Statistical methods for assessing measurement error (reliability) in variables relevant to sports medicine. *Sports Medicine*, 26, 217–238.

Bird, S. and Davison, R. (1997). *Physiological Testing Guidelines*, 3rd edn. Leeds: The British Association of Sport and Exercise Sciences.

CHAPTER 1

HEALTH AND SAFETY

Graham Jarman

INTRODUCTION

Laboratory and field work activities present potential hazards to investigators and participants. The purpose of this chapter is to provide a guide on how to implement a risk assessment approach to health and safety management in these settings.

A DUTY OF CARE

Principal investigators and consultants need to be cognisant of their responsibility to exercise a duty of care to athletes and exercisers, participants in research studies, clients and co-workers. This duty of care is made explicit in the enabling legislation enshrined in the Health and Safety at Work Act 1974. Of particular note are the following sections of the Act:

- Section 2: General duties of employers to their employees.
- Section 3: General duties of employers and the self-employed to persons other than their employees.
- Section 7: General duties of employees at work.

The details of the 1974 Act can be found on the HealthandSafety.co.uk website. The general duties of the Act are qualified by the principle of 'so far as is reasonably practicable', that is, steps to reduce risk need not be taken if they are technically impossible or the time, trouble and cost of measures would be grossly disproportionate to the risk (HSE, 2003). In essence, the law requires that good management and common sense are applied to identify the risks associated with an activity and that sensible measures are implemented to control those risks.

RISK ASSESSMENT

Risk assessment is the cornerstone of health and safety management practice. The Management of Health and Safety Regulations 1999 requires that risk assessments are carried out for *all* activities and that significant findings are documented. Other regulations require that specific types of assessment are made for certain work areas, for example, working with substances (COSHH), noise and manual handling. The Health and Safety Executive (HSE) publication '*A Guide to Risk Assessment Requirements*' (HSE, 1996) examines the common features of the assessments as required by the various regulations and highlights the differences between them.

Risk assessment is essentially an examination of what in the workplace could cause harm to people. It is a structured analysis of what can cause harm, an assessment of the likelihood and impact of something harmful happening and a means to identify measures that can be implemented to mitigate the occurrence of harmful incidents.

APPROACH

The HSE advocate a five-step approach to risk assessment (HSE, 1999):

- Step 1: identify the hazards;
- Step 2: decide who might be harmed and how;
- Step 3: evaluate the risk and decide whether existing precautions are adequate or more should be done;
- Step 4: record significant findings;
- Step 5: review assessment and revise if necessary.

IDENTIFYING HAZARDS

A hazard is something that has the potential to be harmful. The following is an indicative, but not exhaustive list, of typical hazards that are likely to exist in a physiology of exercise laboratory and field-based settings:

Working with equipment:

- Electrical hazard
- Entrapment hazards
- Falls or trips.

Changes in the physiological state of participants:

- Cardio-vascular complications
- Fainting

- Vomiting
- Musculo-skeletal injury.

The administration of pharmacologically active substances and nutritional supplements:

- Overdose or acute effects
- Chronic effects
- Hypersensitivity (allergic responses).

The use of hazardous materials:

- Chemicals or laboratory reagents
- Potentially infectious material (body fluids).

Modifications to the environment:

- Heat stress
- Cold stress
- Hypoxia or hyperoxia
- Other gas mixtures.

Hazard identification could be undertaken as a systematic inspection of the laboratory or could be integral to the design of an experimental protocol.

DECIDING WHO MIGHT BE HARMED

It is important to identify who might be harmed by any activity undertaken in the laboratory. As well as considering investigators, consultants and co-workers it is imperative that participants involved in investigative procedures are adequately protected from harm. When procedures involve participants from the following groups, for example, additional precautions might be needed to reduce the risk to levels that are considered to be acceptable:

- Minors
- The ageing
- Those with learning difficulties
- Those with underlying medical conditions.

Consider also members of the public or visitors to your premises if there is a chance that they could be harmed by your activities.

EVALUATING AND CONTROLLING RISKS

Risk is an appraisal of the likelihood of a hazard causing harm and the consequence of that harm if realised. This can be represented numerically by multiplying a perceived likelihood rating by a perceived consequence rating. Table 1.1 provides an example of how a simple risk rating system could operate and how this could be used as a means to prioritise actions to control and manage risks.

Managing risk is concerned with reducing likelihood and consequence associated with particular hazards to a level at which they can be tolerated. *Control measures* are actions or interventions that reduce risk to an acceptable level. In general, the following principles should be applied in the order given:

- try a less risky option, for example, substitution;
- prevent access to a hazard, for example, by guarding;
- reduce exposure, for example, by organising the work differently;

Table 1.1 An example of a 3 × 3 risk rating system

	Consequence (C)
3	Major (death or severe injury)
2	Serious (injuries requiring three days or more absence from work)
1	Slight (minor injuries requiring no or brief absence from work)
	Likelihood (L)
3	High (event is likely to occur frequently)
2	Medium (event is likely to occur occasionally)
1	Low (event is unlikely to occur)
Risk rating (C × L)	*Action and timescale*
1 (Trivial)	No action is required to deal with trivial risks
2 (Acceptable)	No further preventative action is necessary but consideration should be given to cost-effective solutions or improvements that impose minimal or no additional cost. Monitoring is required to ensure that controls are maintained
4 (Moderate)	Effort should be made to reduce the risk but the cost of prevention should be carefully measured and limited. Risk reduction measures should be implemented within three to six months depending on the number of people exposed to hazard
6 (Substantial)	Work should not be started until the risk can be reduced. Considerable resources may have to be allocated to reduce the risk. Where the risk involves work in progress, the problems should be resolved as quickly as possible
9 (Intolerable)	Work should not be started or continued until the risk level has been reduced. Whilst the control measures should be cost effective, the legal duty to reduce the risk is absolute. This means that if it is not possible to reduce the risk, even with unlimited resources, then the work must not be started or must remain prohibited

- use of personal protective equipment, for example, use of gloves in blood sampling;
- provision of welfare facilities, for example, washing facilities for removal of contamination.

The legislation requires that you must do what is reasonably practicable to make your work and workplace safe. If risks can not be reduced to an acceptable level by applying cost-effective control measures then consideration must be given to whether or not particular activities can be justified.

RECORDING OUTCOMES OF RISK ASSESSMENT

Any risk assessment that is undertaken must be '*suitable and sufficient*'. It is a requirement that all significant findings are recorded and there should be documentary evidence to show that:

- a proper check was made;
- consultation with those affected was undertaken if appropriate;
- all obvious hazards have been dealt with;
- precautions are reasonable and any remaining risk is low (or tolerable).

Documentation must be retained for future use and outcomes should be communicated to any individuals who could be affected by the activity. The documents should be retained as evidence that risk assessments have been undertaken, this is particularly important if any civil liability action is taken as a result of an accident. The outcomes of risk assessments can be incorporated into other laboratory documents such as manuals, codes of practice or standard operating procedures.

REVIEWING ASSESSMENTS

It is good practice to review risk assessments periodically to ensure that control measures are effective and that significant risks are being adequately managed. If there are any material changes that affect the risk assessment and control measures in operation, then a review of the assessment should be undertaken. It is important that any amended documents are version-controlled and that all individuals who are affected are informed of the revisions.

PERSONAL INJURY CLAIMS AND PROFESSIONAL INDEMNITY

Laboratory and field-based activities can never be risk free. Should an incident which causes injury or damage occur, it is important that:

- Documentary evidence can be provided to demonstrate that a duty of care has been exercised.

- That professional indemnity and public liability insurance is in place to cover any legal cost or awards of damages if, for example, a case of negligence is proven.

In the event of personal injury to a client or co-worker there is a formal process by which this is dealt. This 'Pre-action Protocol' is covered in detail on the Department for Constitutional Affairs (DCA) website; of particular interest are the lists of standard disclosure documents given in the annex.

The BASES code of conduct (BASES, 2000) states that: 'members must ensure that suitable insurance indemnity cover is in place for all areas of work that they undertake'. Care must be taken to understand the scope, limitations and exclusions associated with any insurance cover to ensure its adequacy.

OBTAINING INFORMATION AND FURTHER GUIDANCE

The earlier discussion covers the generality of risk assessment as a process. It is recommended that reference is made to the relevant approved codes of practice and related guidance leaflets that are published by the HSE. The HSE website (www.hse.gov.uk) provides a breadth of advice in the form of down-loadable leaflets. Some suggested further reading on some of the specific issues and hazards encountered in the physiology of exercise laboratory are given here.

REFERENCES AND FURTHER READING

British Association of Sport and Exercise Sciences (BASES). (2000). *Code of Conduct*. http://www.bases.org.uk/newsite/pdf/Code%20of%20Conduct.pdf. Accessed 16 December 2005.

Department for Constitutional Affairs. *Pre-Action Protocol for Personal Injury Claims*. http://www.dca.gov.uk/civil/procrules_fin/contents/protocols/prot_pic.htm. Accessed 16 December 2005.

Health and Safety Executive (HSE). (1996). *A Guide to Risk Assessment Risk Assessment Requirements*. http://www.hse.gov.uk/pubns/indg218.pdf. Accessed 11 November 2005.

Health and Safety Executive (HSE). (1999). *Five Steps to Risk Assessment*. http://www.hse.gov.uk/pubns/indg163.pdf. Accessed 11 November 2005.

Health and Safety Executive (HSE). (2001). *Blood-borne Viruses in the Workplace*. http://www.hse.gov.uk/pubns/indg342.pdf. Accessed 16 December 2005.

Health and Safety Executive (HSE). (2003). *Health and Safety Regulation...A Short Guide*. http://www.hse.gov.uk/pubns/hsc13.pdf. Accessed 15 November 2005.

Health and Safety Executive (HSE). (2005). *Coshh: A Brief Guide to the Regulations*. http://www.hse.gov.uk/pubns/indg136.pdf. Accessed 16 December 2005.

Healthandsafety.co.uk. '*[A guide to] The Health and Safety at Work etc Act 1974. (Elizabeth II 1974. Chapter 37)*'. http://www.healthandsafety.co.uk/haswa.htm. Accessed 15 November 2005.

Medicines and Healthcare Products Regulatory Agency. (2003). *Guidance Note 8: A Guide to what is a Medicinal Product*. http://www.mhra.gov.uk/home/groups/commsic/documents/publication/con007544.pdf. Accessed 16 December 2005.

Office of Public Sector Information. *The Management of Health and Safety at Work Regulations*. http://www.opsi.gov.uk/si/si1999/19993242.htm#13. Accessed 17 November 2005.

CHAPTER 2

PSYCHOLOGICAL ISSUES IN EXERCISE TESTING

Craig A. Mahoney

INTRODUCTION

Exercise testing usually serves one of two purposes:

- health screening and diagnosis of disease;
- fitness testing for sport/exercise.

Exercise tests for the general population are normally designed to provide exercise professionals with information on disease diagnosis or prevention, rehabilitation and intensities to commence an exercise programme (de Vries and Housh, 1995). The feedback participants receive can help to establish appropriate intrinsic motivation to achieve goal outcomes. Goal determination will vary depending upon the types of test being completed, the background to the participant, the person commissioning the tests and a range of less tangible factors.

While some participants will self-refer to receive exercise testing as part of a health club membership package, an increasing number of participants do not. The latter are often asked to attend for exercise testing as part of a corporate programme offered to employees (seen by the employers as an employee benefit) though employees might not always be positive about the impact of the assessment, the imposition on their normal lives or their perceived understanding of the purpose of the tests. Many organisations now have minimum health (or fitness) standards required for continued employment; these include the ambulance service, fire service, some police forces, professional football referees and many other occupations that have a measured, objective physical component to their employment base.

For sporting populations, the ongoing assessment of fitness may be related to the process of monitoring training programmes, assessing recovery from injury or medical intervention, or be used as part of a selection process.

Undoubtedly, many athletes approach the regular (often several times per year) assessments with minimum fuss and limited concerns about their ability to perform well, and confidence in the outcomes. However, there is another group who (like employee assessment programmes) have lingering concerns about the purpose and outcomes of exercise testing. These participants often worry for some time before the tests and then present themselves in an over anxious state. This needs to be recognised, reconciled and minimised if the results of tests are to be valid and reliable for coaches, athletes and exercise professionals.

MENTAL ENERGY

Regardless of the types of physical activity in which people engage, whether it is exercise, practice for sport, sports performance or fitness testing, an individual's mental energy to concentrate attention and maintain a positive mental attitude is essential in ensuring optimum physical performances. It is very easy to waste mental energy and therefore physical energy on worry, stress, fretting over distractions and negative thoughts. This will have the combined effect of reducing enjoyment and adversely affecting results. Effective concentration will help to maintain sound technique, for example, during running on the treadmill or when performing shuttles, while enabling participants to conserve energy. Fatigue brought about by physical effort or cognitive stress will result in muscle tiredness and a downward spiral of negative thinking that will exacerbate feelings of pain, fatigue, hopelessness and defeat.

MOTIVATION

Exercise testing has a strong association with intrinsic motivation. There is anecdotal evidence to show that people who score higher than anticipated on exercise tests can become more motivated and more committed, while those who score lower than anticipated can become less motivated and less committed. This seems to be associated with Attribution Theory that attempts to explain behaviours and has been evidenced in medical settings (Rothman et al., 1993).

Athletes and participants from the general population will often involve themselves in exercise testing to identify training needs, screen their health, or evaluate the effectiveness of their training plan or health programme. In the absence of improvement or at the very minimum maintenance, that is, no relapse, participants can lose momentum and the associated intrinsic motivation for training. Though testing should not be considered in isolation from other factors such as; the time of the season for an athlete or lifestyle factors for general population participants, as these additional factors can aid in ensuring training, improvement and personal commitment are all part of an holistic personal development plan for health and fitness benefits. On this basis an understanding of the performance profiling needs and personal goals of all participants should be fully understood by the exercise professional both to tailor the right testing programme and ensure that the designed training programme meets the perceived and actual needs of each participant. It is unlikely that exercise testing will in itself, result in motivation to exercise. Beginners especially, are highly susceptible to positive feedback and vulnerable to negative outcomes. New exercise participants can often display low self-efficacy and are likely to find limited motivation from the experience of exercise testing. Once they have established some skills associated with exercise, results from exercise testing may be a useful form of feedback to aid the motivation to continue.

STRESS AND ANXIETY

Just as psychological preparation for performance in sport is now a recognised part of athlete preparation, so too are psychological aspects of physiological assessment and these should be understood by the exercise professional, and arguably the participant as well. The right mental approach to exercise testing often begins when a date is established for that 'fear inducing' battery of assessments agreed between the coach(es) and the sport scientist(s), since the very thought of exercise testing can create anxiety in many. Others will see this as an opportunity to excel, show why they should be selected above others or merely gain a better understanding of their current physiological status. However, from that point forward, participants will often consciously and subconsciously worry about the types of test, the purpose of the tests, previous experience with the tests and this can result in a range of cognitive concerns, which manifest themselves as fear and trepidation.

Cognitive appraisal by participants is common in testing environments. It is not uncommon for some participants to worry considerably about being tested and the test procedures they will have to follow. Whether this is laboratory- or field-based does not seem to matter. A significant determinant of worry is the basis of the testing, for example, is it part of regular in-season assessment or does it form part of a selection process? Sometimes it can be used in association with employment criteria, for instance, to maintain minimum work standards.

SEQUENCING

The ACSM (2005) has recommended completing tests in a particular sequence, to minimise the effects of tests on one another (Heyward, 1998). The order is broadly:

- Resting blood pressure and heart rate
- Neuromuscular tension and/or stress (if included)
- Body composition
- Muscular fitness
- Cardiorespiratory endurance
- Flexibility.

This order is regardless of whether the tests are field or laboratory based. However, it is often helpful to negotiate with, and agree, particular sequencing in tests. Experience has shown that many older athletes, who have participated in fitness test batteries over many years, are often more comfortable with some tests being completed prior to others.

Some clients might be apprehensive prior to being exercise tested and will demonstrate elevated anxiety about the testing process and particular tests included in the battery. The use of the Multi Stage Fitness Test seems consistently to raise concerns in the minds of athletes prior to its completion. This usually coincides with prior experience in the test. Naive populations, such as school children or first time participants on the test, do not normally present with this anxiety. Another test frequently associated with elevated anxiety is the assessment of maximum oxygen uptake. Endurance tests often raise fear and anxiety in the minds of participants. This can be related to previous experience, but is often underpinned by an absence of training, an existing injury or a general dislike for the test or its protocols. Exercise professionals have an obligation to screen the participant fully and ensure they are in the best physical, but also mental, health to complete the testing programme. Recognition should be given, that varying the intensity of encouragement to participants will also affect results. Due to the motivational basis to the Multi Stage Fitness Test, excessive encouragement can lead to significantly improved results for some participants.

It is incumbent on test administrators to minimise the impact of anxiety in the testing process. Test anxiety can reduce the validity and reliability of test results. To ensure this is not a compounding variable in the efficacy of the test results, clients should always be put at ease upon arrival. Establishing a good rapport between the tester and participant(s) should help to achieve this. Providing a relaxed non-intimidating environment should help to foster a confident but relaxed approach to the battery of tests being used. Ensure the environment is safe, friendly, quiet, private (where appropriate) and comfortable if possible. Careful consideration of temperature and humidity should be taken in consultation with the coaching and support staff and obviously will depend on the purpose of the tests, which may form part of an acclimatisation process. Ensuring appropriate and careful calibration has been completed will also serve to reduce concerns from the group(s) being tested.

HUMAN BEHAVIOUR

How well the exercise professional understands human behaviour and personal mood variables will play an important part in the testing experience of the participant. The importance of understanding basic human behaviour cannot be overstated in an exercise test setting. The British Association of Sport and Exercise Sciences accreditation scheme, established as a gold standard in applied sport and exercise science, has recently acknowledged the need for minimum sport science knowledge from all key sciences prior to the approval of accreditation.

When testing different groups, exercise professionals will build up a strong knowledge of the individuals and the types of test most appropriate for these specialist populations. Paediatric populations are one such specialist group. Because of the dearth of knowledge that such participants might have about laboratory- or field-based testing, it is essential that those working with young people ensure a caring, compassionate and sensitive approach to their work. This is particularly relevant in understanding the goals and motivations that young people will have towards exercise testing, particularly if the procedures are invasive, intense and not fully understood by the participants (Whitehead and Corbin, 1991; Goudas *et al.*, 1994). Notwithstanding the ethical issues that will have had to be approved prior to this, young children will often need more advanced habituation to some testing procedures. If blood samples are being taken, as will often occur in paediatric studies, then demonstration might help, as might anaesthetic creams to minimise pain. The use of mouth pieces is being superseded by face masks in peak oxygen uptake testing, however if mouth pieces are to be used it will often help habituate children and minimise anxiety if they can be given a mouth piece to take home for several days prior to the test.

MODEL OF BEHAVIOUR CHANGE

When prescribing exercise programmes after laboratory- or field-based exercise testing, it is essential that exercise professionals are cognisant of adherence issues, the stages of behaviour change a participant might be demonstrating and willingness on the part of the participant. This will vary between populations. For example, professional and elite athletes would normally present themselves for testing with high intrinsic motivation to succeed and strong personal commitment to prescribed programmes if they possess confidence in the exercise professional and feel the programmes are beneficial to their personal (and usually sporting) potential. This requires the exercise professional to be able to communicate with all exercise test participants on a practical level using language, which is sincere, simple, but not degrading. Most athletes want to be confident that the exercise professional is aware of contemporary issues and training protocols, but they might not want (or need) to know the intricate detail behind the theory.

To this end, establishing clear motives for exercise testing, having an unambiguous understanding of goals that will arise from this and an awareness of the stage, in the Stages of Behaviour Change (Prochaska and DiClemente, 1986), the participant is in will be important (Figure 2.1). These are less relevant concerns when working with elite athletes, since the Governing Body, coaching staff or the athletes have probably established these themselves. However, when assessing participants from the general population who might be part of a corporate testing programme or who have simply made themselves available for testing, or who are part of a research study involving non-elite participants, these covert motives become much more relevant. The concerns arising from this include; will the participant adhere to any exercise programme arising from analysis of the test results; how will adherence

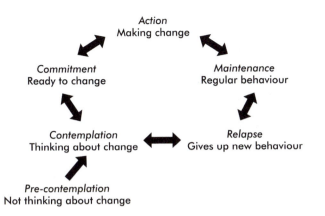

Figure 2.1 Models of the stages of behaviour

be optimised (e.g. social support, priority level, re-assessment, time allocation, SMART goals, monitoring), what factors can help them achieve their goals (or the goals of the programme funder), when re-assessment occurs will this be a positive experience for the participant?

Clearly, the whole area of exercise testing is affected by factors, which exhibit both a cognitive and a behavioural basis. To help minimise their effect on test results, exercise professionals (and participants) should have an understanding of these and be prepared to adapt testing regimens to minimise their impact wherever possible. The following is offered as guidance to participants and exercise professionals.

Guidance for participants

Prior to the day of testing, ensure good rest and sleep where possible. In the same way physical performance can be affected by under or over arousal, so too can exercise test performances. If feeling lethargic then use arousal strategies such as:

- Light exercise
- Warm up routines
- Loud music.

If participants are feeling anxious, then the following relaxation routines might help:

- Calming music
- Relaxation exercises or tape
- Imagery relaxation
- Calm movements, breathing control and tranquil thoughts
- Biofeedback (if previously developed)
- Thought awareness and positive thinking.

Guidance for exercise professionals

- Give time for habituation.
- Allow participants to take home pieces of equipment that may cause anxiety, for example, mouth pieces for VO_{2max} tests, especially children.
- Ensure all participants are supported in an equitable and consistent manner.
- Be sensitive to participant concerns related to some tests, for example, Multi Stage Fitness Test where motivation is highly significant and the test often generates considerable concerns in the minds of performers.

Exercise testing is a complex and multifaceted activity that combines the academic knowledge, practical skills, experiential awareness and personal capacity of the testing team to ensure the participant(s) have the most positive experience achievable. Where possible, exercise professionals must endeavour to ensure that all participants have a positive mental state leading up to and upon engaging in the exercise testing programme. This will increase the potential both for optimal and maximum achievable results. However, regardless of the scientific competency of the exercise physiologist undertaking the testing, sometimes no matter how effective their communication, counselling or relaxation skills, a patient with heart failure that might be progressing to end stage (i.e. death or transplant) and is being evaluated, by, for example, cardiopulmonary exercise testing, for this reason is not going to approach the test with a positive mental state. In such circumstances all we can do is manage those things that are within our control in order to minimise the potential negative experiences of testing.

REFERENCES

American College of Sports Medicine (ACSM). (2005). *ACSM's Guidelines for Exercise Testing and Prescription*, 7th edn. Philadelphia, PA: Lippincott Williams & Wilkins.

de Vries, H.A. and Housh, T.J. (1995). *Physiology of Exercise: for Physical Education, Athletics and Exercise Science*, 5th edn. Dubuque: WC Brown.

Heyward, V.H. (1998). *Advanced Fitness Assessment and Exercise Prescription*. Champaign, IL: Human Kinetics.

Goudas, M., Biddle, S. and Fox, K. (1994). Achievement goal orientations and intrinsic motivation in physical fitness testing with children. *Pediatric Exercise Science*, 6: 159–167.

Prochaska, J.O. and DiClemente, C.C. (1986). Toward a comprehensive model of change. in W.R. Miller and N. Heather (eds) *Addictive Behaviours: Processes of Change*. pp. 3–27. New York: Plenum Press.

Rothman, A., Salovey, P., Turvey, C. and Fishkin, S. (1993). Attributions of responsibility and persuasion: increasing mammography utilisation among women over 40 and with internally oriented message. *Health Psychology*, 12: 39–47.

Whitehead, J.R. and Corbin, C.B. (1991). Youth fitness testing: the effect of percentile – based evaluative feedback on intrinsic motivation. *Research Quarterly for Exercise and Sport*, 62(2): 225–231.

CHAPTER 3

BLOOD SAMPLING

Ron Maughan, Susan M. Shirreffs and
John B. Leiper

INTRODUCTION

The collection of blood samples from human subjects is required in many physiological, biochemical and nutritional investigations. The use to be made of the sample will determine the method of collection, the volume of blood required and the way in which the specimen is handled.

BLOOD SAMPLING AND HANDLING

Many different methods and sites of blood sampling can be used to collect samples for analysis, and the results obtained will be affected by the sampling site and by the procedures used in sample collection. A detailed discussion of the sampling procedures and of the consequences for measurement of various parameters is presented by Maughan *et al.* (2001).

The main sampling procedures involve collection of arterial, venous, arterialised venous or capillary blood. In most routine laboratory investigations of interest to the sports scientist, arterial blood sampling is impractical and unnecessarily invasive, and will not be considered in detail here. Where arterial blood is required, arterial puncture may be used, but in most situations, collection of arterialised venous blood as described later gives an adequate representation of arterial blood.

Venous blood

Venous blood sampling is probably the method of choice for most routine purposes: sampling from a superficial forearm or ante-cubital vein is simple, painless and relatively free from risk of complications. Sampling may be by venous puncture or by an indwelling cannula. Where repeated sampling is necessary at short time intervals, introduction of a cannula is obviously preferred to avoid repeated venous punctures. Either a plastic cannula or a butterfly-type cannula can be used. The latter has obvious limitations if introduced into an ante-cubital vein, as movement of the elbow is severely restricted. However, because it is smaller and therefore less painful for the subject, as well as being very much less expensive, it is often preferable if used in a forearm vein, provided that long-term access is not required. A 21 g cannula is adequate for most purposes, and only where large volumes of blood are required will a larger size be necessary. In most situations where vigorous movements are likely, the forearm site is preferred to the elbow. Clotting of blood in the cannula is easily avoided by flushing with sterile isotonic saline. Where intermittent sampling is performed, the cannula may be flushed with a bolus of saline to which heparin $(10–50 \ IU \cdot ml^{-1}$ of saline) is added, allowing the subject freedom to move around between samples. Alternatively where the subject is to remain static, as in a cycle or treadmill exercise test, a continuous slow infusion (about $0.3 \ ml \cdot min)^{-1}$ of isotonic saline may be used, avoiding the need to add heparin. Collection of samples by venous puncture is not practical in most exercise situations, and increases the risk that samples will be affected by venous occlusion applied during puncture. If repeated venous puncture is used, care must be taken to minimise the duration of any occlusion of blood flow and to ensure that sufficient time is allowed for recovery from interruption of blood flow before samples are collected.

Flow through the superficial forearm veins is very much influenced by skin blood flow, which in turn depends on ambient temperature and the thermoregulatory strain imposed on the individual. In cold conditions, flow to the limbs and to the skin will be low, and venous blood will be highly desaturated. Where sampling occurs over time, therefore, and where the degree of arterialisation of the venous blood will influence the measures to be made, this may cause major problems. For some metabolites which are routinely measured, the difference between arterial and venous concentrations is relatively small and in many cases it may be ignored. Where a difference does occur and is of importance, the effect of a change in arterialisation of the blood at the sampling site may be critical.

Arterialised venous blood

Where arterial blood is required, there is no alternative to arterial puncture, but for most practical purposes, blood collected from a superficial vein on the dorsal surface of a heated hand is indistinguishable from arterial blood. This reflects both the very high flow rate and the opening of arterio-venous shunts in the hand. Sampling can conveniently be achieved by introduction of a butterfly cannula into a suitable vein. The hand is first heated, either by immersion up to the forearm for at least 10 min in hot (about 42°C) water (Forster *et al.,* 1972) or by insertion into a hot air box (McGuire *et al.,* 1976). If hot water immersion is used prior to exercise, arterialisation – as indicated by oxygen saturation – can be maintained for some considerable time by wearing a glove, allowing this technique to be used during exercise studies. This procedure allows large volumes of blood to be collected without problems. Capillary sampling by the fingerprick method cannot guarantee adequate volumes for many procedures.

Capillary blood

Where only small samples of blood are required, capillary blood samples can readily be obtained from a fingertip or earlobe. The use of micromethods for analysis means that the limited sample volume that can be obtained should not necessarily be a problem in metabolic studies. It is possible to make duplicate measurements of the concentrations of glucose, lactate, pyruvate, alanine, free fatty acids, glycerol, acetoacetate and 3-hydroxybutyrate as well as a number of other metabolites on a single 20 μl blood sample using routine laboratory methods (Maughan, 1982).

The sampling site should be arterialised, by immersion of the whole hand in hot (42°C) water in the case of the finger tip, and by the use of a rubefacient in the case of the earlobe. Samples can be obtained without stimulating vasodilatation, but bleeding is slower, the volumes that can be reliably collected are smaller, and the composition of the sample is more variable. It is essential that a free flowing sample is obtained. If pressure is applied, an excess of plasma over red cells will be obtained. Samples are most conveniently collected into graduated glass capillaries where only small volumes are required (typically 10–100 μl). The blood must never be expelled from these tubes by mouth, because of the obvious risks involved. Volumes greater than about 0.5 ml are difficult to obtain.

BLOOD TREATMENT AFTER COLLECTION

Analysis of most metabolites can be carried out using whole blood, plasma or serum, but the differential distribution of most metabolites and substrates between the plasma and the intracellular space may affect values. It is convenient to use whole blood for the measurement of most metabolites. Glucose, glycerol and lactate are commonly measured on either plasma or whole blood, but free fatty acid concentration should be measured using plasma or serum. The differences become significant where there is a concentration difference between the intracellular and extracellular compartments.

If plasma is to be obtained by centrifugation of the sample, a suitable anticoagulant must be added. A variety of agents can be used, depending on the measurements to be made. The potassium salt of EDTA is a convenient anticoagulant, but is clearly inappropriate when plasma potassium is to be measured. Heparin is a suitable alternative in this situation. For serum collection, blood should be added to a plain tube and left for at least 1 h before centrifugation: clotting will take place more rapidly if the sample is left in a warm place. If there is a need to stop glycolysis in serum or plasma samples (e.g. where the concentration of glucose, lactate or other glycolytic intermediates is to be measured), fluoride should be added. Where metabolites of glucose are to be measured on whole blood, the most convenient method is immediate deproteinisation of the sample to inactivate the enzymes which would otherwise alter the concentrations of substances of interest after the sample has been withdrawn.

Control of factors affecting blood and plasma volumes

Blood and plasma volumes are markedly influenced by the physical activity, hydration status and posture of the subject prior to sample collection. The sampling site and method can also affect the haemoglobin concentration, as arterial, capillary and venous samples differ in a number of respects due to fluid exchange between the vascular and extravascular spaces and to differences in the distribution of red blood cells (Harrison, 1985). The venous plasma to red cell ratio is higher than that of arterial blood, although the total body haemoglobin content is clearly not acutely affected by these factors. Haemodynamic changes caused by postural shifts will alter the fluid exchange across the capillary bed, leading to plasma volume changes that will cause changes in the circulating concentration. On going from a supine position to standing, plasma volume falls by about 10% and whole blood volume by about 5% (Harrison, 1985). This corresponds to a change in the measured haemoglobin concentration of about 7 g·l^{-1}. These changes are reversed on going from an upright to a seated or supine position. These changes make in imperative that posture is controlled in studies where haemoglobin changes are to be used as an index of changes in blood and plasma volume over the time course of an experiment. It is, however, common to see studies reported in the literature where samples were collected from subjects resting in a supine position prior to exercise in a seated (cycling or rowing) or upright (treadmill walking or running) position. The changing blood volume not only invalidates any haematological measures made in the early stages of exercise, it also confounds cardiovascular measures as the stroke volume and heart rate will also be affected by the blood volume.

SAFETY ISSUES

Whatever method is used for the collection of blood samples, the safety of the subject and of the investigator is paramount. Strict safety precautions must be followed at all times in the sampling and handling of blood. It is wise to assume that all samples are infected and to treat them accordingly. This means wearing gloves and appropriate protective clothing and following guidelines for handling of samples and disposal of waste material. Appropriate antiseptic procedures must be followed at all times, including ensuring cleanliness of the sampling environment, cleaning of the puncture site and use of clean materials to staunch bleeding after sampling. Blood sampling should be undertaken only by those with appropriate training and insurance cover, and a qualified first-aider should be available at all times. All contaminated materials must be disposed of using appropriate and clearly identified waste containers. Used needles, cannulae and lancets must be disposed off immediately in a suitable sharps bin: resheathing of used needles must never be attempted. Sharps – whether contaminated or not – must always be disposed off in an approved container and must never be mixed with other waste. Any spillage of blood must be treated immediately.

There is clearly a need for appropriate training of all laboratory personnel involved in any aspect of blood sampling and handling. Most major hospitals run courses for the training of phlebotomists, who are often individuals with no medical background. The taking of blood samples is a simple physical skill, and a medical training is not required when expert assistance is at hand. What is essential, though, is the necessary back up if something goes wrong, and a suitable training in first aid and resuscitation should be seen as a necessary part of the training for the sports scientists who collect blood samples outwith a hospital setting.

REFERENCES AND FURTHER READING

Dacie, J.V. and Lewis, S.M. (1968). *Practical Haematology*, 4th edn. London: Churchill, pp. 45–49.

Forster, H.V., Dempsey, J.A., Thomson, J., Vidruk, E. and DoPico, G.A. (1972). Estimation of arterial PO_2, PCO_2, pH and lactate from arterialized venous blood. *Journal of Applied Physiology*, 32: 134–137.

Harrison, M. (1985). Effects of thermal stress and exercise on blood volume in humans. *Physiological Review*, 65: 149–209.

Maughan, R.J. (1982). A simple rapid method for the determination of glucose, lactate, pyruvate, alanine, 3-hydroxybutyrate and acetoacetate on a single 20 µl blood sample. *Clinica Chimica Acta*, 122: 232–240.

Maughan, R.J., Leiper, J.B. and Greaves, M. (2001). Haematology. In R.G. Eston and T.P. Reilly (eds) *Kinanthropometry and Exercise Physiology Laboratory Manual*, 2nd edn. London: Spon Volume 2, pp. 99–115.

McGuire, E.A.H., Helderman, J.H., Tobin, J.D., Andres, R. and Berman, M. (1976). Effects of arterial versus venous sampling on analysis of glucose kinetics in man. *Journal of Applied Physiology*, 41: 565–573.

CHAPTER 4

ETHICS AND PHYSIOLOGICAL TESTING

Steve Olivier

WHAT IS ETHICS?

What is it to behave in an ethical manner as a researcher? The term 'ethics' suggests a set of standards by which behaviour is regulated, and these standards help us to decide what is acceptable in terms of pursuing our aims, as well as helping us to distinguish between right and wrong acts. The principal question of ethics is 'What ought I do?'

Broadly speaking, ethical actions are derived from principles and values, which are in turn derived from ethical theories. The major ethical theories are briefly introduced here for two reasons: to enable researchers to identify where principles are derived from, and to facilitate deeper thought on how potential actions may be justified.

Virtue theory focuses on being a 'good' person, and doing the right thing (e.g. being fair, honest and so on) necessarily flows from being a 'good' person. Utilitarian (consequential) theory attaches primary importance to the consequences of actions – if the 'good' consequences outweigh the 'bad' ones for all concerned by the action, then the action is right and is morally required. Lastly, Deontology holds that primacy is attached to meeting duties and obligations, that the ends do not justify the means, and that an individual's preferences, interests and rights should be respected. It is worth noting that codes of ethics are generally deontological in nature.

There are three basic principles upon which our conception of research ethics is based, namely respect for persons, beneficence (doing good) and justice. Applying these to research contexts involves consideration of autonomy (an individual's right to self-determination), obligations not to harm others (including physical, psychological or social harm), utility (producing a net balance of benefits over harm), justice (distributing benefits and harms fairly), fidelity keeping promises and contracts), privacy, and veracity (truthfulness). More specific ethical considerations would include recognition of cultural factors, preserving participant anonymity (or confidentiality, as appropriate), non-discrimination, sanctions against offenders, compliance with procedures and reports of violations (Olivier, 1995).

INFORMED CONSENT

A central feature of modern biomedical research ethics is the notion of obtaining first person, written, voluntary informed consent from research participants. Given that it is a required element of most projects, researchers need to be aware of what the concept involves.

First, 'informed' implies that potential participants (or their legal representatives) obtain sufficient information about the project. This information must be presented in such a way that it is matched to the appropriate comprehension level (see Olivier and Olivier, 2001; and Cardinal, 2000, for further details on establishing comprehension levels), enabling participants to evaluate and understand the implications of what they are about to agree to. Second, 'consent' implies free, voluntary agreement to participation, without coercion or unfair inducement.[1]

Consent can be considered to be informed when 'it is given in the full, or clear, realization of what the tests involve, including an awareness...of risk attached to what takes place' (Mahon, 1987, p. 203). Further, 'Subjects must be fully informed of the risks, procedures, and potential benefits, and that they are free to end their participation in the study with no penalty whatsoever' (Zelaznik, 1993, p. 63).

Consent is deemed ethically acceptable if the participant receives full disclosure of relevant information, if the implications are understood, if the participant voluntarily agrees to participate, if opportunities to freely ask relevant questions are present throughout the duration of the project and if the participant feels able to withdraw from the procedures at any time.

The informed consent form

The informed consent form, normally signed by the participants, should be tailored to the specific project that it relates to. The document should include the following elements:

- an explanation of the purposes of the project;
- a description of the procedures that will involve participants, including the time commitment;
- identification and description of any risks/discomforts, and potential benefits that can reasonably be foreseen, as well as any arrangements for treatment in the case of injury;
- statements regarding confidentiality, anonymity and privacy;
- identification of an appropriate individual whom the participants can approach regarding any questions about the research;

- a statement that participation is voluntary, that consent has been freely obtained and that participants may withdraw at any time without fear of sanction.

A consent form should not include language that absolves the researcher from blame, or any other waiver of legal rights releasing, or appearing to release any-one from liability (Liehmon, 1979; Veatch, 1989). The consent form should conclude with a statement that the participant has read the document and understands it, and should provide space underneath for a signature and the date. Space should also be provided for signatures of the researcher and an independent witness.

Written consent is considered to be the norm for all but the most minor of research procedures. It can serve to protect participants as well as investiga-tors, and serves as proof that some attention has been paid to the interests of the participants. Written consent is superior to oral in that the form itself can be used as an explanatory tool and as a reference document in the communi-cation process between researchers and participants. Also, presenting informa-tion orally as well as in written form may have the advantage of prompting participants to ask relevant questions. However, when there are doubts about the literacy level of participants, oral information should supplement proxy[2] written consent.

Witnessed consent may be particularly useful when participants are elderly or have intellectual or cultural difficulties in speech or comprehension. In these cases, an independent person, such as a nurse or a community/religious leader, should sign a document stating that the witness was present when the researcher explained the project, and that in the opinion of the witness, the participant understood the implications of the research and consented freely.

Special legal or institutional considerations may apply when the research involves, inter alia, pregnant women, foetuses, prisoners, children, wards of the state or when deception is used. Research requiring deception, or procedures carrying an unusually high risk of harm, will typically require that a researcher satisfies additional conditions.[3]

There is little unanimity concerning the practice of paying research participants, particularly when intrusive procedures are involved. Researchers should be satisfied that payment does not constitute coercion, and remuneration should not adversely affect the judgement of potential participants in respect of risk assessment. Statements on payment to participants should not deflect attention away from the other information in the informed consent form.

Obtaining informed consent at the start of a project may not be sufficient – circumstances may change and new ethical considerations might arise[4] – and researchers should be aware that consent with participants might have to be renegotiated. This might also mean that emergent issues are referred back to the original ethics committee for clearance. It is worth noting that obtaining informed consent does not ensure that a research project is ethical. The research itself must be ethical, and researchers should consider the moral issues that apply to their work.

Children as research participants[5]

When utilising children as research participants, you should consider not only their rights to choose to participate in research (and to withdraw), but also issues such as power differentials, and coercion, in the recruitment process. If you are using a gatekeeper for access (such as a coach, or teacher), that person should not recruit children on your behalf, and should not have access to any individualised data collected. Beware of obtaining proxy consent, as it is unlikely that anyone in a relatively low hierarchical position (such as pupils in a school) will refuse to participate if someone higher up (e.g. a teacher, or Head) gives permission on their behalf (Homan, 2002). You should obtain active rather than passive (assumed) consent. Passive consent involves making the assumption that non-refusal constitutes tacit agreement to participate. While this is a much easier method of recruiting, it may disregard the autonomous wishes (or voluntariness) of participants.

The Medical Research Council (2004) supports the use of children in research as long as the benefits and risks are carefully assessed. Where there is no benefit to child participants, the risk needs to be minimal (see MRC, 2004, pp. 14–15 for categories of risk). Minimal risk activities include questioning, observing and measuring children,[6] and obtaining bodily fluids without invasive intervention. This rules out more invasive procedures such as muscle biopsies.

In England and Wales, anyone who has reached the statutory age of majority (eighteen years) can consent to being a research participant in therapeutic or nontherapeutic[7] studies. For therapeutic research, the Family Reform Act 1969 provides that anyone over 16 can provide consent. Below 16, it is suggested that no one under 12 can provide individual consent (rather than assent, it should be noted), but that children over 12 can provide consent if they are deemed sufficiently mature by the researcher (Nicholson, in Jago and Bailey, 2001). For nontherapeutic work, there is no precise age below 18 at which a child acquires legal capacity, but again, for anyone over 12, an assessment of maturity must be made. The problem with this, of course, is that researchers must 'accept the possibility of prosecution if their interpretation of a child's competence to consent is deemed unacceptable' (Jago and Bailey, 2001, p. 531).

Given that most research by BASES members is nontherapeutic, what should you do? For participants under eighteen, obtain parental consent, first person consent from the participants, and proxy consent from a relevant authority figure if appropriate. If your potential participants are aged 7 to 12, obtain assent (acquiescence, or yea saying) on a simplified form, as well as parental and proxy consent as appropriate. In all cases, the language used on consent and assent forms should be tailored to the participants' comprehension levels (see Olivier and Olivier, 2001).

The ethics review process

The emphasis on research ethics in recent decades is a response to abuses perpetrated on human research participants in the past. This chapter is not the place to enumerate such details (see McNamee *et al.*, 2006), but suffice to say that the regulatory response has been to create a system of ethical review with which investigators must comply.

All funding bodies will insist, as part of the review process, that potential projects are carefully scrutinised with regard to ethical implications. Regulations in the United Kingdom are not as consistently applied as in the United States, but nevertheless, most institutions (e.g. universities, laboratories) will require formal approval of a project before data collection can proceed. Even for unfunded projects, submitting a project for ethical review has benefits for participants (protection of their rights, safety) and for researchers (evidence of compliance with proper procedures, rigour of study design). So, while some researchers view formal ethics review as a bureaucratic impediment to conducting research, it is deemed to be a valuable (if somewhat flawed) process that protects individuals and facilitates good science (Olivier, 2002).

Given that systems of ethics review vary from institution to institution and across funding bodies, it is important for the individual researcher (or team leader) to ascertain what the obligations are with regard to ethics review and compliance. Also, research managers need to be conversant with broader regulatory systems such as the Department of Health Research Governance Framework, NHS Local Research Ethics Committees and the recent introduction into UK law of the European Clinical Trials Directive (see McNamee *et al.*, 2006).

Codes of conduct and accreditation

Codes of conduct and accreditation schemes, such as those administered by BASES, are particularly useful in terms of promoting and maintaining professional competence. A code of conduct though, while promoting ethical behaviour, does not ensure it. This is because rules can conflict, because they are not exhaustive of all moral situations, because they may not take consequences of actions into account, and because they don't consider important contextual issues. Further, if rules are very specific you need an inordinate number to cover all relevant situations, and if they are general then they are likely to be of little practical use. Lastly, and perhaps most importantly, simple rule-following is mechanical, and doesn't promote moral engagement.

Researchers should adhere to the requirements of the BASES Code of Conduct, but should also carefully consider the specific ethical issues that arise from their own projects. It is incumbent on individual researchers, as human agents of moral decision-making, to personally and carefully consider ethical issues inherent in their projects, and to analyse, evaluate, synthesise and apply appropriate principles and values.

Checklist

The checklist below is designed to assist you in preparing your project for ethical review. Remember though that projects are different, and encompass a variety of ethical issues. The checklist is just a start. The challenge for all researchers is to think independently about the ethical issues presented by their work.

- Make sure that you get voluntary, written first-person informed consent. If this is deemed inappropriate, you need to justify the exception.
- Check institutional or legal guidelines about parental consent, and about obtaining a child's assent. In the case of using children as research participants, obtain the necessary parental consent, and the child's assent.
- When using vulnerable populations (e.g. the aged, wards of the state or other agencies), check that you comply with any ethical requirements specific to that group. For example, you may need witnessed consent for cognitively impaired participants.
- Satisfy yourself that participants understand the nature of the project, including any risks or potential benefits. Describing the project to them verbally will often assist in this process.
- Explain to participants that they are free to ask questions at any time, and that they can withdraw from the project whenever they want to.
- Make sure that no coercion occurs during the recruitment process. (Here you need to be clear on issues such as the researcher not being a teacher or assessor of participants' work, for example in the case of students.)
- Allow participants a 'cooling off' period to consider their participation (the time between reading the form and actually agreeing to take part).
- Assess the risk of physical, psychological or social harm to participants.
- Provide medical or other appropriate backup in the event of any potential harm in the categories mentioned earlier.
- Provide medical or other screening, as appropriate.
- Assess the risk of harm to yourself as a researcher, and any assistants (e.g. handling of body fluids, or personal safety in interview situations).
- Provide for the safe conduct of the research if anything has been identified in the preceding point (e.g. correct laboratory procedures; protection in interviews; ability to contact emergency services).
- Assess the impact of any cultural, religious, or gender issues that may pertain to your participants, and/or the dissemination of your findings.
- Provide adequate assurances regarding privacy, confidentiality, anonymity, and how you will securely store and treat your data.
- Satisfy yourself that any payments or inducements offered to participants do not adversely influence their ability to make an informed assessment of the risks and benefits of participation.
- Satisfy yourself that any funding or assistance that you receive with the research will neither result in a conflict of interest, nor compromise your academic integrity.
- If your study involves deception, state the reasons/justification, and indicate how you will debrief the participants about the deception.
- Set measures in place to provide participants with feedback/information on completion of the project.
- And of course, make sure that you have received approval to proceed from the appropriate regional, national or institutional ethics committees.

NOTES

1   I recognise that that this reduction of the concept of informed consent is simplistic, and begs the fallacy of composition (Morgan, 1974), which is the notion that one can break down complex terms into their constituents and then merely add them up as if the sum of the parts was equal to the whole. Nevertheless, it is a useful starting point for the practical application of informed consent procedures.

2   Proxy consent is consent given for an individual, by someone else, for example a parent, religious leader, etc. When seeking proxy consent, particular care should be taken to consider the issues surrounding autonomy and paternalism (see McNamee *et al.*, 2006).

3   For example, justification for deception would include that the research is important, that the results are unobtainable by other methods, that participants are not harmed, and that thorough debriefing occurs if appropriate.

4   Such as the application of new measurement procedures, for example.

5   I would like to thank Malcolm Khan, Senior Lecturer in Law at Northumbria University, for commenting on the legal accuracy of this section.

6   Such activities must be carried out in a sensitive way, with due consideration given to the child's autonomy.

7   I recognise the difficulties with this distinction in terms of describing medical research, but feel that is still useful in terms of much of the research conducted by BASES members.

REFERENCES

Cardinal, B.J. (2000). (Un)Informed consent in exercise and sport science research? A comparison of forms written for two reading levels. *Research Quarterly for Exercise and Sport*, 71(3): 295–301.

Homan, R. (2002). The principles of assumed consent: the ethics of Gatekeeping. In M. McNamee and D. Bridges (eds), *The Ethics of Educational Research*, pp. 23–40. Oxford: Blackwell.

Jago, R. and Bailey, R. (2001). Ethics and paediatric exercise science: issues and making a submission to a local ethics research committee. *Journal of Sports Sciences*, 19: 527–535.

Liehmon, W. (1979). Research involving human subjects. *The Research Quarterly*, 50(2): 157–163.

Mahon, J. (1987). Ethics and drug testing in human beings. In J.D.G. Evans (ed.), *Moral Philosophy and Contemporary Problems*. Cambridge: Press syndicate of the University of Cambridge.

Medical Research Council. (2004). *MRC Ethics Guide: Medical Research Involving Children*. http://www.mrc.ac.uk/pdf-ethics_guide_children.pdf#xml=http://www.mrc.ac.uk/scripts/texis.exe/webinator/search/xml.txt?query=children&pr=mrcall&order=r&cq=&id=422bfe0f2, accessed 7 March 2005.

McNamee, M. Olivier, S. and Wainwright, P. (2006). *Research Ethics in Exercise, Health and Sport Sciences*. Abingdon: Routledge.

Morgan, R. (1974). *Concerns and Values in Physical Education*. London: G Bell and Sons.

Nicholson, R.N. (ed.) (1986). *Medical Research with Children: Ethics, Law and Practice*. Oxford: Oxford University Press. Cited in Jago, R. and Bailey, R. (2001).

Ethics and paediatric exercise science: issues and making a submission to a local ethics research committee. *Journal of Sports Sciences*, 19: 527–535.

Olivier, S. (1995). Ethical considerations in human movement research. *Quest*, 47(2): 135–143.

Olivier, S. (2002). Ethics review of research projects involving human subjects. *Quest*, 54: 194–204.

Olivier, S. and Olivier, A. (2001). Comprehension in the informed consent process. *Sportscience*, 5(3): www.sportsci.org.

Veatch, R.M. (ed.) (1989). *Medical Ethics*. Boston, MA: Jones and Bartlett Publishers.

Zelaznik, H.N. (1993). Ethical issues in conducting and reporting research: a reaction to Kroll, Matt and Safrit. *Quest*, 45(1): 62–68.

PART 2

METHODOLOGICAL ISSUES

CHAPTER 5

METHOD AGREEMENT AND MEASUREMENT ERROR IN THE PHYSIOLOGY OF EXERCISE

Greg Atkinson and Alan M.Nevill

INTRODUCTION

Exercise physiologists need to make an informed choice of the most appropriate measurement tool before they start collecting data from athletes or research participants. The main criteria governing this choice are:

- the appropriate level of invasiveness and convenience of use;
- the available budget;
- the degree of test–retest measurement error;
- the degree of agreement with an alternative method, which is possibly more invasive, less convenient or more expensive.

It is important to note that the most expensive and invasive measurement tool might *not* necessarily be associated with the least test–retest measurement error. Moreover, *all* measurement methods that are employed in order to measure some aspect of human physiology have some degree of test–retest error attributable to natural biological variation. For example, use of the so-called '*gold standard*' Douglas bag method of gas analysis is still associated with substantial test–retest error due to human variability in oxygen consumption kinetics during exercise (Atkinson *et al.*, 2005a). Similarly, whilst it is conventional to compare a new automatic blood pressure monitor with sphygmomanometry, this latter method is, again, associated with substantial test–retest measurement error (Bland and Altman, 1999) that is biological in origin. This ubiquity of biological variability governs several major considerations when analysing the performance characteristics of physiological measurement tools:

- Ideally, an examination of test–retest measurement error should be inherent in any examination of the agreement between measurement tools.

Moreover, *both* measurement tools (not just the more convenient or cheaper alternative) should be appraised for test–retest measurement error. Only through such an analysis can a firm conclusion be made regarding the source of any disagreement between different methods of measurement (Bland and Altman, 2003; Atkinson *et al.*, 2005a).

- Some aspects of least-squares regression (LSR) should be used with caution to examine agreement between measurement methods relevant to exercise physiology. It is likely that both physiological measurement methods show approximately similar degrees of test–retest error due to the major component of this error being ubiquitous and biological in origin. This error, present when using either measurement method, means that an important assumption for LSR might be violated leading to biased estimates of LSR slope and intercept statistics (Ludbrook, 1997; Bland and Altman, 2003; Atkinson *et al.*, 2005a). These statistics are conventionally used to make inferences about systematic differences between methods but the slope and intercept of a LSR line is unbiased only if the 'predictor' method is associated with substantially lower levels of test–retest measurement error than the other method or is in fact a 'fixed' variable. Moreover, the prediction philosophy of regression does not sit well with the fact that most researchers desire to select, *a priori*, the best measurement tool to use throughout their investigations, rather than them aiming to predict measurements using another method as part of their study.

- Like all statistics, those used to describe error and agreement are population specific, since different populations may show different degrees of error due to biological sources. Different individuals sampled from the same population may also show different degrees of error, for example, individuals who record the highest physiological values in general might also show the greatest amount of measurement error. Therefore, whether error might differ for different individuals in the population, and whether the statistical precision of the sample error estimate is adequate, are important considerations.

There are other philosophical issues, which underpin the statistical techniques used to appraise a physiological measurement tool. Our aim is not to discuss these issues, since there are now several comprehensive reviews in which the background to the statistical analysis is explained (Atkinson and Nevill, 1998; Bland and Altman, 1999, 2003; Atkinson *et al.*, 2005a). Alternatively, we aim to summarise the most important aspects of a measurement study in the form of a checklist for exercise physiologists.

A METHOD AGREEMENT AND MEASUREMENT ERROR CHECKLIST

We present a checklist, which may be useful to exercise physiologists interested in appraising a measurement tool, either if they are performing a measurement study themselves or if they are reading a relevant paper already published in a scientific journal. By 'measurement study', we mean an investigation into either the agreement between measurement methods (a method comparison study) or test–retest measurement error (a repeatability or reliability study). We have categorised the various important points into (1) Delimitations, (2) Systematic error examination, (3) Random error examination and (4) Statistical precision.

1   Delimitations

- Ideally, the measurement study should involve at least 40 participants. If there are less than 40 participants, then scrutiny of confidence limits for the error statistics becomes even more important (see Section 4), since error estimates calculated on a small sample can be imprecise (Atkinson, 2003).
- Try to match the characteristics of the measurement study to planned uses of the measurement tool, that is, a similar population, a similar time between repeated measurements (for investigations into test–retest error), a similar exercise protocol as well as comparable resting conditions during measurements.
- Select *a priori* an amount of error that is deemed acceptable between the methods or repeated tests. This delimitation may depend on whether one wishes to use the measurement tool predominantly for research purposes (i.e. on a sample of participants) or for making measurements on individuals (e.g. for health screening purposes or for sports science support work). Atkinson and Nevill (1998) termed these considerations 'analytical goals'.
- For research purposes, the analytical goal for measurement error is best set via a statistical power calculation. One could delimit an amount of test–retest error (described by the standard deviation of the differences, for example) on the basis of an acceptable statistical power to detect a given difference between groups or treatments with a feasible sample size (Atkinson and Nevill, 2001). If a relatively large sample is feasible for future research, then a given amount of measurement error should have less impact on use of the measurement tool, and *vice versa* (Figure 5.1).
- For use of the measurement tool on individuals, one might delimit the acceptable amount of error on the basis of the 'worst scenario' individual difference, which would be allowable. This delimitation is related to the 95% limits of agreement (LOA) statistic (Bland and Altman, 1999, 2003) as well as applications of the standard error of measurement (SEM) statistic (Harvill, 1991). For example, a difference as large as 5 beat·min^{-1} between two repeated measurements of heart rate during exercise would probably still make little difference to the prescription of heart rate training 'zones' to individuals.
- The use of arbitrary 'rules of thumb' such as accepting adequate agreement between methods or tests on the basis of a correlation coefficient being above 0.9 or a coefficient of variation (CV) being below 10% is discouraged, since no relation is made between error and real uses of the measurement tool with such generalisations (Atkinson, 2003). Nevertheless, reviews (e.g. Hopkins, 2000), in

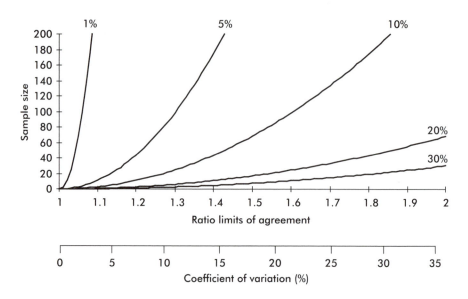

Figure 5.1 A nomogram to estimate the effects of measurement repeatability error on whether 'analytical goals' are attainable or not in exercise physiology research. Statistical power is 90%. The different lines represent different worthwhile changes of 1%, 5%, 10%, 20% and 30% due to some hypothetical intervention. The measurement error statistics, which can be utilised are the LOA and CV. For example, a physiological measurement tool, which has a repeatability CV of 5% would allow detection of a 5% change in a pre-post design experiment (using a paired *t*-test for analysis of data) and with a feasible sample size (~20 participants)

Source: Batterham, A.M. and Aktinson, G. (2005). How big does my sample need to be? A primer on the murky world of sample size estimation. *Physical Therapy in Sport 6*, 153–163.

which error statistics are cited for various measurements may be useful in establishing a 'typical' degree of acceptable error to select.[1]

2   Systematic error examination

- Compare the mean difference between methods/tests with the *a priori* defined acceptable level of agreement (see Figure 5.2 and Section 4 below on the use of confidence limits for interpretation of this mean difference).
- If systematic error is present between repeated tests using the same method (i.e. a repeatability study) and if no performance test has been administered, then be suspicious about the design of the repeatability study. Perhaps, there have been carry-over effects from previous measurements being obtained too close in time to subsequent measurements. Such a scenario could occur with measurements of intra-aural temperature, for example (Atkinson *et al.*, 2005b).
- If a performance test *is* incorporated in the protocol, then systematic differences between test and retest(s) in a repeatability study may occur due to learning effects, for example. Such information is important for advising future researchers how many familiarisation sessions might be required prior to the formal recording of physiological values.

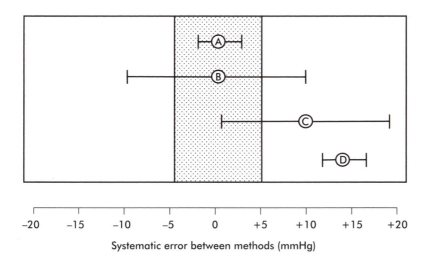

-20   -15   -10   -5   0   +5   +10   +15   +20

Systematic error between methods (mmHg)

Figure 5.2 Using a confidence interval and 'region of equivalence' (shaded area) for the mean difference between methods/tests. Using the agreement between two blood pressure monitors as an example, if both of the 95% CI fall inside an *a priori* selected 'region of equivalence' of ±5 mmHg for the mean difference between methods, as is the case with point A, then we can be reasonably certain that the true systematic difference between methods is not clinically or practically important. For point D, even the lower 95% confidence limit of 10.6 mmHg for the systematic difference between methods does not lie within the designated region of equivalence, so we can be reasonably certain that the degree of systematic bias would have practical impact. The width of the CIs for points B and C suggest that the population mean bias might be practically important, but one would need more cases in the measurement study to be reasonably certain of the true magnitude, or in the case of point B, even the direction of the systematic bias for the population

- Examine whether the degree of systematic error alters over the measurement range. One can consult the Bland–Altman plot for this information (Figure 5.3). The presence of 'proportional' bias would be indicated if the points on the Bland–Altman plot show a pronounced downward or upward trend over the measurement range. If this characteristic is present, it means that the systematic difference between methods/tests differs for individuals at the low and high ends of the measurement range. Bland and Altman (1999) and Atkinson *et al.* (2005a) discuss how this proportional bias can be explored and modelled.

3   Random error examination

- Scrutinise the degree of random error between methods/tests that is present (Figure 5.3). Popular statistics used to describe random error include the SEM (Harvill, 1991), which is also known as the within-subjects standard deviation, CV, LOA and standard deviation of the differences. Each of these statistics will differ, since they are based on different underlying philosophies. For example, the LOA statistic is rooted in clinical work and can be viewed as representing the 'worst scenario' error that one might observe for an

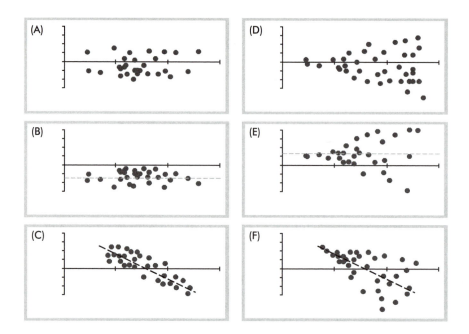

Figure 5.3 Various examples of relationships between systematic and random error and the size of the measured value as shown on a Bland-Altman plot. (A) Proportional random error. (B) systematic error present, which is uniform in nature, random errors also uniform. (C) Systematic error present, which is proportional to size of measured value, random errors uniform. (D) Proportional random error with no systematic error. (E) Uniform systematic error present with proportion random error. (F) Proportional systematic and random error

Source: Atkinson, G., Davison, R.C.R. and Nevill, A.M. (2005). Performance characteristics of gas analysis systems: what we know and what we need to know. *International Journal of Sports Medicine*, 26 (Suppl. 1): S2–S10.

individual person. The SEM statistic is popular amongst psychologists who conceptualise an average of many repeated measurements; the 'true score' for an individual (Harvill, 1991).

- Explore the relationship between random error and magnitude of measured value (Atkinson and Nevill, 1998; Bland and Altman, 1999). If the random error does increase in proportion to the size of the measured value, then a ratio statistic should be employed to describe the measurement error (e.g. CV) (Nevill and Atkinson, 1997; Bland and Altman, 1999). If the error is 'homoscedastic' (uniform over measurement range), than measurement error can be described in the particular units of measurement by calculating the LOA or SEM. Complicated relationships between magnitude of error and measured value can be analysed using non-parametric methods according to Bland and Altman (1999) or by calculating measurement error statistics for separate sub-samples within a population (Lord, 1984).

- In a repeatability study, which involves multiple retests, examine whether random error changes between separate test and retest(s). The researcher could explore whether random error reduces as more tests are administered (Nevill and Atkinson, 1998). If this is so, and in keeping with the advice above for exploration of systematic error, the researcher should communicate this learning effect on random error so that future users know exactly how many familiarisation sessions are required for total error variance to be minimised for their research.

4   Statistical precision

- Calculate the 95% confidence interval (CI) for the mean difference between methods/tests (Jones *et al.*, 1996) and compare this CI to a 'region of equivalence' for the two methods of measurement (Figure 5.2). This CI is *not* the same as the LOA statistic. Scrutiny of the lower and upper limits of this CI should not change the conclusion that has been arrived at regarding the acceptability of systematic error between methods/tests (Figure 5.2). For example, one might observe a mean difference of 10 mmHg between blood pressure measuring devices but the 95% CI might be −4 to 24 mmHg. This means that the population mean difference between methods could be as much as 4 mmHg in one direction or as much as 24 mmHg in the other direction. Only a narrower CI, mediated mostly by a greater sample size, would allow one to make a more conclusive statement regarding systematic error. Atkinson and Nevill (2001) and Atkinson *et al.* (2005a) discuss further the use of CI's and limits.
- Calculate confidence limits for the random error statistics. As above, scrutiny of the CI should not change the decision that has been made about acceptability of random error. For example, a physiological measurement method with a repeatability CV of 30% and associated CI of 25–35% indicates poor repeatability, even if the lower limit of the CI is taken into account. Bland and Altman (1999) provide details relevant to limits of agreement. Hopkins (2000) shows how to calculate confidence limits for CV and Morrow and Jackson (1993) provide details for intraclass correlation.

SUMMARY

We have presented a checklist for exercise physiologists, who are interested in examining the performance characteristics of a particular measurement tool. The most important issues to consider generally are the specific application of measurement tool (research or individual), degree of systematic and random error between methods or repeated tests, and adequate statistical precision of error estimates. All these issues cannot be encapsulated into a single statistic. Therefore, the exercise physiologist should be aware of the several statistics, which are used to measure agreement and error, especially in view of the impact error has on the validity of eventual study conclusions.

NOTE

1   One point worth noting is that several professional bodies (e.g. The British
     Hypertension Society) have their own evidence-based guidelines on acceptable lev-
     els of method agreement and measurement error involving the mean and standard
     deviation of differences statistics (O'Brien, 1998). Unfortunately, such agreed
     standards are rare in exercise physiology but they would be helpful.

REFERENCES

Atkinson, G. (2003). What is this thing called measurement error? In T. Reilly,
    M. Marfell-Jones (eds), *Kinanthropometry VIII: Proceedings of the 8th International
    Conference of the International Society for the Advancement of Kinanthropometry
    (ISAK)* pp. 3–14. London: Taylor and Francis.

Atkinson, G. and Nevill, A.M. (1998). Statistical methods in assessing measurement error
    (reliability) in variables relevant to sports medicine. *Sports Medicine*, 26: 217–238.

Atkinson, G. and Nevill, A.M. (2001). Selected issues in the design and analysis of sport
    performance research. *Journal of Sports Sciences*, 19: 811–827.

Atkinson, G., Davison, R.C.R. and Nevill, A.M. (2005a). Performance characteristics of
    gas analysis systems: what we know and what we need to know. *International
    Journal of Sports Medicine*, 26 (Suppl. 1): S2–S10.

Atkinson, G., Todd, C., Reilly, T. and Waterhouse, J.M. (2005b). Diurnal variation in
    cycling performance: influence of warm-up. *Journal of Sports Sciences*, 23(3):
    321–329.

Bland, J.M. and Altman, D.G. (1999). Measuring agreement in method comparison
    studies. *Statistical Methods in Medical Research*, 8: 135–160.

Bland, J.M. and Altman, D.G. (2003). Applying the right statistics: analyses of
    measurement studies. *Ultrasound and Obstetrics in Gynecology*, 22: 85–93.

Harvill, L.M. (1991). An NCME instructional module on standard error of measurement.
    Educational Measurement: Issues and Practice, 10: 33–41.

Hopkins, W. (2000). Measures of reliability in sports medicine and science. *Sports
    Medicine*, 30: 1–15.

Jones, B. *et al.* (1996). Trials to assess equivalence: the importance of rigorous methods.
    British Medical Journal, 313: 36–39.

Lord, F.M. (1984). Standard errors of measurement at different ability levels. *Journal of
    Educational Measurement*, 21: 239–243.

Ludbrook, J. (1997). Comparing methods of measurement. *Clinical and Experimental
    Pharmacology and Physiology*, 24: 193–203.

Morrow, J.R. and Jackson, A.W. (1993). How 'significant' is your reliability? *Research
    Quarterly for Exercise and Sport*, 64: 352–355.

Nevill, A.M. and Atkinson, G. (1997). Assessing agreement between measurements
    recorded on a ratio scale in sports medicine and sports science. *British Journal of
    Sports Medicine*, 31: 314–318.

Nevill, A.M. and Atkinson, G. (1998). Assessing measurement agreement (repeatability)
    between 3 or more trials. *Journal of Sports Sciences*, 16: 29.

O'Brien, E. (1998). Automated blood pressure measurement: state of the market in 1998
    and the need for an international validation protocol for blood pressure measuring
    devices. *Blood Pressure Monitoring*, 3: 205–211.

CHAPTER 6

SCALING: ADJUSTING PHYSIOLOGICAL AND PERFORMANCE MEASURES FOR DIFFERENCES IN BODY SIZE

Edward M. Winter

INTRODUCTION

It is well established that measures of performance and physiological characteristics are influenced by the size of the body as a whole or of its exercising segments in particular (Schmidt-Nielsen, 1984; Åstrand and Rodahl, 1986). Consequently, if the qualitative properties of tissues are to be explored meaningfully, differences in size have to be partitioned out by adjusting scores. Scaling is the technique that is used to make these adjustments and there has been a revival of interest in this area which impacts on those with interests in the physiology of exercise.

It has been suggested that in sport and exercise physiology there are four main uses of scaling techniques (Winter, 1992):

1   To compare an individual against standards for the purpose of assessment.
2   To compare groups.
3   In longitudinal studies that investigate the effects of growth or training.
4   To explore possible relationships between physiological characteristics and performance.

There is enthusiastic debate about when scaling might be appropriate and in particular, how it should be done. In heavyweight rowing for instance, in which body weight is supported, absolute measures either of performance or physiological characteristics are key and hence, do not require adjustment. Conversely, in activities such as running where body mass is unsupported and has to be carried, some form of scaling might be informative.

However, there is an intuitive attraction to adjust measures so as to develop insight into underlying mechanisms. It is at this point that serious consideration has to be given to possible methods.

RATIO STANDARDS

Traditionally, physiological characteristics such as oxygen uptake ($\dot{V}O_2$) have been scaled simply by dividing them by an anthropometric variable, for instance body mass (BM). This produces a ratio standard and the particular standard $\dot{V}O_2$/BM expressed as ml·kg^{-1}·min^{-1} is probably the most widely used value in the physiology of exercise. However, it was suggested nearly 60 years ago by Tanner (1949) and confirmed by Packard and Boardman (1987) and Winter et al. (1991) that these standards can be misleading. Tanner (1949) stated that the ratio standard should be applied only when a 'special circumstance' has been satisfied.

For an outcome measure y and a predictor variable x, the special circumstance that allows the legitimate use of a ratio standard is given by:

$$v_x/v_y = r$$

where: v_x = coefficient of variation of x, that is (SD$x/\bar{x}$) $\times$ 100
v_y = coefficient of variation of y, that is (SD$y/\bar{y}$) $\times$ 100
r = Pearson's product–moment correlation coefficient.

Rarely is this special circumstance tested and arguably it is even rarer for it to be satisfied. As the disparity between each side of the equation increases, the ratio standard becomes increasingly unstable and distorts measures under consideration.

An effect of the unchallenged use of ratio standards is an apparent favourable economy in submaximal exercise in large individuals compared with those who are diminutive, whereas for maximal responses the opposite occurs. This latter observation has bedevilled researchers in the field of growth and development who see children's endurance performance capabilities increase during adolescence while simultaneously, their aerobic capabilities seemingly deteriorate.

ALLOMETRY

The preferred form of scaling is non-linear allometric modelling (Schmidt-Nielsen, 1984; Nevill et al., 1992). This modelling is based on the relationship:

$$y = ax^b$$

where: y = a performance or physiological outcome measure
x = an anthropometric predictor variable
a = the constant multiplier
b = the exponent.

The terms a and b can be identified by taking natural logarithms (ln) of both the predictor variable and outcome measure and then regressing ln y on ln x (Schmidt-Nielsen, 1984; Winter and Nevill, 2001). Groups can be compared either by analysis of covariance on the log–log regression lines or via power function ratios, that is, y/x^b. These types of ratio are created first, by raising x to the power b to create a power function and then second, by dividing y by this power function. The power function ratio presents y independent of x. As a note of caution, it should be acknowledged that this simple type of regression is not without its problems and Ricker (1973) provides a useful introduction to some of the vagaries of linear modelling.

THE SURFACE LAW

The surface area of a body is related to its volume raised to the power 0.67 and this relationship illustrates what is called the *surface law* (Schmidt-Nielsen, 1984). This means that as a body increases in mass and hence volume, there is a disproportionate reduction in the body's surface area. Conversely, as a body reduces in mass, its surface area becomes relatively greater. This is a fundamental principle which underpins for instance, the action of enzymes during digestion and partly explains differences in thermoregulation in children and adults. Heat exchange with the environment occurs at the surface of a body so thermogenesis and hence energy expenditure must occur to replace heat lost. The precise rate of thermogenesis is dependant on the temperature differences involved. For bodies that are isometric, that is, they increase proportionally, surface area increases as volume raised to the power two-thirds.

It has been suggested (Åstrand and Rodahl, 1986) and demonstrated (Nevill 1995; Welsman *et al.*, 1996; Nevill *et al.*, 2003) that maximal oxygen uptake ($\dot{V}O_{2max}$) and related measures of energy expenditure can be scaled for differences in body mass by means of the surface law; body mass can be raised to the two-thirds power and then divided into absolute values of $\dot{V}O_2$. This produces a power function which describes the aerobic capabilities of a performer with units of ml·kg$^{-0.67}$·min^{-1}. Typical values for elite athletes are presented by Nevill *et al.* (2003). They range from (mean $\pm$ SD) 192 ± 19 ml·kg$^{-0.67}$·min^{-1} for women badminton players to 310 ± 31 ml·kg$^{-0.67}$·min^{-1} for elite standard heavyweight men rowers. When their aerobic capabilities are expressed as ratio standards, the characteristics of the heavyweight men rowers appear considerably more modest yet their event demands high aerobic capability.

ELASTIC SIMILARITY

An alternative approach has been to use the power three quarters. This is based on McMahon's (1973) model of elastic similarity which acknowledges that growth in most living things is not isometric; body segments and limbs grow at different rates and hence, relative proportions change. In addition, buckling loads and other elastic properties for instance of tendons, are not accounted for in a simple surface-law approach. Moreover, in inter-species studies, animals that differ markedly in size seem to be described by a body mass exponent that approximates 0.75.

ALLOMETRIC CASCADE

However, yet another approach has recently been advanced: the allometric cascade model for metabolic rate (Darveau *et al.*, 2002). This model acknowledges two important considerations: first, the non-isometric changes in the body's segments that accompany growth and development and training induced hypertrophy; and second, the tripartite nature of $\dot{V}O_2$ and in particular, $\dot{V}O_{2max}$. The $\dot{V}O_{2max}$ is the global outcome of the rate at which the body can extract oxygen from the atmosphere via the cardiopulmonary system, transport it via the cardiovascular system and use it in skeletal muscle. The ability to release energy is as strong as the weakest part of this three-link chain.

Darveau *et al.* (2002) ascribed a weighting to each of these three facets and predicted an exponent for maximal and submaximal metabolic rate. For the former the exponent was between 0.82 and 0.92. For the latter, equivalent values were 0.76–0.79. Seemingly successful attempts have been made to validate these exponents in exercising humans (Batterham and Jackson, 2003).

RECOMMENDATIONS

In the light of these considerations and the possible confusion they create, how should the results of exercise tests be expressed? To report the results of laboratory and field-based tests which meaningfully reflect the performance and physiological status of athletes and exercisers, investigators should:

- Report absolute values of performance measures and physiological characteristics.
- Report ratio standards only when Tanner's special circumstance has been satisfied.
- For expediency, use the surface law exponent of 0.67 to scale $\dot{V}O_2$ or other related assessments of energy expenditure for differences in body mass or the size of exercising segments.
- Verify the choice of a particular exponent but acknowledge that because of sampling errors comparisons between groups might be compromised.
- For $\dot{V}O_2$ and $\dot{V}O_{2max}$ consider applying the allometric cascade model.

REFERENCES

Åstrand, P.-O. and Rodahl, K. (1986). *Textbook of Work Physiology*, 3rd edn. New York: McGraw-Hill.

Batterham, A.M. and Jackson, A.S. (2003). Validity of the allometric cascade model at submaximal and maximal metabolic rates in men. *Respiratory Physiology and Neurobiology*, 135: 103–106.

Darveau, C.-A., Suarez, R.K., Andrews, R.D. and Hochachka, P.W. (2002). Allometric cascade as a unifying principle of body mass effects on metabolism. *Nature*, 417: 166–170.

McMahon, T. (1973). Size and shape in biology. *Science*, 179: 1201–1204.

Nevill, A.M. (1995). The need to scale for differences in body size and mass: an explanation of Kleiber's 0.75 mass exponent. *Journal of Applied Physiology*, 77: 2870–2873.

Nevill, A.M., Ramsbottom, R. and Williams, C. (1992). Scaling physiological measurements for individuals of different body size. *European Journal of Applied Physiology*, 65: 110–117.

Nevill, A.M., Brown, D., Godfrey, R., Johnson, P.J., Romer, L., Stewart, A.D. and Winter, E.M. (2003). Modelling maximum oxygen uptake of elite endurance athletes. *Medicine and Science in Sports and Exercise*, 35: 488–494.

Packard, G.C. and Boardman, T.J. (1987). The misuse of ratios to scale physiological data that vary allometrically with body size. In M.E. Feder, A.F. Bennett, W.W. Burggren and R.B. Huey (eds), *New Directions in Ecological Physiology*. pp. 216–236. Cambridge: Cambridge University Press.

Ricker, W.E. (1973). Linear regressions in fishery research. *Journal of Fisheries Research Board*, Canada, 30: 409–434.

Schmidt-Nielsen, K. (1984). *Scaling: Why is Animal Size so Important?* Cambridge: Cambridge University Press.

Tanner, J.M. (1949). Fallacy of per-weight and per-surface area standards and their relation to spurious correlation. *Journal of Applied Physiology*, 2: 1–15.

Welsman, J., Armstrong, N., Nevill, A., Winter, E. and Kirby, B. (1996). Scaling peak O$_2$ for differences in body size. *Medicine and Science in Sports and Exercise*, 28: 259–265.

Winter, E.M. (1992). Scaling: partitioning out differences in size. *Pediatric Exercise Science*, 4: 296–301.

Winter, E.M. and Nevill, A.M. (2001). Scaling: adjusting for differences in body size. In: R. Eston and T. Reilly (eds), *Kinanthropometry and Exercise Physiology Laboratory Manual: Tests, Procedures and Data*, 2nd edn. *Volume 1: Anthropometry*, pp. 321–335. London: Routledge.

Winter, E.M., Brookes, F.B.C. and Hamley, E.J. (1991). Maximal exercise performance and lean leg volume in men and women. *Journal of Sports Sciences*, 9: 3–13.

CHAPTER 7

CIRCADIAN RHYTHMS

Thomas Reilly

CHRONOBIOLOGICAL BACKGROUND

Chronobiology is the science of biological rhythms. Circadian rhythms refer to cyclical fluctuations that recur regularly each solar day. The term is based on the Latin words circa (about) and dies (a day), reflecting that the endogenous rhythm (determined in constant conditions in an isolation unit) exceeds 24 h but is fine-tuned to a 24-h period by exogenous factors. These include light, temperature, habitual activity and social influences.

The circadian rhythm can be stylised by cosinor analysis. The period is predetermined as 24 h, the acrophase refers to the time when the peak occurs and the amplitude is half the distance between the highest and lowest values on the cosine curve. The trough occurs 12 h after the acrophase and after 24 h the next cycle commences. The hourly changes in core body temperature provide an example of a typical cosine function, with an acrophase around 17.50 h. Diurnal variation refers to changes within the normal daylight hours and nychthemeral conditions apply during normal habitual experiences.

The endogenous component of circadian rhythms, that is the body clock, is located in the suprachiasmatic nuclei within the hypothalamus. These nerve cells have receptors for melatonin, the hormone secreted from the pineal gland. This substance has circadian timekeeping functions due to its direct effects on the suprachiasmatic nuclei. These cells have a direct neural pathway from the retina and another input pathway through the intergeniculate leaflet. The visual receptors that enable light signals to synchronise the body clock and the environment act to assess the time of dawn and dusk according to several aspects of the quality and quantity of light.

Melatonin is secreted as darkness falls and is inhibited by light. The hormone has vasodilatory properties, causing body temperature to fall in the evening. Metabolic and other physiological functions slow down as the body prepares itself for sleep. The circadian rhythm in synthesis and release of serotonin, a substrate for melatonin and a brain neurotransmitter, is implicated in the sleep–wakefulness cycle. Whilst body temperature is regarded as a fundamental variable with which many human performance measures co-vary, the sleep–activity cycle reflects the circadian rhythm in the body's arousal system. Cells with timekeeping roles have also been located in peripheral tissues. The overall result is that the environmental light–dark cycle, the human activity–sleep cycle and the circadian system are integrated with respect to equipping the body to operate best over each day.

RHYTHMS IN PERFORMANCE

Field tests

All-out efforts such as time trials in cycling, swimming and rowing demonstrate circadian rhythms closely in phase with changes in body temperature. The evidence points to an endogenous component that combines with exogenous factors to influence the outcome (Drust et al., 2005). When such tests are conducted in applied settings, time-of-day effects should be considered.

Muscle strength

The maximal capability of muscle to exert force may be measured under isometric and dynamic conditions. Traditionally isometric measures were used, the maximal voluntary contraction being recorded at a specific joint angle for purposes of replication and comparison with others. Portable dynamometers have been used for assessment of grip, back and leg strength in field conditions. Circadian rhythms have been identified for grip strength, elbow flexion, knee extension and back extension (Table 7.1). The peak time usually coincides with the acrophase in body temperature, the amplitude is 5–10% of the mean value. Observations of the quadriceps muscle after electrical stimulation suggest that peripheral more than central mechanisms are implicated in the rhythm in isometric force.

   The measurement error in maximal voluntary contraction due to time of day may be corrected, provided the cosine function of the muscle group in question is known. The corrected value (MVCcorr) can be estimated by the equation:-

$$MVCcorr = \frac{MVCt}{1 + A \times \cos°(15t + 15p)}$$

where t is the time of day in decimal clock hours at which the test is performed, A is the amplitude of MVC as a per cent of the mean divided by 100, and p is the acrophase. Whilst the correction was originally designed for clinical assessments (Taylor et al., 1994), the equation could be used for other strength tests that display time-of-day effects.

   It is now more common to measure dynamic muscle strength in laboratory assessments rather than isometric force. Peak torque is measured under concentric and eccentric modes of muscle action and at different angular velocities using isokinetic dynamometry. Comprehensive familiarisation of subjects is required and test–retest variation may be high initially at fast angular velocities. Circadian rhythms have been reported for concentric peak torque of the knee extensors, the amplitudes and peak times being close to those reported for isometric force (see Table 7.1).

Anaerobic performance

Measures of anaerobic performance range from single, so-called explosive actions to formal measurement of maximal power output and its decline as

Table 7.1 Circadian variation in muscle strength and power from various sources. Only those publications where at least six measures have been recorded to characterise the rhythms have been cited

Muscle performance	Peak time (decimal clock hours)	Amplitude (% mean value)	Reference
Isometric strength			
Grip strength			
Left	18.00	6.4	Atkinson et al., 1993
Left	17.80	6.5	Atkinson et al., 1994
Right	17.90	4.7	Atkinson et al., 1994
Leg strength	18.20	9.0	Coldwells et al., 1994
	18.25	7.6	Atkinson et al., 1994
(90° extension)	17.80	7.1	Taylor et al., 1994
Back strength	16.88	10.6	Coldwells et al., 1994
	18.30	6.9	Atkinson et al., 1994
Dynamic strength			
(Concentric mode)			
Knee extensors			
1.05 rad·s^{-1}	15.47	3.7	Bambaeichi et al., 2004
1.05 rad·s^{-1}	18.64	6.2	Atkinson et al., 1995
1.57 rad·s^{-1}	18.00	4.6	Atkinson and Reilly, 1996
3.14 rad·s^{-1}	17.86	8.2	Atkinson et al., 1995
Knee flexors			
3.14 rad·s^{-1}	19.76	7.2	Atkinson et al., 1995
Anaerobic power			
Broad jump	17.75	3.4	Reilly and Down, 1986
Stair run	17.26	2.1	Reilly and Down, 1992
Flight time	20.30	2.4	Atkinson et al., 1994

exercise is sustained. The Wingate test entails exercise on a cycle ergometer for 30 s, allowing peak anaerobic power, anaerobic capacity and a 'fatigue index' to be recorded. Peak power and mean power over the 30 s have been reported to be 8% higher in the evening (15.00 and 21.00 h) compared with night-time (03.00 h) (Hill and Smith, 1991). A higher circadian amplitude in peak and mean power output was found when the test was adapted for use on a swim bench (Reilly and Marshall, 1991). The large amplitude was attributed to the complex simulated swimming action compared to the grosser movement engaged in arm cranking. These rhythms are evident after prior activity so a systematic warm-up does not eliminate the circadian effect on anaerobic performance.

Power production can also be monitored in a stair-run and in jump tests (Atkinson and Reilly, 1996). The circadian rhythm in the standard stair-run test peaked at 17.26 h, the amplitude being 2.1% of the 24-h mean (see Table 7.1). Similar findings apply to standing broad jump (amplitude 3.4%) and flight time in a vertical jump (2.4%). Bernard et al. (1998) showed that flight time and jump power ($W \cdot kg^{-1}$) were greater in the afternoon and evening (14.00 and 18.00 h) than in the morning (09.00 h), the difference between means amounting to 7.0% and 2.6%, respectively. Such variations can have pronounced effects on global performance in training or competition, highlighting the need to reduce measurement error to a minimum when anaerobic performance is assessed.

PHYSIOLOGICAL RESPONSES

Rest

Circadian rhythms are evident in a range of endocrine, respiratory, digestive and renal functions. There is close correspondence between the circadian rhythm in core temperature and that in oxygen consumption ($\dot{V}O_2$) and minute ventilation ($\dot{V}E$), the change in temperature accounting for 37% and 24% of the variation in these metabolic measures, respectively (Reilly and Brooks, 1982). The amplitude of the rhythm in $\dot{V}E$ is greater than that of $\dot{V}O_2$; over and above the reduced requirement for oxygen at night-time, bronchoconstriction decreases the flow of air through the respiratory passages. Resting values are recorded over 10 min in order to reduce measurement error and, if a pre-exercise resting value is needed, it is acceptable to have the subject on the ergometer to be used, for example sitting motionless and comfortable on a cycle or rowing ergometer.

Heart rate at rest tends to be recorded in assessments of athletes, notably in endurance specialists whose training regimens lead to low resting values. The rhythm in heart rate tends to occur earlier in the afternoon than does that of $\dot{V}O_2$ or $\dot{V}CO_2$, this phase lead being attributed in part to changes in catecholamines whose peaks occur around 13.00 h. Adrenaline and noradrenaline have been linked with diurnal variations in alertness rather than enslaved to the rhythm in body temperature, although some dependence is likely (see Reilly *et al.*, 1997).

Submaximal exercise

In the main, the circadian rhythms evident at rest persist during light and moderate exercise. The rhythm in $\dot{V}O_2$ parallels that in $\dot{V}CO_2$, indicating stability in the respiratory exchange ratio. A standard light snack is recommended, at least 3 h prior to testing, to avoid circadian influences in substrate utilisation. It seems also that the energy cost of locomotion and the net mechanical efficiency are constant with time of day. When running economy is employed, the resting $\dot{V}O_2$ value should be subtracted, otherwise 'economy' would appear to be improved at night-time.

The rhythms in $\dot{V}O_2$ and $\dot{V}CO_2$ tend to fade as exercise is intensified. In contrast the rhythm in $\dot{V}E$ is accentuated and is reflected in a circadian rhythm in the ventilation equivalent of oxygen (Reilly and Brooks, 1990). The rhythm in $\dot{V}E$ may partly explain the mild dyspnoea sometimes associated with exercising in the early morning and the elevated perceived exertion noted at this time. The rhythm in heart rate persists for both arm and leg exercise, but decreases as exercise approaches maximal effort. Psychophysical methods also display circadian rhythmicity, expressed in the self-chosen work-rate. This value determines the pace individuals set for sustaining continuous exercise.

The 'anaerobic threshold' is used as a submaximal index of aerobic capacity. Forsyth and Reilly (2004) used the Dmax method to indicate 'lactate threshold' in rowers and reported a circadian rhythm for $\dot{V}O_2$ and heart rate at the threshold; the higher values for both variables were in phase with the rectal temperature data. When lactate threshold is used as a marker of performance change, tests should be conducted at the same time of day to eliminate circadian influences.

Maximal responses

The amplitude of the resting rhythm in $\dot{V}O_2$ would represent $<0.3\%$ of the $\dot{V}O_{2max}$ in a typical endurance athlete; a variation of this magnitude at maximal exercise is hard to detect. When subjects exercise to voluntary exhaustion, the highest $\dot{V}O_2$ value is referred to as peak rather than maximal if standard physiological criteria are not fulfilled. Arm exercise does not generally yield a plateau in $\dot{V}O_2$ before subjects desist in an incremental test to voluntary exhaustion, so a circadian rhythm reflects the influence of the total work done rather than innate physiological capacity. When subjects failing to demonstrate a plateau in $\dot{V}O_2$ during leg exercise to exhaustion in an incremental test were recalled for repeat testing, $\dot{V}O_{2max}$ was found to be stable (Reilly and Brooks, 1990).

The rhythm in submaximal heart rate is evident at exhaustion, albeit reduced in amplitude. The lowered values at night-time may be attributed to a decreased sympathetic drive. The variation is insufficient to affect cardiac output which, like $\dot{V}O_{2max}$, is a stable function. Whilst field performance tests display circadian variation, the effect cannot be explained by fluctuations in the transport or delivery of oxygen to the active muscles.

OVERVIEW

The evidence that circadian rhythms influence many physical fitness and performance measures is comprehensive. Therefore, serial tests on an individual athlete should be conducted at the same time of day for results to be compared.

The influence of individual differences on human circadian rhythms seems to be small. Lifestyle factors, such as morning or evening types, have no major effects on rhythm characteristics, nor has personality type. The phasing of the rhythm is relatively advanced with ageing, shifting towards a more morning-type profile. Fitness does not affect the acrophase of circadian rhythms but may increase their amplitude by means of a lowered trough. The rhythm is influenced by menstrual cycle phase, the decreased amplitude in muscle performance during the luteal compared to the follicular phase being linked to fluctuations in reproductive steroid hormones (Bambaeichi *et al.*, 2004).

Sports scientists must consider the time of day when planning and conducting fitness tests. This recommendation applies to both laboratory and field measures. Such care should be part of an overall preparation for administering test protocols that commence with familiarising the individual with the test procedures. Reduction in measurement error is paramount if changes between tests are to be identified and interpreted properly. This attention to detail is an essential part of quality control.

REFERENCES

Atkinson, G. and Reilly, T. (1996). Circadian variation in sports performance. *Sports Medicine*, 21: 292–312.

Atkinson, G., Coldwells, A. and Reilly, T. (1993). A comparison of circadian rhythms in work performance between physically active and inactive subjects. *Ergonomics*, 36: 273–281.

Atkinson, G., Coldwells, A., Reilly, T. and Waterhouse, J. (1994). An age-comparison of circadian rhythms in physical performance measures. In S. Harris, H. Suominen, P. Era and W.S. Harris (eds), *Towards Healthy Aging: International Perspectives Part 1. Physical and Biomedical Aspects Volume 3, Physical Activity, Aging and Sports*, pp. 205–216, Albany, NY: Center for Study of Aging.

Atkinson, G., Greeves, J., Reilly, T. and Cable, N.T. (1995). Day-to-day and circadian variability of leg strength measured with the LIDO isokinetic dynamometer. *Journal of Sports Sciences*, 13: 18–19.

Bambaeichi, E., Reilly, T., Cable, N.T. and Giacomoni, M. (2004). The isolated and combined effects of menstrual phase and time-of-day on muscle strength of eumenorrheic women. *Chronobiology International*, 21: 645–660.

Bernard, T., Giacomoni, M., Gavarry, O., Seymat, M. and Falgairette, G. (1998). Time-of-day effects in maximal anaerobic leg exercise. *European Journal of Applied Physiology*, 77: 133–138.

Coldwells, A., Atkinson, G. and Reilly, T. (1994). Sources of variation in back and leg dynamometry. *Ergonomics*, 37: 79–86.

Drust, B., Waterhouse, J., Atkinson, G., Edwards, B. and Reilly, T. (2005). Circadian rhythms in sports performance: an update. *Chronobiology International*, 22: 21–44.

Forsyth, J.J. and Reilly, T. (2004). Circadian rhythms in blood lactate concentration during incremental ergometer rowing. *European Journal of Applied Physiology*, 92: 69–74.

Hill, D.W. and Smith, J.C. (1991). Circadian rhythms in anaerobic power and capacity. *Canadian Journal of Sports Science*, 16: 30–32.

Reilly, T. and Brooks, G.A. (1982). Investigation of circadian rhythms in metabolic responses to exercise. *Ergonomics*, 25: 1093–1107.

Reilly, T. and Brooks, G.A. (1990). Selective persistence of circadian rhythms in physiological responses to exercise. *Chronobiology International*, 7: 59–67.

Reilly, T. and Down, A. (1986). Circadian variation in the standing broad jump. *Perceptual and Motor Skills*, 62: 830.

Reilly, T. and Down, A. (1992). Investigation of circadian rhythms in anaerobic power and capacity of the legs. *Journal of Sports Medicine and Physical Fitness*, 32: 342–347.

Reilly, T. and Marshall, S. (1991). Circadian rhythms in power output on a swim bench. *Journal of Swimming Research*, 7: 11–13.

Reilly, T., Atkinson, G. and Waterhouse, J. (1997). *Biological Rhythms and Exercise*. Oxford: Oxford University Press.

Taylor, D., Gibson, H., Edwards, R.H.T. and Reilly, T. (1994). Correction of isometric strength tests for time of day. *European Journal of Experimental Musculoskeletal Research*, 3: 25–27.

PART 3

GENERAL PROCEDURES

CHAPTER 8

LUNG AND RESPIRATORY MUSCLE FUNCTION

Alison McConnell

INTRODUCTION

The following section will describe briefly the structure and function of the healthy respiratory system, as well as considering why the assessment of lung and respiratory muscle function is relevant to sport and exercise science. The final section will describe the equipment and procedures for undertaking basic lung function and respiratory muscle assessments.

PHYSIOLOGY OF BREATHING

A detailed description of the physiology of the respiratory system is beyond the scope of this section, and the reader is referred to West (1999) for this information. However, in order to place the assessment of the respiratory system into context, it is necessary to provide a very brief overview of the act of breathing.

The principal function of the respiratory system is the exchange of the respiratory gases, oxygen and carbon dioxide. The movement of air into and out of the lungs is brought about by the contraction of skeletal muscles, which are activated by both automatic and conscious control mechanisms. The structure of the lungs provides for a huge interface between air and capillary blood; it has been estimated that the combined alveolar surface area of both adult lungs is equivalent to that of half a tennis court. Each alveolus is surrounded by a dense network of capillaries. The large surface area of the gas/blood interface, combined with the high affinity of haemoglobin for oxygen, and the sigmoid shape of its dissociation curve, ensure the complete equilibration of the respiratory gases across the respiratory membrane. Accordingly, arterial oxygen saturation remains around 97%, even during heavy exercise (see later for exceptions), and oxygen transport in healthy human beings at sea level is not generally considered to be limited by the diffusing capacity of their lungs.

The precise mechanisms that control the level of breathing (minute ventilation, $\dot{V}_E$) in response to changing metabolic demand remain relatively poorly understood. However, it is known that the control is more closely linked to carbon dioxide production than to oxygen uptake (Wasserman et al., 1978). As well as ensuring the maintenance of oxygen delivery during exercise, the respiratory system plays a crucial role in acid–base homeostasis. Stimulation of the carotid chemoreceptors by hydrogen ions drives up $\dot{V}_E$, and facilitates the removal of carbon dioxide (in excess of metabolic demand), which increases pH (Wasserman et al., 1975). The ventilatory compensation for a metabolic acidosis ensures that exercise can be sustained above the lactate threshold for much longer than would otherwise be the case.

WHY ASSESS LUNG AND RESPIRATORY MUSCLE FUNCTION?

Minute ventilation displays a more than 10-fold increase between rest and peak exercise, with typical resting values of 8–10 $l \cdot min^{-1}$ and values approaching 150–200 $l \cdot min^{-1}$ during maximal exercise. The highest values for $\dot{V}_E$ are recorded in athletes such as rowers, where it is not uncommon for $\dot{V}_E$ to reach 250 $l \cdot min^{-1}$ at peak exercise in elite, open-class oarsmen.

The relevance of lung function to elite endurance performance remains a topic of debate, since it is well known that there is no ventilatory (diffusion) limitation to performance in healthy human beings at sea level. The exceptions to this received wisdom are elite endurance trained individuals; 40–50% of this group show arterial oxygen desaturation at peak exercise, which is indicative of a diffusion limitation to oxygen transport (Powers et al., 1993). However, the aetiology is multifactoral, and the phenomenon is not explained totally by mechanical constraints upon breathing.

Notwithstanding these observations of diffusion limitation in endurance athletes, the apparent excess capacity of the ventilatory system has led to the assumption that there is no ventilatory limitation to exercise performance. However, it is a common observation that endurance athletes tend to have large lung volumes, even when body size is taken into account. Other evidence from untrained individuals also points to a relationship between lung function and maximal oxygen uptake that cannot be explained by body size (Nevill and Holder, 1999). The reasons for these observations are currently unknown.

Other evidence also points to a potential ventilatory limitation to exercise performance. Breathing is brought about by the action of muscles, which can demand as much as 16% of oxygen uptake during maximal exercise (Harms, 2000). The inspiratory muscles (which undertake the majority of the mechanical work of breathing) have been shown in numerous studies to exhibit fatigue after both short, high intensity bouts of exercise (Johnson et al., 1993; Babcock et al., 1996; McConnell et al., 1997; Volianitis et al., 2001a; Romer et al., 2002a, 2004; Lomax and McConnell, 2003), and prolonged moderate intensity exercise such as marathon running (Loke *et al.*, 1982; Hill *et al.*, 1991). This is suggestive of a system that is working at the limits of its capacity. The fact that pre-fatigue of the respiratory muscles impairs performance (Mador and Acevedo, 1991), and that specific inspiratory muscle training improves performance (Volianitis *et al.*, 2001a; Romer *et al.*, 2002a,b) adds further weight to the argument that the ventilatory system exerts a limitation to exercise performance.

Whilst it is debatable whether superior lung function is associated with superior endurance performance, it is well recognised that impaired lung function has a detrimental influence upon exercise performance (Aliverti and Macklem, 2001). Although impairment of lung function may not necessarily result in a compromise to gas exchange, studies on people with lung disease demonstrate that the breathlessness associated with lung function impairment becomes an exercise-limiting factor (Hamilton *et al.*, 1996). Similarly, high levels of respiratory muscle work and inspiratory muscle fatigue have been implicated in impairment of exercise performance due to blood flow 'stealing' by the respiratory muscles (Harms, 2000).

Accordingly, the routine assessment of lung function and respiratory muscle function in athletes is worthwhile, and essential in any athlete who reports inappropriate levels of breathlessness during training or competition. The source of inappropriate breathlessness is most likely to be exercise-induced asthma. Data from the GB team that competed in the Athens Olympics indicated that 21% of the squad had exercise-induced asthma that qualified for treatment under International Olympic Committee criteria (Dickinson *et al.*, 2005 (in press). Prevalence rates were highest in the sports of swimming and cycling (over 40%). The prevalence rate in Team GB as a whole was more than twice that in the UK general population (8%).

ROUTINE ASSESSMENT AND INTERPRETATION OF LUNG FUNCTION

The guidance below is based upon a variety of sources, but principally the recommendations of the American Thoracic Society (American Thoracic Society, 1995) and European Respiratory Society (Quanjer *et al.*, 1993), as well as extensive practical experience.

A 'classic', global test of breathing capacity is the maximum voluntary ventilation (MVV) test. The capacity to move air in and out of the lungs is influenced by the participants' physical size, age, gender and race. All other things being equal (e.g. age, gender, etc.), the outcome of an MVV test is also influenced by the condition of the respiratory muscles (weakness and susceptibility to fatigue), narrowing of the airways (e.g. asthma), loss of lung elastic recoil (e.g. emphysema), as well as the distensability of the lungs and thoracic cage (e.g. scoliosis). The MVV is therefore a somewhat 'blunt instrument' that should lead on to more specific tests in the presence of a relatively poor performance.

The MVV requires the participant to breathe in and out as hard as possible for a predetermined time, usually 15 s (MVV_{15}). The test is most easily performed using an electronic spirometer that measures flow rate directly, and most proprietary spirometry systems have a function that permits MVV testing. It is important that the equipment has a low resistance to airflow, as a back pressure will impair the validity of the measurements. The test can also be performed for longer durations (e.g. 4 min, discussed later), but then requires supplemental carbon dioxide to prevent severe hypocapnia. The participant requires strong encouragement throughout the test, which shows a task learning effect. During serial assessments of MVV_{15} (repeated to obtain a reliable value (two values should be within 10% or 20 $l·min^{-1}$)), at least 3 min should be allowed between tests. Because the manoeuvre results is some hypocapnia it is also helpful to instruct the participant to hold their breath at the end of the test in order to allow normocapnia to be restored more rapidly.

The 4-min MVV ($MVV_{4\,min}$, also known as the maximum sustained ventilation) gives an index of the fatigue resistance of the respiratory muscles, since the progressive decline in flow rate is due to muscle fatigue. There is also a task learning effect in the assessment of $MVV_{4\,min}$, which should not be repeated for at least an hour; visual feedback of a target $\dot{V}_E$ is also helpful. Most healthy untrained people can sustain 60–70% of their MVV_{15} for 4 min, and trained individuals over 80% (Anholm et al., 1989), that is, $MVV_{4\,min}$ is 60–80% of MVV_{15}. At peak exercise, healthy people achieve a $\dot{V}_E$ of 70–80% of their MVV_{15} (Hesser et al., 1981); thus, $MVV_{4\,min}$ and peak exercise $\dot{V}_E$ are broadly equivalent.

STATIC LUNG FUNCTION

After the MVV, the most basic assessment of lung function involves the measurement of lung volumes (static lung volumes), which are measured in litres and expressed under BTPS conditions. Figure 8.1 illustrates the static lung volumes, definitions of which are provided below:

- *Total lung capacity (TLC).* The volume of air in the lungs at full inspiration. This cannot be measured without access to specialised equipment.
- *Vital capacity (VC).* The maximum volume that can be exhaled/inhaled between the lungs being completely inflated and the end of a full expiration. VC can be measured during either a 'forced' (with maximal effort; FVC) or relaxed manoeuvre (VC). The relaxed manoeuvre is more appropriate for patients with lung disease whose airways tend to collapse during a forced manoeuvre.
- *Residual volume (RV).* The volume of air remaining in the lungs at the end of a full expiration. This cannot be measured without access to specialised equipment.
- *Functional residual capacity (FRC).* The volume of air remaining in the lungs after a resting tidal breath. This changes during exercise, when it becomes known as end expiratory lung volume (EELV).
- *Expiratory and inspiratory reserve volumes (ERV/IRV).* The volumes available between the beginning or end of tidal breath and TLC and RV, respectively.

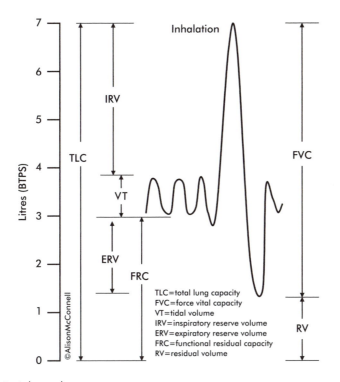

Figure 8.1 Static lung volumes

Lung function is influenced by a number of physiological and demographic factors, as well as by the presence of disease. For example, there is a strong influence of body size, gender and age, as well as ethnicity. For this reason, there are population-specific prediction equations that assist in the interpretation of measured values (see Lung Function Reference Values in the Reference section). Generally, lung volumes are greater in larger individuals, are lower in women, and decrease with age. A component of the effect of gender appears to be independent of the effect of body size (Becklake, 1986).

A description of the breathing manoeuvres required to assess lung volumes is given in the next section.

DYNAMIC LUNG FUNCTION

The condition of the airways (as distinct from the measurement of lung volumes) can be assessed using a technique known as spirometry (dynamic lung function). Obstructive lung diseases such as asthma are diagnosed by measuring the rate of expiratory airflow during forced expiratory manoeuvres. By plotting either volume against time (Figure 8.2 (A) or flow against volume (by integration of the flow signal) (Figure 8.2 (B)), a 'spirogram' is constructed. Figure 8.2 (A) and (B) illustrate each of these approaches and identifies a number of parameters that provide information about airway function (see legend for details). The most commonly used index of airway calibre is the forced expiratory volume in 1 s (FEV_1), which can be assessed using either a bellows (also known as wedge) spirometer, or using electronic spirometry. Electronic spirometers allow the construction of so-called flow volume loops (Figure 8.2 (B)).

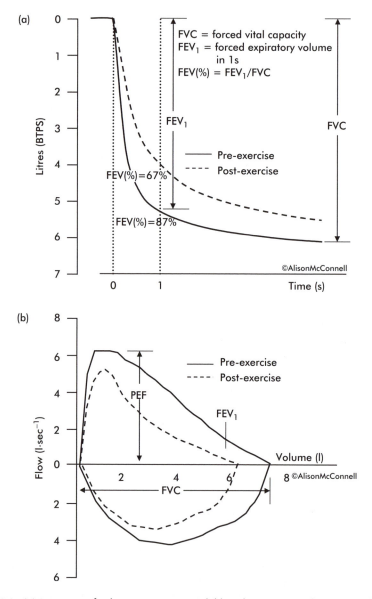

Figure 8.2 (a) Spirogram of volume against time. Solid line depicts a normal tracing. Dashed line depicts the response of an individual with EIA after an exercise challenge (bronchoconstriction). Note the decline in the ratio of FEV_1 to FVC in the presence of bronchoconstriction. (b) Spirogram of volume against flow. Line coding as above

A method that many asthma patients use to self-monitor their airway function is peak expiratory flow (PEF) measurement. Whilst this is easily assessed, using very inexpensive equipment (e.g. Mini Wright Peak Flow Meter), it is highly effort dependent and its reliability is poor. Accordingly, PEF is satisfactory for patient self-monitoring, but not for diagnostic testing (Quanjer *et al.*, 1997).

Because FEV_1 is influenced by vital capacity, it is expressed as a fraction of vital capacity (FEV%). In the presence of normal airways, FEV% should exceed 80% for individuals under 30 years, and 75% up to late middle age.

Conducting a dynamic lung function test

The description here is for conducting a forced flow volume loop, but the basic principles are the same for static lung volume assessment and FEV_1 measured using a bellows spirometer.

1   Ensure that your equipment is calibrated and working properly (e.g. check for leaks in hoses).
2   Ensure that all equipment that will come in contact with the participant (e.g. mouthpiece), or that he/she will inhale through (tubing), is sterile, and/or protected by a disposable viral filter.
3   Complete any necessary consent documentation.
4   Measure the particpant's stature.
5   Explain to the participant exactly what you wish them to do before starting the test.
6   Measurements can be made seated or standing, but ensure that no clothing restricts the thorax, and that the neck is slightly extended.
7   For a manouvre, explain that the manoeuvre must be performed 'forcefully'. Then fit the nose clip, instruct the participant to go onto the mouthpiece, and to inhale 'until [your] lungs are as full as they can possibly be'.
8   When it is clear that they have achieved this (don't hesitate at this point), instruct them to 'blow out as hard and fast as [you] can...keep going, out, out, squeeze out'. Encourage them to keep going until they cannot squeeze any more air from their lungs. During this latter phase, encourage them to keep breathing 'out, out, squeeze out'.
9   As soon as it is clear that their lungs are empty instruct them to 'breathe in as fast as [you] can...big deep breath, keep going'. You can describe this manoeuvre as being like a huge gasp. If you are measuring FEV_1 using a bellows spirometer, there is no requirement to undertake the inspiratory part of the manoeuvre (stop at the end of 8).
10 Take the mouthpiece out of the participant's mouth (having a piece of tissue ready for saliva). Leave the nose clip in place and allow around 15–30 s rest before repeating the procedure.
11 If there were deficiencies in the quality of the manoeuvre, for example, he/she didn't exhale fully, explain what went wrong and what can be done to improve things on the next attempt.

12   Once you have three measurements of FVC and FEV_1 that were technically satisfactory and are within 5% (or 100 ml) of each other you can stop (it may take as many as eight attempts to achieve this).
13   Remove the mouthpiece (having a piece of tissue ready for saliva) and nose clip.
14   Report the largest of the three technically satisfactory measurements (best of three).
15   If eight manoeuvres are performed without achieving the 5% criterion, then record the highest value measured.

Common faults

- Incomplete inspiration or expiration;
- initiation of the manoeuvre before the participant is attached to the mouthpiece;
- leakage of air around the lip/mouthpiece interface;
- coughing.

Most modern electronic spirometers will store the data within a built-in database, and will also calculate 'percent predicted values' by referencing the participant's measured values to appropriate prediction equations (see Lung Function Reference Values).

It is relatively rare to encounter an otherwise healthy physically active person who has clear evidence of abnormal lung function in a rested state. However, it is possible to have normal lung function in a rested state, and to have an airway responsiveness to exercise, or exercise-induced asthma (EIA). People with EIA generally complain of breathlessness following exercise, or during repeated bouts of exercise that are interspersed with rest. If EIA is suspected, then lung function should be assessed post-exercise, preferably after an activity that typically provokes symptoms. Measurement should be repeated 3, 5, 10, 15, 20 and 30 min after exercise has ceased (this is to take account of the individual variation in the time of the nadir of the response). A fall in FEV_1 of >10% is indicative of EIA. Specialist EIA testing facilities are available at the Olympic Medical Institute and some English Institute of Sport Centres. If lung function assessment identifies an individual with apparently abnormal lung function, they should be referred to their GP with a copy of their test results for confirmation and treatment, as appropriate.

One might imagine that it is very rare to identify an athlete with EIA, who was unaware that they had the condition. However, routine EIA screening of Team GB prior to the Athens Olympics identified seven athletes with no previous diagnosis of EIA (Dickinson *et al.*, 2005 (in press)). This represented ~10% of the symptomatic athletes who were referred for testing, and 2.6% of the entire squad. Measurement of lung function is therefore recommended as a routine part of athlete profiling (Dickinson *et al.*, 2005 (in press)).

ASSESSMENT AND INTERPRETATION OF RESPIRATORY MUSCLE FUNCTION

Unlike lung function, which can be predicted with reasonable accuracy on the basis of stature, gender and age, respiratory muscle function shows much less predictability. This makes interpretation of one-off measurements largely meaningless, unless there is gross weakness. However, the observation that some forms of exercise are associated with inspiratory muscle fatigue (IMF), and that specific inspiratory muscle training (IMT) abolishes IMF and improves performance, has led to an interest in this method of assessment. The role of the expiratory muscles in exercise limitation currently remains unknown.

Thus, assessment of respiratory muscle function is recommended for diagnosis of exercise induced IMF, as well as for monitoring the influence of IMT. In 2002, the American Thoracic Society and European Respiratory Society published an extensive set of guidelines for respiratory muscle assessment (ATS/ERS, 2002). Readers wishing to know more about respiratory muscle assessment in general are referred to this source. Below is a practical guide to assessing just one aspect of muscle function, namely, respiratory muscle strength.

For obvious reasons, it is not possible to obtain a direct measurement of respiratory muscle force output. Accordingly, surrogate measurements are used to provide an index of global respiratory muscle strength. The maximal respiratory pressures measured at the mouth provide simple indices that are very reliable when performed by competent 'technicians' (Romer and McConnell, 2004). Respiratory pressures are measured against an occluded airway (incorporating a 1 mm diameter leak to maintain an open glottis) at prescribed lung volumes (discussed later). The equipment required is portable and hand-held (mouth pressure meter).

Because maximum respiratory pressures are indices of maximal strength, they are highly effort-dependent and require well-motivated participants. Efforts must be sustained for at least 1.5 s in order that an average pressure over 1 s can be calculated (by the measuring instrument). This averaging enhances the reliability of the measurement. The measurements must also be made at predetermined lung volumes. This is because of the length–tension relationship of the respiratory muscles. Maximal inspiratory pressure (MIP) is measured at residual volume and maximal expiratory pressure (MEP) at total lung capacity.

Care must also be taken to ensure that any task learning and other effects are expressed fully before recording measured values. It has been shown that there is a considerable effect of repeated measurement upon MIP, even in experienced participants. This effect is large enough to mask changes in MIP due to the effects of inspiratory muscle fatigue. After 18 repeated trials, MIP was 11.4% higher than the best of the first three measurements made (Volianitis *et al.*, 2001b). However, this learning effect can be overcome to a large extent by a bout of sub-maximal inspiratory loading prior to the assessment of MIP (two sets of 30 breaths against an inspiratory threshold load equivalent to

40% of the best MIP measured during the first three efforts). Following this prior loading, the difference between the best of the first three efforts and the 18th measurement was only 3%. Thus, the time taken to obtain reliable measurements of MIP can be curtailed considerably by implementing a bout of prior loading.

For reasons given earlier, it is difficult to offer typical values for MIP and MEP, and the reader is cautioned against using reference values within the literature, because their predictive power and functional relevance is questionable. Notwithstanding this, it is possible to offer some 'ball park' estimates of values that should be expected in healthy young people; males range MIP = 110–140 cmH_2O; females range MIP = 90–120 cmH_2O. Values for MEP are typically 30% higher than MIP.

After strenuous exercise, MIP can fall by between 10% and 30%. (McConnell *et al.*, 1997; Volianitis *et al.*, 2001a; Romer *et al.*, 2002a; Lomax and McConnell, 2003), depending upon the intensity of exercise and its modality (swimming appears to be a very potent stimulus to IMF, see Lomax and McConnell, 2003). Following a 4–6 week programme of inspiratory muscle training (IMT), improvements in MIP in the order of 25–35% could be observed (Volianitis *et al.*, 2001a; Romer *et al.*, 2002a,b).

Conducting a MIP and MEP test

"Contraindications – MIP and MEP efforts produce large changes in thoracic, upper airway, middle ear and sinus pressures. This is contraindicated in people with a history of spontaneous pneumothorax, recent trauma to the rib cage, a recently perforated eardrum (or other middle ear pathology), or acute sinustis (until the condition has resolved). Urinary incontinence is also a contraindication for MEP, but not MIP testing."

1   Ensure that your equipment is calibrated and working properly.
2   Ensure that all equipment that will come in contact with the participant (e.g. mouthpiece), or that he/she will inhale through, is sterile, and/or protected by a disposable viral filter.
3   Complete any necessary consent documentation.
4   Explain to the participant exactly what you wish them to do before starting the test. During the measurement of MIP and MEP the participant will be unable to generate any airflow against the mouth pressure meter, which contains only a small (1 mm) leak; they must be prepared for this.
5   Measurements can be made seated or standing, but ensure that no clothing restricts the thorax, and that a nose-clip is in place.
6   Because of the length–tension relationship of the respiratory muscles, inspiratory pressures (MIP) are measured at residual volume and expiratory pressures (MEP) at total lung capacity.
7   Ideally, MIP measurements should be preceded by a bout of prior loading to reduce the effect of repeated measurement (discussed earlier, and Volianitis *et al.*, 2001b). If suitable equipment is not available for this procedure, up to 18 repeated trials may be necessary to establish reliable data (Volianitis *et al.*, 2001b). However, a pragmatic compromise between rigour and time constraints is to record up to 10 efforts.

8   For MIP assessment, ensure that the participant 'squeezes out slowly' to residual volume.

9   Then instruct them to 'breathe in hard…pull, pull, pull' holding the effort for at least 2 s, and no more than 3 s, and maintaining encouragement throughout. Then instruct the participant to 'relax and come off the mouthpiece' (having a piece of tissue ready for saliva).

10   Take the meter from the participant and record the measured value.

11   Leave the nose-clip in place and allow at least 30 s rest before repeating.

12   If there were deficiencies in the quality of the manoeuvre, for example, he/she did not sustain the effort for long enough, explain what went wrong and what can be done to improve things on the next attempt.

13   For MEP assessment, ensure that the participant breathes in fully to total lung capacity.

14   Then instruct them to 'breathe out hard…push, push, push' holding the effort for no more than 3 s (then as in 9 earlier). During the expiratory effort there is a tendency for air to leak around the lips/mouthpiece. Leaks can be prevented if the participant pinches their lips in place around the mouthpiece by encircling them with the thumb and forefinger.

15   Serial measurements of MIP or MEP should not differ by more than 10%, or 10 cmH$_2$O (which ever is the smallest), and should be repeated until three measurements meet the criterion. At least five measurements should be made.

16   Report the largest of the three technically satisfactory measurements (best of three).

17   If 10 manoeuvres are performed without achieving the 10% criterion, then record the highest value measured.

Common faults

- Incomplete inspiration or expiration
- Not maintaining the effort for long enough.

ACKNOWLEDGEMENT

I am grateful to Dr Lee Romer for his advice during the preparation of this manuscript. I also declare a beneficial interest in the POWERbreathe® inspiratory muscle trainer (royalty share of licence income).

REFERENCES

Aliverti, A. and Macklem, P.T. (2001). How and why exercise is impaired in COPD. *Respiration*, 68(3): 229–239.

American Thoracic Society. (1995). ATS statement – standardization of spirometry. 1994 update. *American Review of Respiratory Disease*, 152(5): 1107–1136.

American Thoracic Society/European Respiratory Society. (2002). ATS/ERS Statement on respiratory muscle testing. *American Journal of Respiratory and Critical Care Medicine*, 166(4): 518–624.

Anholm, J.D., Stray-Gundersen, J., Ramanathan, M. and Johnson, R.L. Jr (1989). Sustained maximal ventilation after endurance exercise in athletes. *Journal of Applied Physiology*, 67(5): 1759–1763.

Babcock M.A., Pegelow D.F., Johnson B.D., Dempsey J.A. (1996). Aerobic fitness effects on exercise-induced low-frequency diaphragm fatigue. *Journal of Applied Physiology*, 81(5): 2156–2164.

Becklake, M.R. (1986). Concepts of normality applied to the measurement of lung function. *American Journal of Medicine*, 80: 1158–1164.

Casaburi, R., Whipp, B.J., Wasserman, K. and Stremel, R.W. (1978). Ventilatory control characteristics of the exercise hyperpnea as discerned from dynamic forcing techniques. *Chest*, 73 (Suppl. 2): 280–283.

Dickinson, J.W., Whyte, G.P., McConnell, A.K. and Harries, M.G. (2005). The impact of changes in the IOC-MC asthma criteria: a British perspective. *Thorax* 60(8): 629–632.

Hamilton, A.L., Killian, K.J., Summers, E. and Jones, N.L. (1996). Symptom intensity and subjective limitation to exercise in patients with cardiorespiratory disorders. *Chest*, 110(5): 1255–1263.

Harms, C.A. (2000). Effect of skeletal muscle demand on cardiovascular function. *Medicine and Science in Sports and Exercise*, 32(1): 94–99.

Hesser, C.M., Linnarsson, D. and Fagraeus, L. (1981). Pulmonary mechanisms and work of breathing at maximal ventilation and raised air pressure. *Journal of Applied Physiology*, 50(4): 747–753.

Hill, N.S., Jacoby, C. and Farber, H.W. (1991). Effect of an endurance triathlon on pulmonary function. *Medicine and Science in Sports and Exercise*, 23(11): 1260–1264.

Johnson, B.D., Babcock, M.A., Suman, O.E. and Dempsey, J.A. (1993). Exercise-induced diaphragmatic fatigue in healthy humans. *Journal of Physiology*, 460: 385–405.

Loke, J., Mahler, D.A. and Virgulto, J.A. (1982). Respiratory muscle fatigue after marathon running. *Journal of Applied Physiology*, 52(4): 821–824.

Lomax, M.E. and McConnell, A.K. (2003). Inspiratory muscle fatigue in swimmers after a single 200 m swim. *Journal of Sports Science*, 21(8): 659–664.

McConnell, A.K., Caine, M.P. and Sharpe, G.R. (1997). Inspiratory muscle fatigue following running to volitional fatigue: the influence of baseline strength. *International Journal of Sports Medicine*, 18(3): 169–173.

Mador, M.J. and Acevedo, F.A. (1991). Effect of respiratory muscle fatigue on subsequent exercise performance. *Journal of Applied Physiology*, 70(5): 2059–2065.

Nevill, A.M. and Holder, R.L. (1999). Identifying population differences in lung function: results from the Allied Dunbar national fitness survey. *Annals of Human Biology*, 26(3): 267–285.

Powers, S.K., Martin, D. and Dodd, S. (1993). Exercise-induced hypoxaemia in elite endurance athletes. Incidence, causes and impact on VO_{2max}. *Sports Medicine*, 16(1): 14–22.

Quanjer, P.H., Tammeling, G.J., Cotes, J.E., Pedersen, O.F., Peslin, R. and Yernault, J.C. (1993). Lung volume and forced ventilatory flows. Report Working Party Standardization of lung function tests; Official Statement European Respiratory Society. *European Respiratory Journal*, 6 (Suppl. 16): 5–40.

Quanjer, P.H., Lebowitz, M.D., Gregg, I., Miller, M.R. and Pedersen, O.F. (1997). Peak expiratory flow: conclusions and recommendations of a Working Party of the European Respiratory Society. *European Respiratory Journal* (Suppl. 24): 2S–8S.

Romer, L.M. and McConnell, A.K. (2004). Inter-test reliability for non-invasive measures of respiratory muscle function in healthy humans. *European Journal of Applied Physiology*, 91(2–3): 167–176.

Romer, L.M., McConnell, A.K. and Jones, D.A. (2000a) Inspiratory muscle fatigue in trained cyclists: effects of inspiratory muscle training. *Medicine and Science in Sports and Exercise*, 34(5): 785–792.

Romer, L.M., McConnell, A.K. and Jones, D.A. (2002b). Effects of inspiratory muscle training upon recovery time during high intensity, repetitive sprint activity. *International Journal of Sports Medicine*, 23(5): 353–360.

Romer, L.M., Bridge, M.W., McConnell, A.K. and Jones, D.A. (2004). Influence of environmental temperature on exercise-induced inspiratory muscle fatigue. *European Journal of Applied Physiology*, 91(5–6): 656–663.

Volianitis, S., McConnell, A.K., Koutedakis, Y., McNaughton, L., Backx, K. and Jones, D.A. (2001a). Inspiratory muscle training improves rowing performance. *Medicine Science Sports Exercise*, 33(5): 803–809.

Volianitis, S., McConnell, A.K. and Jones, D.A. (2001b). Assessment of maximum inspiratory pressure. Prior submaximal respiratory muscle activity ('warm-up') enhances maximum inspiratory activity and attenuates the learning effect of repeated measurement. *Respiration*, 68(1): 22–27.

Wasserman, K. (1978). Breathing during exercise. *New England Journal of Medicine*, 298(14): 780–785.

Wasserman, K., Whipp, B.J., Koyal, S.N. and Cleary, M.G. (1975). Effect of carotid body resection on ventilatory and acid-base control during exercise. *Journal of Applied Physiology*, 39(3): 354–358.

West, J.B. (1999). Respiratory Physiology, 6th edn. London: Lippincott, Williams & Wilkins.

LUNG FUNCTION REFERENCE VALUES

Adults

Crapo, R.O., Morris, A.H. and Gardner, R.M. (1981). Reference spirometric values using techniques and equipment that meet ATS recommendations. *American Review of Respiratory Disease*, 123: 659–664.

Knudson, R.J., Lebowitz, M.D., Holberg, C.J. and Burrows, B. (1983). Changes in the normal maximal expiratory flow-volume curve with growth and aging. *American Review of Respiratory Disease*, 127: 725–734.

Quanjer, P.H., Tammeling, G.J., Cotes, J.E., Pedersen, O.F., Peslin, R. and Yernault, J.C. (1993). Lung volume and forced ventilatory flows. Report Working Party Standardization of lung function tests; Official Statement European Respiratory Society. *European Respiratory Journal*, 6 (Suppl. 16): 5–40.

Children/adolescents

Quanjer, P.H., Borsboom, G.J.J.M., Brunekreef, B., Zach, M., Forche, G., Cotes, J.E., Sanchis, J. and Paoletti, P. (1995). Spirometric reference values for white European children and adolescents: Polgar revisited. *Pediatric Pulmonology*, 19: 135–142.

Wang, X., Dockery, D.W., Wypij, D., Fay, M.E. and Ferris, B.G. (1993). Pulmonary function between 6 and 18 years of age. *Pediatric Pulmonology*, 15: 75–88.

CHAPTER 9

SURFACE ANTHROPOMETRY

Arthur D. Stewart and Roger Eston

INTRODUCTION

Anthropometry is defined as 'measurement of the human body'. Surface anthropometry may therefore be defined as the science of acquiring and utilising surface dimensional measurements which describe the human phenotype. Measurements of mass, stature, skeletal breadths, segment lengths, girths and skinfolds are used, either as raw data or derived ratios or predicted values to describe human size, proportions, shape, composition and symmetry. Historically, anthropometry draws from diverse disciplines including anatomy, physiology, nutrition and medicine, and the multiplicity of methodologies which prevail have caused some confusion for the exercise scientist in practice today.

Previous attempts to standardise surface anthropometric measures did not achieve widespread recognition (Lohman *et al.*, 1988; Reilly *et al.*, 1996). Nevertheless, the recommendation by Reilly *et al.* (1996) to ensure inclusion of the thigh measurement with the four commonly used upper body skinfolds (biceps, triceps, subscapular and iliac crest) (Durnin and Womersley, 1974) to provide a more valid estimate of body fat has recently been confirmed in healthy young men and women (Eston *et al.*, 2005). The publication of 'Anthropometrica' (Norton and Olds, 1996) was a significant advance in the anthropometric sciences, particularly for the application of surface anthropometry techniques. This text has formed the basis of the content of the accreditation courses approved by the International Society for the Advancement of Kinanthropometry (ISAK). The general procedures and location of the various sites are also described and illustrated by Hawes and Martin (Hawes and Martin, 2001), however the current definitive guide for all anthropometric procedures is ISAK's standards manual (ISAK, 2001) (revised 2006). The purpose of this chapter is to summarise key principles and methods for measuring the most commonly used skinfolds and girths.

MEASUREMENT PRE-REQUISITES

For all measurements, subjects require appropriate information in advance, and informed written consent should be obtained. Anthropometry requires a spacious (minimum 3 m × 3 m) well-illuminated area, affording privacy. Subjects should present for measurement in suitable apparel, recovered from previous exercise, fully hydrated and voided. Clothing should conform to the natural contours of the skin and allow easy access for landmarking and measurement. For males, running shorts or swimwear is ideal, and for females, either a two-piece swimming costume, or running shorts and a sports top which exposes the shoulders and abdominal area, are suitable. (One-piece swimwear, rowing suits or leotards are *not* suitable.) Some subjects may prefer a loose fitting shirt which can be lifted to access measurement sites. All measurements (except hip girth, which is measured over close fitting clothing for reasons of modesty) are performed on clean, dry unbroken skin. Cultural differences may preclude the acquisition of some or all measurements in some subjects. Measurement of females or children by male anthropometrists requires particular sensitivity and the individual's entitlement to a chaperone. It is always advisable to have another adult (preferably female) present in such circumstances.

RECOMMENDED EQUIPMENT

Stadiometer – (e.g. Holtain, Crosswell, Crymych, UK) mounted on wall or stand with sliding headboard and accurate to 0.1 cm.

Weighing scales – calibrated and graduated to 100 g suggested range to be up to 150 kg (e.g. SECA, Birmingham, UK).

Skinfold calipers – Harpenden (British Indicators, c/o Assist Creative Resources, Wrexham, UK) calibrated to 10 g·mm^{-2}, scale to 80 mm in new models, 40 mm in old ones, which can be read to 0.1 mm by interpolation. Holtain (Crosswell, Crymych, UK) calipers are of similar quality and can be used with equal precision.

Anthropometric tape – Metal, with a stub extending several centimetres beyond the zero line. The Rosscraft anthropometric tape (Rosscraft Innovations Inc, Vancouver, Canada) is a modified version of the Lufkin W606PM (Cooper Industries, USA). Both can be read to 0.1 cm, and are recommended.

Segmometer – A flexible metal tape with rigid sliding branches for identifying lengths and landmark locations (Rosscraft Innovations Inc, Vancouver, Canada) read to 0.1 cm.

Anthropometric box – These are not commercially available, but should be made from plywood or a strong fibre-board equivalent capable of supporting an individual who may weigh 150 kg. The box should be 30 cm × 40 cm × 50 cm, to facilitate ease of measuring subjects of differing size.

PROCEDURES

Stature is measured to 0.1 cm without footwear and with the head in the Frankfort plane (orbitale and tragion are horizontally aligned). The heels are together and touching the scale of the stadiometer. The subject inspires for measurement, and the recorder brings down the headboard to compress the hair.

Body mass is measured to 0.1 kg. The subject wears exercise apparel or light clothing but no footwear. If nude mass is required, clothing could be weighed separately.

Landmarking. Skeletal Landmarks (bony locations defining measurement sites) are located via palpation of overlying soft tissue. Because some measurements vary considerably over a short distance, landmarking the correct site is essential for reproducible measurements. Landmarks should be located generally and then released. They should then be re-located specifically before marking, as the skin can move several centimetres in relation to underlying bone. Skinfold locations are marked with a cross, with two lines intersecting at right angles. A longer line should represent the orientation of the skinfold, and the shorter line should define the finger and thumb placement. Bony edges are commonly marked with a short (0.5 cm) line, while points (e.g. the inferior tip of the scapula) are marked with a dot, from which linear measurements are made.

Protocol. Measurements should be made on the *right* side of the body. Left-handed subjects may have greater muscle mass on the left limb, in which case girths on both sides can be recorded. Subjects are encouraged to relax their muscles before measurement to reduce discomfort and improve reproducibility. Measurements should be made in series – moving from one site to the next until the entire protocol is complete.

Skinfolds. Ensure the skin is dry and unbroken, and the landmark is clearly visible. The anthropometrist's left hand approaches the subject's skin surface at 90°. The skinfold is raised at the marked site, with the shorter line visible at the edge of the anthropometrist's forefinger and thumb. The fold is grasped firmly in the required orientation, following natural cleavage lines of the skin and raised far enough (but no further) so the fold has parallel sides. Palpation helps avoid incorporating underlying muscle into the grasp. The near edge of the caliper blades are applied to the raised fold 1 cm away from the thumb and forefinger, at a depth of mid-fingernail (see Figure 9.1).

The calipers are held at 90° to the skinfold, the spring pressure is released and the measurement value recorded 2 s afterwards. In the case of large skinfolds, the needle is likely to be moving at this time, but the value is recorded nonetheless. The calipers are removed before the skinfold is released.

Skinfold locations are illustrated in Figure 9.2. and described in Table 9.1.

Girths. A cross-handed technique is used with the stub held in the left hand, and the case in the right hand. Approaching from the side of the subject, the stub is passed around the body segment, grasped by the right hand, and then passed back to the left hand which pulls it to the appropriate tension. The middle fingers of both hands can then be used for 'pinning' the tape, and moving it a short distance up or down and maintaining its orientation 90° to the long

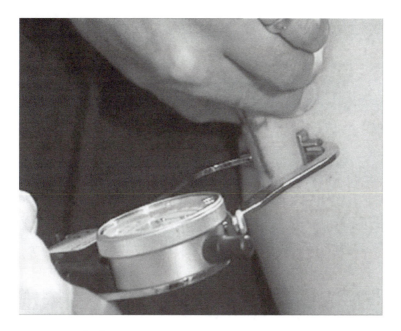

Figure 9.1   A triceps skindfold measurement illustrating appropriate technique

axis of the segment. There should be no visible indentation of the skin at the measurement. In the case of maximal measurements it is necessary to measure lesser measurements superior and inferior to the final measurement site. If the skin surface is concave, the tape spans the concavity in a straight line. For torso sites, measurements should be made at the end of a normal expiration (Table 9.2).

MEASUREMENT PROFORMA

The *mean* of duplicate or the *median* of triplicate measures (when the first two measures differ by more than 5% for skinfolds and 1% for other measures) is recommended. In some situations only a single set of measures is possible, and the error of the measurer needs to be quantified as this governs the meaning and implication of the data (Pederson and Gore, 1996). This should be in the form of *Technical Error of Measurement (TEM)*, and expressed as a percentage of the measurement value.

$$\text{TEM} = [\Sigma(x_2 - x_1)^2] \cdot 2n^{-1}$$

$$\% \text{ TEM} = 100 \cdot \text{TEM} \cdot m^{-1}$$

where x_1 and x_2 are replicate pairs of measures, n is the number of pairs and m is the mean value for that measure across the sample.

Table 9.1   Skinfold measurements

Skinfold	Location and landmarking	Orientation	Body position for measurement
Triceps[a]	Mid-point of a straight line between the acromiale and the radiale on the posterior aspect of the arm	Vertical	Standing Shoulder slightly externally rotated
Subscapular	2 cm lateral and 2 cm inferior to the inferior angle of the scapula	Oblique – ~45° dipping laterally	Standing
Biceps[a]	Mid-point level of a straight line between the acromiale and the radiale on the Anterior aspect of the arm	Vertical	Standing Shoulder slightly externally rotated
Iliac crest	Immediately superior to the crest of the ilium, on the ilio-axilla line	Near horizontal	Standing Right arm placed across torso
Supraspinale	The intersection of a horizontal line drawn from the crest of the ilium, with a line joining the anterior superior iliac spine and the anterior axillary fold	Oblique	Standing
Abdominal	5 cm lateral of the midpoint of the umbilicus	Vertical	Standing
Thigh[a]	Mid-point of the perpendicular distance between the inguinal crease at the mid-line of the thigh and the mid-point of the posterior border of the patella when seated with the knee flexed to 90°	Longitudinal	Sitting with leg extended and foot supported, the subject extends the knee and clasps hands under hamstrings and lifts gently for measurement
Medial calf	The most medial aspect of the calf, at the level of maximum girth, with subject standing and weight evenly distributed	Vertical	Standing, foot on box, with knee at 90°

Note
a These sites ideally require a wide-spreading caliper or segmometer to locate, because curvature of the skin surface affects site location if a tape is used

Error magnitude varies with the recorder, the measurement type and site. For serial measurements, a statistical basis for detecting real change should be included. Because the TEM equates to the standard error of a single measurement, then overlapping standard errors indicate no significant change in serial measures – either at the 68% (for 1SE) or 95% (for 2SE) level. Clearly, experienced anthropometrists with low TEMs are several times more likely to detect real change than others.

The conversion of raw data into indices may be justified in terms of fat patterning (Stewart, 2003b; Eston *et al.*, 2005) (skinfold ratios) corrected girths (Martin *et al.*, 1990), proportions (the ratio of segment lengths or

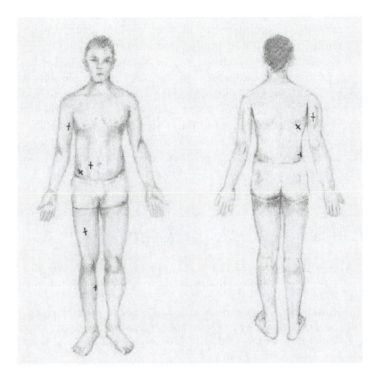

Figure 9.2   Skinfold locations
Source: M. Svensen

Table 9.2   Girth measurements

Girth	Location	Body position	Notes
Chest	At level of mid-sternum	Arms abducted slightly	Measure at the end of a normal expiration
Waist	Narrowest circumference between thorax and pelvis	Arms folded	Mid-point between iliac crest and 10th rib, if no obvious narrowing
Hip	At the level of maximum posterior protuberance of buttocks	Relaxed, feet together	Measure from the side, over clothing
Upper arm	Mid acromiale-radiale	Arm abducted slightly, elbow extended	
Forearm	Maximum	Shoulder slightly flexed, elbow extended	
Mid-thigh	Mid trochanterion – tibiale laterale level	Weight equally distributed	
Calf	Maximum	Weight equally distributed	

anthropometric somatotype (Heath and Carter, 1967). Corrected girths involve subtracting the skinfold multiplied by pi from the limb girth, and are a useful surrogate for muscularity. Predicting tissue masses of fat (Sinning *et al.*, 1985; Stewart, 2003c) or muscle (Martin *et al.*, 1990; Stewart, 2003a) has obvious appeal but is problematic. Numerous methodological assumptions govern the conversion of linear surface measurements into tissue mass, and sample-specificity restricts the utility of many equations. If used, they should be accompanied by the standard error of the estimate or confidence limits, as well as total error of prediction equations (Stewart and Hannan, 2000), although the use of raw anthropometric data is becoming more accepted and is to be encouraged.

REFERENCES

Durnin, J.V.G.A. and Womersley, J. (1974). Body fat assessment from total body density and its estimation from skinfold thickness: measurements on 481 men and women aged from 16 to 72 years. *British Journal of Nutrition*, 32: 77–97.

Eston, R.G., Rowlands, A.V., Charlesworth, S., Davies, A. and Hoppitt, T. (2005). Prediction of DXA-determined whole body fat from skinfolds: importance of including skinfolds from the thigh and calf in young, healthy men and women. *European Journal of Clinical Nutrition*, 59: 695–702.

Hawes, M. and Martin, A. (2001). Human body composition. In Eston, R.G. and Reilly, T. (eds), *Kinanthropometry and Exercise Physiology Laboratory Manual: Tests, Procedures and Data. Volume 1: Anthropometry*. Routledge, London, pp. 7–46.

Heath, B.H. and Carter, J.E.L. (1967). A modified somatotype method. *American Journal of Physical Anthropology*, 27: 57–74.

International Society for the Advancement of Kinanthropometry. (2001). International standards for anthropometric assessment. North West University (Potchefstroom Campus), Potchefstroom 2520, South Africa: ISAK (revised 2006).

Lohman, T.G., Roche, A.F. and Martorell, R. (eds) (1988). *Anthropometric Standardization Reference Manual*. Champaign, IL. Human Kinetics.

Martin, A.D., Spenst, L.F., Drinkwater, D.T. and Clarys, J.P. (1990). Anthropometric estimation of muscle mass in men. *Medicine and Science in Sports and Exercise*, 22: 729–733.

Norton, K. and Olds, T. (eds) (1996). *Anthropometrica*. Sydney: University of New South Wales Press, pp. 77–96.

Pederson, D. and Gore, C. (1996). Anthropometry measurement error. In K. Norton and T. Olds (eds), *Anthropometrica*. Sydney: University of New South Wales Press, pp. 77–96.

Reilly, T., Maughan, R.J. and Hardy, L. (1996). Body fat consensus statement of the steering groups of the British Olympic Association. *Sports Exercise and Injury*, 2: 46–49.

Sinning, W.E., Dolny, D.G., Little, K.D., Cunningham, L.N., Racaniello, A., Siconolfi, S.F. and Sholes, J.L. (1985). Validity of 'generalised' equations for body composition analysis in male athletes. *Medicine and Science in Sports and Exercise*, 17: 124–130.

Stewart, A.D. (2003a). Fat patterning – indicators and implications. *Nutrition*, 19: 568–569.

Stewart, A.D. (2003b). Anthropometric fat patterning in male and female subjects. In T. Reilly and M. Marfell-Jones (eds), *Kinanthropometry VIII*. London, Routledge, pp. 195–202.

Stewart, A.D. (2003c). Mass fractionation in male and female athletes. In T. Reilly and M. Marfell-Jones (eds), *Kinanthropometry VIII*. London, Routledge, pp. 203–210.

Stewart, A.D. and Hannan, W.J. (2000). Body composition prediction in male athletes using dual X-ray absorptiometry as the reference method. *Journal of Sports Sciences*, 18: 263–274.

CHAPTER 10

MEASURING FLEXIBILITY

Nicola Phillips

Flexibility has been defined as 'the intrinsic property of body tissues, which determines the range of motion achievable without injury at a joint or group of joints' (Holt *et al.*, 1996, p. 172). However the term flexibility has historically involved some confusion or contention. Inconsistencies in terminology used by varying disciplines has been a major factor, where the term often means different things to different disciplines. For example, Kisner (2002) defined flexibility as the ability of a muscle to relax and yield to stretch. This definition emphasises the contractile component of soft tissue structures around a joint rather than the movement available at a specific joint or joints.

Before considering appropriate measures of flexibility it is therefore important to clarify what is meant by flexibility, which type of flexibility you want to measure and whether a test is appropriate for that measure. Accuracy and reliability of testing has been discussed in general in previous chapters but the type of flexibility being measured will have a major impact on the validity of any specific test. When measuring flexibility, it should not be thought of as a whole body component but as a joint or body segment specific issue. Flexibility will often be joint-specific in different sports and measurement should therefore reflect those variations.

STATIC AND DYNAMIC TESTS

Static flexibility is a measure of range of movement, usually passive, around one or more joints in a body segment. Static flexibility is thought to be primarily limited by an individual's ability to tolerate stretch and could therefore be affected by factors such as varying tolerance of discomfort or state of relaxation. Physiotherapists and other health professionals might also assess limits of motion through 'end-feel' of the movement (Norkin and White, 1995), which is a subjective measure of resistance to the limits of movement. Although a fairly sensitive measure when performed by an experienced individual, the subjective nature with limited scope to attribute a figure to the outcome poses definite limitations to this technique.

Dynamic flexibility is considered by some disciplines as the range of motion usually achieved through active movement involving muscle contraction. Following this definition, test movements would be made specific to functional movements required in various sporting activities (MacDougall *et al.*, 1991). The measures would thus be the same as for passive stretching but would follow a different strategy for achieving the range of movement. However other disciplines would regard dynamic flexibility as something entirely different. Gleim and McHugh (1997) discuss dynamic flexibility in terms of measuring increasing stiffness in a muscle as range is increased and it is put on a stretch, either actively or passively. It has been argued that this is the more objective way of measuring flexibility. In view of the lack of consensus in current literature regarding dynamic flexibility, this chapter will be restricted to measuring range of movement to that reflecting passive flexibility.

EQUIPMENT

There is a wide variety of measuring tools available and some will be more applicable in certain sports than others. They also vary in complexity of use and cost. The simplest and cheapest would probably be the standard tape measure, whereas the more costly would be use of a digital camera and appropriate software for angular measurement. Goniometers or digital inclinometers are also used to measure joint angles. The procedures described in the regional sections of this chapter could be used interchangeably with most of the equipment listed. However, it is important to decide on a specific piece of equipment for each test and to use it consistently for flexibility measures to be meaningful.

PROCEDURES

The following procedures are essential for flexibility measurement as each of the conditions below has been shown to affect flexibility.

- The environment should be standardised, especially regarding temperature and whether measurements are made inside or outside.
- Any warm up should also be standardised as this could have major effects on muscle extensibility.
- Starting positions should be recorded carefully to allow repetition on subsequent occasions for meaningful comparison.
- Instructions should also be standardised, particularly if there is a likelihood of different testers taking measurements over a training/competitive season.

- The actual protocol should be the same each time, including the number of attempts as there is likely to be a learning as well as a warm up effect with many of the tests. It would be usual to decide on a mean or the best of three attempts.
- An appropriate battery of tests should be selected for an individual sport as different sports will have very different flexibility requirements and will be joint or region specific.

SPECIFIC MEASURES

The following sections describe commonly used tests of flexibility. Some tests measure single joint movement, whereas others measure multi-joint segments. It is by no means a comprehensive list but provides a sufficient battery of tests to be able to assess most frequently measured segments. There are also numerous variations of the tests described which have been modified for sport specificity. It is beyond the scope of this chapter to discuss the myriad adaptations, therefore the tests described will provide a standard starting point from which to develop a sports-specific testing protocol as appropriate.

UPPER LIMB

Shoulder flexion

- The subject lies supine with knees bent and back flattened (Figure 10.1).
- The arm is raised above head with elbow straight.
- The angle between the humerus and trunk is then measured using a goniometer, inclinometer or motion analysis software package.

Tip – make sure that the lumbar spine does not come away from the support surface giving an appearance of additional shoulder range of movement.

Some sports, for example, gymnastics, swimming, racket or throwing activities involve greater range of movement than this test allows. An alternative test could be used for these sports:

- The subject lies prone with chin or forehead resting on support surface and arms stretched above head.
- The arm is lifted from the support surface.
- The distance of the arm from the support surface can then be measured using a tape measure.

Tip – ensure that the head remains in contact with the support surface to standardise range of movement measured.

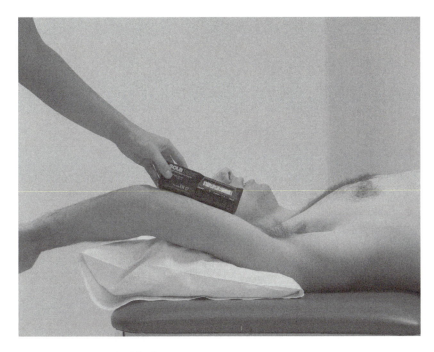

Figure 10.1 Measuring shoulder flexion in supine

Reach behind back – combined rotation, adduction and extension

Part 1

- The subject stands with arms by their side (Figure 10.2).
- The subject is instructed to raise one arm behind their head and reach down their spine as far as possible.
- The distance of the middle finger from the 7th Cervical vertebra is measured with a tape measure.
- The distance can then be compared to the contralateral side.

Part 2

- The subject stands with arms by their side.
- The subject is instructed to take their arm behind their back and reach up as far as possible.
- The distance of the middle finger from the 7th Cervical vertebra is measured with a tape measure.

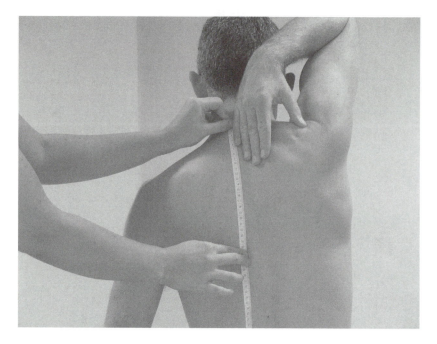

Figure 10.2 Measuring combined elevation and external rotation (back scratch)

Shoulder rotation

Although the reach tests incorporate rotation, isolated internal or external rotation are important measures in some sports. For instance, Tyler *et al.* (1999) reported a significant relationship between shoulder internal rotation and posterior shoulder tightness in baseball pitchers. The following tests are options for more specific measures.

Internal rotation

- The subject lies supine with knees bent and back flattened and with shoulder to be tested held at 90° abduction and elbow in 90° flexion (Figure 10.3).
- The tester fixes the scapula by placing the hand over the acromion.
- The subject can be asked to actively internally rotate or moved into the range passively, depending on the testing protocol chosen. (The choice of procedure should be recorded to ensure accurate repetition on subsequent testing.)
- The angle of rotation of the forearm can be measured with a goniometer, inclinometer or motion analysis software.

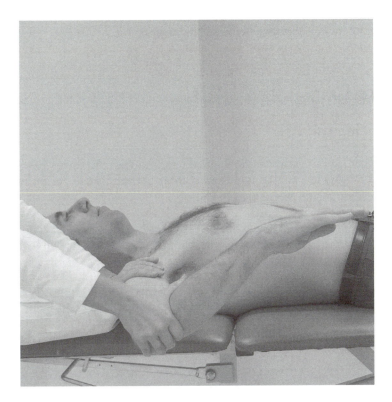

Figure 10.3 Measuring shoulder internal rotation at 90° abduction in supine

External rotation

- The subject lies supine with knees bent and back flattened and with shoulder to be tested held at 90° abduction and elbow in 90° flexion.
- The tester fixes the scapula by placing the hand over the acromion.
- The subject can be asked to actively internally rotate or moved into the range passively, depending on the testing protocol chosen. (The choice of procedure should be recorded to ensure accurate repetition on subsequent testing.)
- The angle of rotation of the forearm can be measured with a goniometer, inclinometer or motion analysis software.

Tip – make sure the trunk or the scapula does not lift from the support surface and that elbow flexion/extension remains constant to avoid giving an appearance of additional shoulder range of movement.

It should be noted that less range of movement would be expected when the scapula is fixed as described earlier than when the subject is asked to freely move into internal or external rotation. The technique described earlier is designed to control accessory scapulothoracic motion and is therefore thought to be more representative of glenohumeral movement (Awan *et al.*, 2002). However, there is a learning element for the measurer in this test, which could affect standardisation on subsequent testing.

LOWER LIMB

Straight leg raise

The straight leg raise is a commonly used test, although there are varied reports about its validity for hamstring flexibility measurement because of the influence of concurrent sciatic nerve stretch during the test. Varied recommendations have been made by previous authors as to whether the contralateral leg should remain straight or flexed during the test (Gajdosik, 1991; Kendall *et al.*, 1997). The test described here uses a straight leg but providing testing remains consistent, either could be used.

- The subject lies supine, arms at side and legs straight (Figure 10.4).
- The tester lifts the leg to be measured, keeping the knee straight.
- Maximum movement is measured as an angle between the leg and the support surface using a goniometer, inclinometer or motion analysis software.

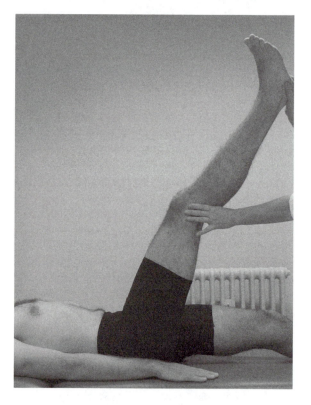

Figure 10.4 Measuring hamstring flexibility using the straight leg raise

Active knee extension test/passive knee extension test

This test is considered to be more specific to hamstrings as opposed to the straight leg raise where neural structures are often a limiting factor (Sullivan *et al.*, 1992; de Weijer *et al.*, 2003). The active knee extension (AKE) was proposed by Gajdosk and Lusin (1983) as a modification of the straight leg raise but there has been some argument about its inter- tester reliability (Worrell *et al.*, 1991). Consequently the test has since been modified to a passive version and both intra-and inter-tester reliability has been reported elsewhere (Gajdosik *et al.*, 1993).

- The subject to be tested lies supine with to be tested held at 90° hip flexion and 90° knee flexion. The contralateral leg is straight.
- AKE – the subject actively extends the knee whilst maintaining the hip at 90° or
- Passive knee extension (PKE) – the subject's knee is passively extended whilst maintaining the hip at 90°.
- Range of movement is measured by the angle of knee flexion using goniometer, inclinometer or motion analysis software.

Hip abduction – adductors/groin

- The subject lies supine with legs straight.
- The tester fixes the opposite hip over the pelvis.
- The subject then slides their leg out to the side as far as possible whilst keeping their toes pointed upwards.
- The angle of abduction can then be measured from the midline using a goniometer, inclinometer or motion analysis software.

Tip: Rotation of the pelvis or the hip can alter the range of movement significantly and therefore need to be standardised between tests.

Hip adduction – tensor fascia latae and iliotibial band

The iliotibial band is a structure that can become shortened in a variety of sports involving running. Adduction of the hip is therefore a useful movement to assess the length of tensor fascia latae and the iliotibial band.

- The subject lies on their side with the leg to be tested uppermost and with their back close to the edge of the testing surface.
- The upper leg is dropped off the edge of the surface, behind the body.
- The angle of drop can be measured using a goniometer, inclinometer or motion analysis software.

Tip: Anything less than 10° of movement is generally regarded as abnormal (Kendall *et al.*, 1997).

Thomas test

The Thomas test is a specific test for hip flexor flexibility as described by Kendall *et al.* (1997). It can be modified through altering knee flexion to include or exclude rectus femoris as a component in the range. Maintaining a straight knee excludes rectus femoris. Adding knee flexion in the test position will assess the length of rectus femoris following assessment of the hip flexors.

- The subject leans back against the edge of a treatment couch or table.
- The non-test thigh is held firmly as closely to the chest as possible.
- The subject lowers into a supine position on the treatment couch or table, whilst maintaining the hip position of the non-test leg.
- The angle of the test thigh to the floor can be measured using a goniometer, inclinometer or motion analysis software (Figure 10.5).
- Rectus femoris range can then be measured by passively flexing the knee.
- The knee angle can then be measured using a goniometer, inclinometer or motion analysis software.

Note that the Thomas test is a sensitive test but requires an appropriate testing surface that is not always possible when testing in the field. The following test also assesses hip flexor movement.

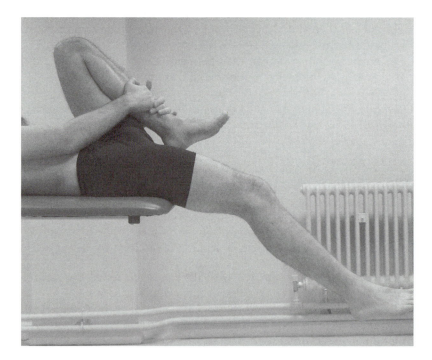

Figure 10.5 Measuring hip flexor length using the Thomas test

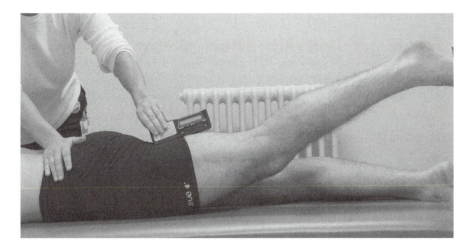

Figure 10.6 Measuring hip flexor length in prone lying

Prone hip extension

- The subject lies prone on the test surface.
- The test leg is lifted whilst keeping the knee in extension.
- The angle of the leg to the test surface can then be measured using a goniometer, inclinometer or motion analysis software (Figure 10.6).
- Adding knee flexion to this movement will introduce rectus femoris as an additional component in the test.

Tip: Ensure that the pelvis is kept in contact with the test surface to prevent the subject from rotating the trunk in order to compensate for any lack of movement. Typical range of movement would be 10°.

Hip rotation

Isolated hip rotation is a useful measure for some sports, particularly when links have been made with limited hip movement through the kinetic chain to upper limb injuries, particularly in sports involving throwing (Kibler, 1995, Kraemer, *et al.*, 1995). The following test is useful for isolating hip movement easily but a limitation is that it will be most transferable to sporting activities that happen in some flexion, which is not the case for throwing activities. However, limitations of hip rotation in a throwing position are often still highlighted by testing in this position but adaptation of the test to be performed in supine might need to be considered for some screening situations.

Internal rotation

- The subject lies supine with the test hip and knee at 90° and the foot relaxed.
- The hip is moved into internal rotation (foot away from midline of body).
- The angle of rotation can be measured using a goniometer, inclinometer or motion analysis software.

External rotation

- The subject lies supine with the test hip and knee at 90° and the foot relaxed.
- The hip is moved into external rotation (foot towards midline of body).
- The angle of rotation can be measured using a goniometer, inclinometer or motion analysis software.

Knee extension

Although knee flexion is frequently measured as an outcome measure following injury and also during function to assess efficacy in activities such as landing, it is not usually regarded as a measure of flexibility and knee measurement has therefore been restricted to extension for these purposes.

Knee extension, or more importantly hyperextension, is a commonly used measure of general flexibility in athletes.

- The subject lies supine with knees straight.
- Extension, or hyperextension is then performed passively or actively, depending on the required protocol.
- The angle is then measured from the tibia to the horizontal using a goniometer, inclinometer or motion analysis software.

Tip: This test can also be performed in standing but care should be taken to standardise hip and ankle positions, which can confound the readings.

Ankle dorsiflexion

The ankle joint has two major plantar flexors, gastrocnemius and soleus. The former is a two joint muscle, extending over the knee, whilst the latter is a single joint muscle originating from the tibia. It is therefore important that both the following dorsiflexion tests are completed to ensure assessment of the flexibility of both muscles.

Ankle dorsiflexion with straight knee – gastrocnemius bias

- The subject is in stride standing and leans forward onto arms (Figure 10.7).

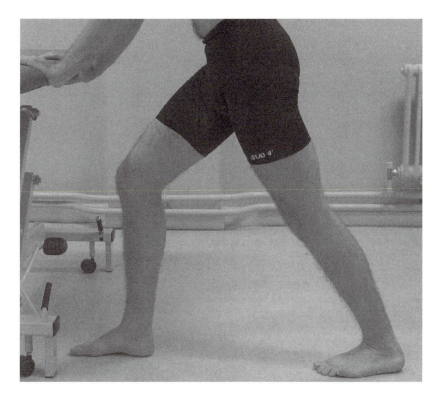

Figure 10.7 Measuring gastrocnemius length in stride standing

- The pelvis is kept in posterior tilt, the hip and knee in extension.
- The rear foot is taken as far back as possible whilst still keeping the heel on the floor.
- The angle of the lower leg to the foot is then measured from the horizontal or the dorsum of the foot, using a goniometer, inclinometer or motion analysis software.

Ankle dorsiflexion with bent knee – soleus bias

- The subject is in stride standing and leans forward onto arms (Figure 10.8).
- The pelvis is kept in posterior tilt, the hip in extension.
- The rear foot is taken as far back as possible whilst still keeping the heel on the floor additional dorsiflexion can then be achieved through knee flexion.
- The angle of the lower leg to the foot is then measured from the horizontal or the dorsum of the foot, using a goniometer, inclinometer or motion analysis software.

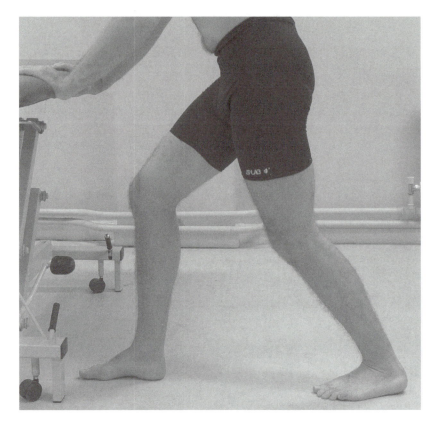

Figure 10.8 Measuring soleus length in stride standing

Ankle dorsiflexion with bent knee – soleus bias – alternative method

- The subject is in stride standing and leans forward onto arms.
- The test leg is placed up against a wall.
- The foot is then placed as far from the wall as possible whilst still being able to maintain the knee in contact with the wall and the heel on the floor.
- The maximal distance where this position can be achieved is then recorded with a tape measure.

Ankle plantar flexion

- The subject lies supine with knees straight. A rolled towel may need to be put under the calf or the feet placed over the edge of a bed to allow heel movement.
- The toes are pointed down, either actively or passively depending on the chosen protocol.

- The angle of the dorsum of the foot to plantigrade (90° to the tibia) using a goniometer, inclinometer or motion analysis software.

TRUNK

Sit and reach

The sit and reach test has traditionally been used as a field test for hamstring flexibility as it requires minimal equipment. However, it is significantly limited by lumbar spine and scapulothoracic flexibility and has therefore been placed in the trunk section. It has been established that anthropometric differences have a significant effect on sit and reach scores (Shephard *et al.*, 1990) making the test inappropriate for comparing between individuals, although it might still be useful for intra-subject comparison.

- The subject is positioned in long sitting, with knees straight and feet placed up against the sit and reach box (Figure 10.9).
- The arms are stretched out, elbows extended, palms down onto the top surface of the box.
- The bar on the sit and reach scale is pushed forward as far as possible by leaning the trunk, whilst maintaining the straight leg position. The movement should be a slow steady stretch with no bouncing permitted.
- The end position is held for 3 s and the scale recorded in centimetres from the sit and reach box.

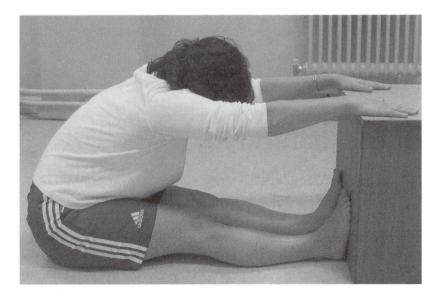

Figure 10.9 Measuring hamstring flexibility using the sit and reach test

Lumbar spine flexion

Lumbar spine flexion in standing is another test frequently used as a measure of general flexibility. There are a variety of tests used but two are described later.

Hands to floor

- The subject starts in a standing position, knees straight, arms by their side.
- The subject leans forward, letting arms drop towards the floor with fingers extended.
- The distance between floor and finger tips is measured with a tape measure.

Tip: In sports requiring higher degrees of flexibility, this test can be performed on a box and the distance from the base of support down to the fingertips below can be measured with a tape measure.

The earlier test is easily conducted but only provides a very gross measure of trunk flexibility as it is impossible to tell whether range has been achieved through lumbar spine of hamstring flexibility, much like the sit and reach test.

The following test provides a more specific measure of lumbar spine flexion (Modified Schrobers).

- The subject starts in a standing position, knees straight, arms by their side.
- The base of a tape measure is held over the 1st sacral vertebra and extended to the 1st lumbar vertebra and the length recorded.
- The subject leans forward, letting arms drop towards the floor.
- The increase in length between the two points described earlier is used as the measure of lumbar spine movement.

Tip: Surfacing marking the skin over the above points will help accuracy. Note that this test requires an element of palpation skill in highlighting fixed anatomical points for measurement.

Lumbar spine extension

This movement is even more difficult to standardise and is likely to need to be adapted for sports that require a very high degree of lumbar extension, such as gymnastics. The following test is more appropriate for general use.

- The subject starts in a standing position, knees straight, arms by their side.
- The subject is instructed to slide their hands down the back of their legs whilst arching backwards.
- The distance from fingertips to the floor can be measured with a tape measure.

Tip: Side flexion of the lumbar spine is less frequently measured in general screening as opposed to situations of injury. It can be measured in a similar way by asking the subject to slide one hand down the side of the same leg and measuring from the floor as above.

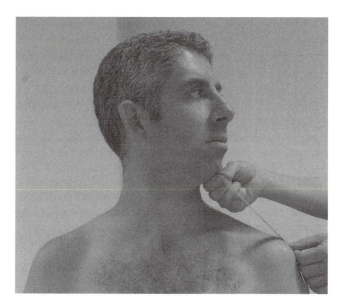

Figure 10.10 Measuring cervical spine rotation in sitting

Cervical spine flexion

- The subject starts in upright sitting with shoulders back.
- The subject drops their head forward and looks down.
- The distance between the chin and the sternal notch can be measured with a tape measure.

Tip: Extension can be measured in a similar way by recording the increase in distance from chin to sternal notch or alternatively the distance between occiput and 7th cervical spine (Figure 10.10).

Cervical spine rotation

Rotation is more likely to be useful as a routine cervical spine movement measure than flexion and extension, particularly in sports such as swimming.

- The subject starts in upright sitting with shoulders back.
- The subject then turns their head to look over one shoulder.
- The distance from the chin to acromion can be measured with a tape measure.

ACKNOWLEDGEMENTS

Thanks to Matthew Townsend and Michelle Evans for their assistance in producing the figures.

REFERENCES

Awan, R., Smith J. and Boon, A.J. (2002). Measuring shoulder internal rotation range of motion: a comparison of 3 techniques. *Archives of Physical Medicine and Rehabilitation*, 83(9): 1229–1234.

de Weijer, V.C., Gorniak, G.C. and Shamus, E. (2003). The effect of static stretch and warm-up exercise on hamstring length over the course of 24 hours. *Journal of Orthopedics and Sports Physical Therapists*, 33(12): 727–733.

Gajdosik, R.L. (1991). Passive compliance and length of clinically short hamstring muscles of healthy men. *Clinical Biomechanics*, 6: 239–244.

Gajdosik, R. Lusin (1983). G. Hamstring muscle tightness. Reliability of an active-knee-extension test. *Physical Therapy*, 7: 1085–1090.

Gajdosik, R.L., Rieck, M.A., Sullivan, D.K. and Wightman, S.E. (1993). Comparison of four clinical tests for assessing hamstring muscle length. *Journal of Orthopedics and Sports Physical Therapists*, 18: 614–618.

Gleim, G.W. and McHugh, M.P. (1997). Flexibility and its effects on sports injury and performance. *Sports Medicine*, 24(5): 289–299.

Holt, J., Holt, L.E. and Pelham, T.W. (1996). Flexibility redefined. In T. Bauer (ed.), *Biomechanics in Sports XIII*, pp. 170–174. Thunder Bay, Ontario: Lakehead University.

Kendall, F.P., McCreary, E.K. and Provance, P.G. (1997). Muscle testing and function, 5th edn. London: Lipincott, William and Wilkins.

Kibler, W.B. (1995). Biomechanical analysis of the shoulder during tennis activities. *Clinical Sports Medicine*, 14(1): 1–8.

Kisner, C. and Colby, L.A. (2002). Therapeutic exercise: Foundation and techniques, 4th edn. F.A. Davis Company.

Kraemer, W.J., Triplett, N.T. and Fry, A.C. (1995). An in depth sports medicine profile of women collegiate tennis players. *Journal of Sports Rehabilitation*, 4: 79–88.

MacDougall, J.D., Wenger, H.A. and Green, H.J. (eds) (1991). Physiological testing of the high-performance athlete, 2nd edn. IL: Human Kinetics Books.

Magnusson, S.P., Simonsen, E.B., Aagaard, P., Buesen, J., Johannson, F. and Kjaer, M. (1997). Determinants of musculoskeletal flexibility: viscoelastic properties, cross-sectional area, EMG and stretch tolerance. *Scandinavian Journal of Medicine, Science and Sports*, 7: 195–202.

Magnusson, S.P., Simonsen, E.B., Aagaard, P., Sorensen, H. and Kajer, M. (1996). A mechanism for altered flexibility in human skeletal muscle. *Journal of Physiology*, 487: 291–298.

Norkin, C.C. and White, D.J. (1995). Measurement of joint motion: a guide to goniometry, 2nd edn. Philadelphia, PA: F.A. Davis.

Shephard, R.J., Berridge, M. and Montelpare, W. (1990). On the generality of the 'sit and reach' test: an analysis of flexibility data for an aging population. *Research Quarterly for Exercise and Sport*, 61(4): 326–330.

Sullivan, M.K., Dejulia, J.J., and Worrell, T.W. (1992). Effect of pelvic position and stretching on hamstring muscle flexibility. *Medicine and Science in Sports and Exercise*, 24(12): 1383–1389.

Tyler, T.F., Roy, T., Nicholas, S.J. and Gleim, G.W. (1999). Reliability and validity of a new method of measuring posterior shoulder tightness. *Journal of Orthopedics and Sports Physical Therapy*, 29(5): 262–269.

Worrell, T.W., Perrin, D.H., Gansneder, B.M. and Gieck, J.H. (1991). Comparison of isokinetic strength and flexibility measures between hamstring injured and non-injured athletes. *Journal of Orthopedics and Sports Physical Therapy*, 13: 118–125.

CHAPTER 11

PULMONARY GAS EXCHANGE

David V.B. James, Leigh E. Sandals,
Dan M. Wood and Andrew M. Jones

INTRODUCTION

Pulmonary gas exchange (PGE) variables include oxygen uptake ($\dot{V}O_2$) and carbon dioxide output ($\dot{V}CO_2$). However, variables representing ventilation, for example expired minute ventilation ($\dot{V}_E$), and derived variables, for example respiratory exchange ratio, will also be considered in this chapter. Gas exchange variables are routinely measured at rest or during exercise to determine the following:

- resting metabolic rate
- work efficiency or economy
- maximal oxygen uptake
- gas exchange thresholds
- gas exchange kinetics.

Such information is routinely used in an exercise context to:

- determine performance potential
- recommend exercise training intensities
- examine the effect of exercise training
- establish causes for exercise intolerance.

With the increased availability of semi-automated PGE measurement systems, and software to determine PGE parameters, it may be tempting to avoid asking too many questions about the quality of the data. However, in this measurement area the quality of the data should always be questioned, not least because measurement errors may be compounded through necessary calculations. This has been nicely summed up previously by Haldane (1912, preface) who in his paper stated that, 'the descriptions are given in considerable detail, as attention to small matters of detail is often of much importance'.

Despite the proliferation of semi-automated on-line measurement systems, the traditional 'off-line' Douglas bag based approach is still recognised by many as the 'gold standard' in the measurement of PGE. An advantage of the Douglas bag method derives from the transparency of the steps in determining PGE variables, so this approach is very useful when attempting to systematically identify potential sources of error. Once such errors have been quantified and minimised, other measurement systems may be compared to the gold standard (for further discussion see Lamarra and Whipp, 1995).

In the first part of this chapter we consider key potential measurement errors when using the Douglas bag technique in normal ambient conditions (i.e. 20.9% O_2), and approaches to quantify and minimise the systematic errors (accuracy) and random errors (precision). The range of applications of the Douglas bag technique will be considered, raising the question about suitable approaches when PGE is changing rapidly. With the increased availability of rapidly responding gas analysers and flow measurement, breath-by-breath determination of PGE is possible. We have chosen not to consider on-line systems involving mixing chambers due to their decreasing popularity, and numerous, often problematic, assumptions (Atkinson *et al.*, 2005). Throughout the chapter we suggest an approach that is uncompromising, and might be considered best practice.

The basic calculation of $\dot{V}O_2$ is:

$$\dot{V}O_2 = \dot{V}_I \times F_IO_2 - \dot{V}_E \times F_EO_2$$

where $\dot{V}_I$ and $\dot{V}_E$ are the rate at which air is inspired and expired respectively, and F_IO_2 and F_EO_2 are the fractions of oxygen in the inspired and expired air, respectively.

The basic calculation of $\dot{V}CO_2$ is:

$$\dot{V}CO_2 = \dot{V}_E \times F_ECO_2 - \dot{V}_I \times F_ICO_2$$

where F_ECO_2 and F_ICO_2 are the fractions of carbon dioxide in the expired and inspired air, respectively. However, due to the small quantity of CO_2 in the inspired gas under normal atmospheric conditions, without introducing any meaningful systematic error, the equation may be rewritten as:

$$\dot{V}CO_2 = \dot{V}_E \times F_ECO_2$$

The volume of a gas varies depending on its temperature (Charles' Law), pressure (Boyle's Law) and content of water vapour. Further calculations to standardise $\dot{V}_I$ and $\dot{V}_E$ are therefore necessary in order that comparisons can be made between data collected in different circumstances. For example, expirate collected in a Douglas bag, which is then allowed to cool to room temperature, is in the form Ambient Temperature and Pressure Saturated (ATPS). Further calculations may be used to convert this volume of expirate into Body Temperature and Pressure Saturated (BTPS) or Standard Temperature and Pressure Dry (STPD) forms.

Through the following section, where relevant, procedures to minimise potential sources of measurement error are presented for each term in the earlier calculations. In addition, the resulting precision (95% confidence limits) of the measurement, and the impact on the precision of $\dot{V}O_2$ itself, is presented. To determine the influence of measurement precision on $\dot{V}O_2$ precision, certain assumptions were necessary. Assumptions include a 45 s collection of expirate during *heavy* intensity exercise, where the following values are used in the calculation: $F_EO_2 = 0.165$; $F_ECO_2 = 0.041$; $\dot{V}_E$ (BTPS) $= 80.0$ $l\cdot min^{-1}$; $\dot{V}O_2 = 3.616$ $l\cdot min^{-1}$; $\dot{V}CO_2 = 3.226$ $l\cdot min^{-1}$.

Commonly, neither off-line (Douglas bag) nor commercially available on-line PGE systems directly measure $\dot{V}_I$. Instead, $\dot{V}_I$ is determined using a nitrogen correction factor that 'converts' $\dot{V}_E$ into $\dot{V}_I$. The assumption on which this correction factor is based is that nitrogen (N_2) is metabolically inert, such that the volume of N_2 expired and the volume of N_2 inspired is equal. This can be represented by the following equation:

$$\dot{V}_I \times F_IN_2 = \dot{V}_E \times F_EN_2$$

where F_IN_2 and F_EN_2 are the fractions of N_2 in inspired and expired air, respectively.

This equation can then be rearranged to calculate $\dot{V}_I$ from $\dot{V}_E$. Although neither F_IN_2 nor F_EN_2 are typically measured when the Douglas bag method is used, it is assumed that inspired air is composed only of O_2, CO_2 and N_2, so with values for O_2 and CO_2, N_2 may be determined:

$$F_IN_2 = 1 - F_IO_2 - F_ICO_2$$

This assumption is valid because the trace gases (i.e. argon, neon, helium, etc.) that comprise ~0.93% of inspired air are metabolically inert and can, therefore, be combined with N_2. Errors may be made if it is assumed that inspired gas fractions are equivalent to outside gas fractions when working in a laboratory, even when the laboratory is well ventilated. Sandals (2003) found the mean (95% confidence limits) for F_IO_2 and F_ICO_2 in a well-ventilated laboratory (with one exercising subject and two experimenters) to be 0.20915 (± 0.00035) and 0.0007 (± 0.0003), respectively over 38 tests. For heavy intensity exercise, the difference between actual inspired and assumed inspired (i.e. outside) gas fractions translates into a systematic error of 0.18% and 0.99% for $\dot{V}O_2$ and $\dot{V}CO_2$, respectively. The only way to correct for such systematic errors is to determine average inspired gas fractions over a series of exercise tests in the normal laboratory conditions. The precision of the $\dot{V}O_2$ and $\dot{V}CO_2$ determination is $\pm 0.88\%$ and $\pm 0.74\%$, respectively. The only way to improve precision further is to determine inspired gas fractions during *each* exercise test.

If it is assumed that expired air is composed only of O_2, CO_2 and N_2, an expression for F_EN_2 that involves F_EO_2 and F_ECO_2, both of which are typically measured, can be used:

$$F_EN_2 = 1 - F_EO_2 - F_ECO_2$$

On-line systems, based on mass spectrometers for determination of gas fractions, directly measure all inspired (F_IN_2, F_IO_2, F_ICO_2) and expired (F_EN_2, F_EO_2, F_ECO_2) gas fractions, thereby reducing the number of necessary assumptions for these systems. It is of relevance to both on-line and Douglas bag methods that although the assumption that N_2 is inert has been challenged in the past (Dudka *et al.*, 1971; Cissik *et al.*, 1972), the assumption is generally considered appropriate, particularly during exercise when minute ventilation is elevated (Wilmore and Costill, 1973).

DOUGLAS BAG METHOD

The discrete nature, and often prolonged duration, of expired gas collections into a Douglas bag suggests that this technique is most suited to determining steady state gas exchange. The technique may also appropriately be employed when PGE variables are changing systematically in a predictable way, for example, during exercise with a progressively increasing work rate (or speed).

The required duration for each discrete collection of expired gas into a Douglas bag should be determined by how quickly the bag becomes filled, which in turn depends on the minute ventilation. It is this requirement that determines sampling frequency, and therefore limits the utility of this technique in examining PGE kinetics. The need to 'fill' the bag is partly related to the need to average over a number of breaths, so it is important to only include a *whole* number of breaths in each collection of expirate. However, the need to adequately fill the bag is also related to minimising several potential sources or error. An obvious source of error is inaccurate timing. One approach to ensure accurate timing is the addition of timer switches to the Douglas bag valves. Before considering the other sources of error, it is perhaps worth pointing out that any leak in the collection apparatus will induce an error, and that it is therefore important to check for leaks in the Douglas bag itself, the two-way valve, the connecting tubing and the three-way breathing valve.

The accuracy and precision of the determined volume of expired gas (expirate) may be affected by the volume measuring device, the determination of ambient pressure, the determination of gas temperature and the volume of any sample removed for determination of gas fractions. Volume is often determined by evacuating collected expirate through a dry gas meter in off-line systems, although traditionally various spirometers have been used. Calibration of the dry gas meter ensures accurate gas volume determination, and calibration is normally undertaken by exposing the dry gas meter to a range of known volumes. To ensure a valid calibration approach, these known volumes should be delivered to the dry gas meter in exactly the same way that expirate would normally be delivered (for further information, see Hart and Withers, 1996). In this regard, a syringe is normally used to pass known volumes of gas into a Douglas bag through the normal collection assembly. The known gas volumes are then passed through the dry gas meter with the help of a vacuum pump. The flow rate of the vacuum pump must be set to that used when evacuating expirate to ensure a valid calibration. A regression equation is then produced to correct the volume meter reading to the actual volume. Sandals (2003) has determined the precision of such a calibration approach to be ±0.057 l. When this level of precision is considered for $\dot{V}O_2$ determination, a value for $\dot{V}O_2$ precision of ±0.86% is calculated.

Ambient pressure is normally determined with a mercury barometer via a Vernier scale. Barometers should be regularly checked for their accuracy. This is possible by applying the following equation (WMO, 1996):

$$P_{B_{LAB}} = P_{B_{SL}}/(H/29.27T_{LAB})$$

where $P_{B_{SL}}$ is the barometric pressure (P_B) at sea-level, $P_{B_{LAB}}$ is the P_B for the laboratory (and both pressures are in hPa where 1 mm Hg = 1.33 hPa), H is the laboratory elevation in metres (from ordinance survey map), and T_{LAB} is the laboratory temperature in Kelvin.

A good quality barometer will normally have a resolution of 0.05 mmHg, and it can be assumed that measurement can be accurately made to within ±0.2 mmHg. When this level of precision is considered for $\dot{V}O_2$ determination, a value for $\dot{V}O_2$ precision of ±0.03% is calculated (Sandals, 2003).

Expired gas temperature is normally determined at the inlet port of the dry gas meter with the use of a thermistor probe. Commonly available probes have a resolution of 0.1°C, and are normally factory calibrated with maximum accuracy in the region of ±0.2°C. Errors in the measurement of expired gas temperature may have a cumulative effect on $\dot{V}_E$ translation from ATPS to STPD (and therefore $\dot{V}O_2$ determination) via both the determination of gas temperature itself, and the use of gas temperature in the determination of the partial pressure of water vapour in the gas (P_{H_2O}). Taking a maximum possible error of ±0.2°C in the determination of gas temperature, this translates into a precision of ±0.07% in the conversion of $\dot{V}_E$ from ATPS to STPD. This is further compounded by a ±0.2 mmHg error in P_{H_2O} determination, resulting in an accumulated precision of ±0.1% in the conversion of $\dot{V}_E$ from ATPS to STPD and hence $\dot{V}O_2$ determination (Sandals, 2003).

The sample volume removed for determination of gas fractions should be accurately determined. This is normally done by removing gas from the Douglas bag at a known flow rate. The accuracy of the flow rate may be checked by filling a bag with a known gas volume, and timing the emptying of the bag. Sandals (2003) has noted significant discrepancies in the actual flow rate and that at which flow controllers are set. Once flow rate is known, the precision of determination of sample volume has been shown to be ±0.007 l·min^{-1}. When this level of precision is considered for $\dot{V}O_2$ determination, a value for $\dot{V}O_2$ precision of ±0.11% is calculated (Sandals, 2003).

Having considered factors that might influence the accuracy and precision of the determined volume of expirate, factors that influence the accuracy and precision of the gas fractions are now considered. Due to the influence of gas fraction determination on $\dot{V}O_2$ and $\dot{V}CO_2$, the importance of accurate and precise determination of expired fractions of oxygen (F_EO_2) and carbon dioxide (F_ECO_2) cannot be overemphasised. Table 11.1, produced by Sandals (2003), demonstrates just how influential errors in F_EO_2 and F_ECO_2 can be in the determination of $\dot{V}O_2$ and $\dot{V}CO_2$. For example, in the heavy exercise intensity

Table 11.1 Effect of a 1% increase in F_EO_2 and F_ECO_2 on the error incurred in the calculation of $\dot{V}O_2$ and $\dot{V}CO_2$ at three levels of exercise intensity

Exercise intensity	1% increase in F_EO_2		1% increase in F_ECO_2	
	% error in $\dot{V}O_2$	% error in $\dot{V}CO_2$	% error in $\dot{V}O_2$	% error in $\dot{V}CO_2$
Moderate	−3.07	0.00	−0.21	+1.01
Heavy	−4.61	0.00	−0.24	+1.01
Severe	−7.94	0.00	−0.30	+1.01

domain, a 1% overestimation for F_EO_2 converts to a 4.61% underestimation of $\dot{V}O_2$. This is because the F_EO_2 variable is used twice in the calculation of $\dot{V}O_2$ and the error incurred at the first stage of the calculation is in the same direction as that which is introduced at the second stage. In the calculation of $\dot{V}CO_2$ no variable is used twice so this amplification effect does not occur. However, since similar factors will influence the determination of $\dot{V}O_2$ and $\dot{V}CO_2$, errors in both calculations should be minimised.

A potential source of error that is often ignored is the contamination of expirate with any residual gas in the bag after evacuation. Whilst vacuum pumps are commonly employed to thoroughly evacuate Douglas bags, residual gas remains mainly in the non-compressible part of the bag (the neck) between the bag and the two-way valve. Minimising the volume of the 'neck' of the bag is an important part of minimising this potential error. However, it is also possible to quantify the volume of the 'neck' of the bags, and correct for the contamination effect of residual gas (Prieur *et al.*, 1998). A correction may be performed by knowing the volume and the concentration of the gas contained in the 'neck' of the bag following evacuation. Flushing the bags with room air of known concentration prior to evacuation allows for such a correction. If this procedure is followed, one may have increased confidence in measurement of the oxygen fraction (F_EO_2), and carbon dioxide fraction (F_ECO_2) in the expirate. Sandals (2003) has calculated that with such a correction the precision is ±0.031 l for the residual volume, which translates to ±0.05% for $\dot{V}O_2$ determination.

When using partial pressure analysers, water vapour partial pressure presents a gas fraction diluting effect, which may lead to a further source of error (Beaver, 1973; Norton and Wilmore, 1975). The water vapour is evident (as water droplets) as the expirate cools to room temperature in the bag. Partial pressure analysers are commonly used in off-line PGE systems, whereas on-line systems are increasingly incorporating mass spectrometry to determine gas fractions. Mass spectrometers are not influenced by water vapour partial pressure. When using partial pressure gas analysers, water vapour should be dealt with consistently when calibrating the analysers (when using dry bottled gases and moist air) and when analysing the expirate (when using saturated air). A possible approach is to first saturate all gas presented to the analysers using Nafian tubing (e.g. MH Series Humidier; Perma Pure Inc, New Jersey, USA) immersed in water, and then cool and dry the gas using a condenser (e.g. Buhler PKE3; Paterson Instruments, Leighton Buzzard, UK) to a consistent saturated water vapour pressure (e.g. 6.47 mmHg at 5.0°C; see Draper *et al.*, 2003). When adopting this approach, the calibration of the analysers and the resulting accuracy of the expired gas fraction measurement are improved.

A two-point calibration (zero and span) is commonly used for the O_2 and the CO_2 analyser. In each case, adjusting the zero setting is equivalent to altering the intercept of a linear function relating the analyser reading to the output from the sample cell, while adjusting the span is equivalent to altering the slope of this relationship. For both analysers, the zero setting is adjusted to ensure that the reading on the analyser is zero when gas from a cylinder of N_2 is passed through the analyser (the zero gas). For the O_2 analyser the span setting is adjusted to ensure that the reading on the analyser is 0.2095 when outside air is passed through the analyser. For the CO_2 analyser the span setting is adjusted to ensure that the reading on the analyser is the same as the gas fraction of a gravimetrically prepared cylinder (normally 0.0400 CO_2). Precise measurements of the atmospheric O_2 fraction since 1915 have been in the range of 0.20945–0.20952 (Machta and Hughes, 1970) and recent data suggest that a realistic current value for the CO_2 fraction would be ~0.00036 (Keeling *et al.*, 1995). It is not clear why the 0.2093 and 0.0003 values for F_IO_2 and F_ICO_2, respectively, have been so widely adopted in the physiological literature. However, it is plausible that they arose from Haldane's investigations of mine air at the start of the twentieth century and have been assumed to be constant over time (Haldane, 1912). The data from the meteorological literature show that the O_2 fraction in fresh outside (atmospheric) air is relatively constant, varying by ~0.00002 over a year (Keeling and Shertz, 1992). The precision of the gravimetrically prepared gas mixtures is reported to be within ±0.0001 of the actual nominal gas fraction (BOC Gases, New Jersey, USA). Taking the worst-case scenario, a precision of ±0.0001 for the measured expired gas fractions, translates into a precision of ±0.34% for $\dot{V}O_2$.

Table 11.2 presents an overview showing that after careful consideration of potential sources of error, and taking action to minimise each potential error, the degree of precision may be quantified. Sandals (2003) has calculated that when each source of uncertainty is combined in the calculation of $\dot{V}O_2$, an overall value for precision may be derived. Whilst we have presented precision for each potential source of error based on certain assumptions (e.g. heavy intensity exercise and 45 s collection of expirate), Sandals (2003)

Table 11.2 Effect of measurement precision on the determined $\dot{V}O_2$ for heavy intensity exercise with 45 s expirate collections

Measurement	Precision in $\dot{V}O_2$ (%)
Volume (with dry gas meter)	0.86
Ambient pressure (with barometer)	0.03
Expired gas temperature (with thermistor probe)	0.10
Sample volume (with flow controller)	0.11
Residual volume	0.05
Expired gas fractions (gas analysers)	0.34

Table 11.3 Total precision in the calculation of $\dot{V}O_2$ and $\dot{V}CO_2$ at three levels of exercise intensity and for four collection periods

Exercise intensity	Total % precision in $\dot{V}O_2$				Total % precision in $\dot{V}CO_2$			
	Collection period (s)				Collection period (s)			
	15	30	45	60	15	30	45	60
Moderate	6.0	3.3	2.4	2.0	6.2	3.5	2.6	2.1
Heavy	3.2	1.8	1.4	1.1	3.4	1.9	1.4	1.2
Severe	1.9	1.2	0.9	0.8	1.8	1.1	0.8	0.7

has determined overall precision for a range of assumptions (see Table 11.3). For the assumptions made earlier in this chapter, the overall precision of $\dot{V}O_2$ determination is calculated to be 1.4%. Of particular note is the finding that the degree of precision is increased as exercise intensity increases (for a given collection period) or the expirate collection duration increases (for a given exercise intensity).

BREATH-BY-BREATH METHOD

Although the Douglas bag method is still regarded by many as the 'gold standard' in the measurement of PGE, the requirement for collection of expired air over relatively lengthy periods (typically 30–60 s) essentially limits its use to steady-state exercise conditions. However, a great deal of important information concerning the integrated pulmonary–cardiovascular–muscle metabolic response to exercise can be gleaned under non-steady-state conditions. For example, the breath-by-breath measurement of PGE during ramp incremental exercise tests in which the external work rate is increased rapidly until the participant reaches exhaustion (typically in 10–12 min) permits not only the determination of the peak $\dot{V}O_2$, but also estimates of 'delta efficiency' (from the slope of the relationship between $\dot{V}O_2$ and work rate) and the lactate threshold (from the associated non-linear responses of $\dot{V}CO_2$ and $\dot{V}_E$) (Whipp *et al.*, 1981). These and other derived PGE variables are useful not only in the evaluation of exercise capacity in athletes and healthy volunteers but also in defining the physiological limitations to exercise performance in disease (Wasserman *et al.*, 1999). Furthermore, the breath-by-breath measurement of PGE in the abrupt transition from rest (or, more often, very light exercise) to a higher constant work rate can provide important information on the dynamic adjustment of oxidative metabolism following a 'step' increase in metabolic demand. The rate at which $\dot{V}O_2$ rises (i.e. the $\dot{V}O_2$ kinetics) during such exercise is another parameter of aerobic fitness, which is relevant both in health and disease and which can also be used to differentiate central versus peripheral limitations to exercise performance (Jones and Poole, 2005).

The principles of measuring PGE continuously (i.e. breath-by-breath) are essentially the same as those outlined earlier except that both flow and the concentration of the expirate are sampled continuously and the necessary calculations are performed 'on-line' by a microcomputer. Typically, commercial metabolic carts integrate a measurement of flow (by directing the expirate through a turbine or ultrasonic flowmeter) with a measurement of the gas concentration profiles (by directing samples of the expirate at the mouth into the gas analysers). One important consideration here is the accurate time-alignment of the gas concentration with the flow signals: the determination of the gas concentration will be delayed relative to that of flow owing to the transit time for the gas sampled at the mouth to arrive at the gas analysers and the subsequent response dynamics of those analysers. This delay is typically measured during the semi-automated calibration procedures of most commercially available systems and 'corrected' during subsequent exercise testing so that the concentration and flow signals measured at the mouth are appropriately aligned. However, greater care must be taken if exercise testing is to be conducted under conditions of hypoxia or hyperoxia because changes in gas viscosity will alter the transit time of the gas from the sample probe to the analysers.

The validity of PGE measurements with metabolic carts is ideally checked by simultaneous measurement of PGE using the Douglas bag technique (assuming that error in the latter is both known and minimised; see earlier). Ideally, PGE should be measured with both systems in the 'steady-state' over a wide range of exercise intensities (from rest to maximum) in a variety of subjects. A difference of less than approximately 5% in measurements of PGE and $\dot{V}_E$ is generally considered acceptable in such comparisons (Lamarra and Whipp, 1995). The accuracy of PGE and $\dot{V}_E$ measurement in the non-steady-state can be ascertained in a similar fashion by comparing the average values obtained by the two systems over the first 2 or 3 min following a step change in work rate. It is recommended that these checks be carried out at least every few weeks.

Despite careful attention to calibration and system maintenance, breath-by-breath PGE measurements are inherently 'noisy'; that is, there is considerable breath-to-breath variability in measures of PGE even in the steady-state. One solution to this is to average the PGE values over 10-s or 15-s periods, and this approach is very effective during incremental exercise tests and in situations where only steady-state PGE values are of interest. During exercise transients, however, such an approach is likely to obscure important events such as the Phase I–Phase II transition. Therefore, in the study of PGE kinetics, it is customary to reduce breath-to-breath noise by converting breath-by-breath values into second-by-second values and then averaging together an appropriate number of like-transitions. Just how many such transitions are required to sufficiently reduce breath-to-breath noise and increase confidence in the parameters derived from subsequent curve-fitting procedures depends on the extent of the noise (this can vary substantially from subject to subject), the amplitude of the PGE response above baseline (which will depend upon the imposed work rate and therefore, to some extent, the fitness of the subject), and the desired confidence level (Lamarra et al., 1987; Lamarra, 1990).

A portion of the breath-to-breath variability in PGE can be attributed to changes in the lung gas stores so that PGE measured at the mouth will not necessarily represent alveolar gas exchange for any given breath. This will be especially true during the first few seconds of a transition from a lower to a higher metabolic rate. Several algorithms have been developed to enable 'correction' of PGE measurements made at the mouth for changes in lung gas stores (Aunchincloss *et al.*, 1966; Wessel *et al.*, 1979; Beaver *et al.*, 1981). Cautero *et al.* (2003) have recently suggested that the methods of Gronlund (1984), in which the respiratory cycle is defined as the time elapsing between two equal O_2 fractions in two subsequent breaths, are more appropriate for the determination of 'true' alveolar gas exchange. However, as recently summarised by Whipp *et al.* (2005), this approach presents as many problems as it solves.

CONCLUSIONS

The measurement and interpretation of the PGE response to exercise is an essential component of the physiological evaluation of subjects across the entire spectrum of fitness and physical activity (from elite athletes to patients with a variety of disease states). Determination of one or more of the parameters of aerobic fitness ($\dot{V}O_2$ peak, work efficiency, lactate/gas exchange threshold and gas exchange kinetics) is likely to be of value in work with sportspeople, except for those who participate exclusively in very short-duration (<60 s) sprint or power events. The same aerobic fitness parameters will determine tolerance to the activities of daily living in elderly and patient populations. However, the measurement of PGE is fraught with the potential for serious error, irrespective of whether 'off-line' or 'on-line' systems are used, and this can lead to flawed, and potentially dangerous, data interpretation. It is important, therefore, that investigators work diligently first to identify, and then to minimise, the errors associated with their measurement systems. PGE data can only be interpreted appropriately in the light of the known error margins.

REFERENCES

Atkinson, G., Davison, R.C.R. and Nevill, A.M. (2005). Performance characteristics of gas analysis systems: what we know and what we need to know. *International Journal of Sports Medicine*, 26 (suppl. 1): S2–S10.

Auchincloss, J.H., Gilbert, R. and Baule, G.H. (1966). Effect of ventilation on oxygen transfer during exercise. *Journal of Applied Physiology*, 21: 810.

Beaver, W.L. (1973). Water vapour corrections in oxygen consumption calculations. *Journal of Applied Physiology*, 35: 928–931.

Beaver, W.L., Lamarra, N. and Wasserman, K. (1981). Breath-by-breath measurement of true alveolar gas exchange. *Journal of Applied Physiology*, 51: 1662–1675.

Cautero, M., di Prampero, P.E. and Capelli, C. (2003). New acquisitions in the assessment of breath-by-breath alveolar gas transfer in humans. *European Journal of Applied Physiology*, 90: 231–241.

Cissik, J.H., Johnson, R.E. and Rokosch, D.K. (1972). Production of gaseous nitrogen in human steady-state conditions. *Journal of Applied Physiology*, 32: 155–159.

Draper, S., Wood, D.M. and Fallowfield, J.L. (2003). The $\dot{V}O_2$ response to exhaustive square wave exercise: influence of exercise intensity and the mode. *European Journal of Applied Physiology*, 90: 92–99.

Dudka, L.T., Inglis, H.J., Johnson, R.E., Pechinski, J.M. and Plowman, S. (1971). Inequality of inspired and expired gaseous nitrogen in man. *Nature*, 232: 265–268.

Gronlund, J. (1984). A new method for breath-to-breath determination of oxygen flux across the alveolar membrane. *European Journal of Applied Physiology and Occupational Physiology*, 52: 167–172.

Haldane, J.S. (1912). *Methods of Air Analysis*. London, Griffin and Co.

Hart, J.D. and Withers, R.T. (1996). The calibration of gas volume measuring devices at continuous and pulsatile flows. *Australian Journal of Science and Medicine in Sport*, 28: 61–65.

Jones, A.M. and Poole, D.C. (eds). (2005). Oxygen Uptake Kinetics in Sport, Exercise and Medicine, London and New York: Routledge.

Keeling, C.D., Whorf, T.P., Wahlen, M. and van der Plicht, J. (1995). Interannual extremes in the rate of rise of atmospheric carbon dioxide since 1980. *Nature*, 375: 666–670.

Keeling, R.F. and Shertz, S.R. (1992). Seasonal and interannual variations in atmospheric oxygen and implications for the global carbon cycle. *Nature*, 358: 723–727.

Lamarra, N. (1990). Variables, constants and parameters: clarifying the system structure, *Medicine and Science in Sports and Exercise*, 22: 88–95.

Lamarra, N. and Whipp, B.J. (1995). Measurement of pulmonary gas exchange, in P.J. Maud, and C. Foster (eds). *The Physiological Assessment of Human Fitness*, Champaign, IL: Human Kinetics, pp. 19–35.

Lamarra, N., Whipp, B.J., Ward, S.A. and Wasserman, K. (1987). Effect of interbreath fluctuations on characterizing exercise gas exchange kinetics. *Journal of Applied Physiology*, 62: 2003–2012.

Machta, L. and Hughes, E. (1970). Atmospheric oxygen in 1967 and 1970. *Science*, 168: 1582–1584.

Norton, A.C. and Wilmore, J.H. (1975). Effects of water vapour on respiratory gas measurements and calculations. *The NSCPT Analyser*, 9: 6–9.

Prieur, F., Busso, T., Castells, J., Bonnefoy, R., Bennoit, H., Geyssant, A. and Denis, C. (1998). Validity of oxygen uptake measurements during exercise under moderate hyperoxia. *Medicine and Science in Sports and Exercise*, 30: 958–962.

Sandals, L.E. (2003). Oxygen uptake during middle-distance running. Unpublished PhD thesis, University of Gloucestershire.

Wasserman, K., Hansen, J.E., Sue, D.Y., Casaburi, R. and Whipp, B.J. (1999). *Principles of Exercise Testing and Interpretation. Maryland*, 3rd edn. USA: Lippincott, Williams and Wilkins.

Wessel, H.U., Stout, R.L., Bastanier, C.K. and Paul, M.H. (1979). Breath-by-breath variation of FRC: effect on VO_2 and VCO_2 measured at the mouth. *Journal of Applied Physiology*, 46: 1122–1126.

Whipp, B.J., Davis, J.A., Torres, F. and Wasserman, K. (1981). A test to determine parameters of aerobic function during exercise. *Journal of Applied Physiology*, 50: 217–221.

Whipp, B.J., Ward, S.A. and Rossiter, H.B. (2005). Pulmonary O_2 uptake during exercise: conflating muscular and cardiovascular responses. *Medicine and Science in Sports and Exercise*, 37: 1574–1585.

Wilmore, J.H. and Costill, D.L. (1973). Adequacy of the Haldane transformation in the computation of exercise $\dot{V}O_2$ in man. *Journal of Applied Physiology*, 35: 85–89.

WMO: World Meteorological Organisation. (1996). Guide to meteorological instruments and methods of observation. *WMO Publication*, 8: 1–21.

CHAPTER 12

LACTATE TESTING

Neil Spurway and Andrew M. Jones

THEORETICAL BACKGROUND

Lactate production by muscle during sub-maximal exercise was traditionally ascribed to a shortfall in oxygen supply, forcing vigorously active muscles to resort in part to anaerobic metabolism for their ATP requirements. However, work in the past 15–20 years (review: Spurway, 1992) has shown that fully aerobic muscle produces lactate when operating above ~50% maximal work rates. Connett *et al.* (1990) propose that this lactate production may be explained by three steps:

1 NADH builds up in mitochondria to drive their electron transport chains faster;
2 in turn, this build-up is reflected in the cytoplasmic NADH pool;
3 elevated cytoplasmic NADH increases the rate of reduction of pyruvate to lactate.

Other suggestions about the mechanism have been made, particularly in terms of changes in the activity of pyruvate dehydrogenase. On all accounts, however, the muscle's ATP production remains essentially the result of *aerobic* glycolysis, so the old concept of a major resort to anaerobic metabolism must be dropped. Even on a conservative view, therefore, the lactate production must be considered to represent an upper bound estimate of the total anaerobic activity, and it is almost certainly a very considerable over-estimate.

Nevertheless, increases in muscle oxidative capacity as a result of training diminish the need for NADH build-up to drive electron transport, and so result in reduced lactate production at any given work rate (Holloszy and Coyle, 1984; Jones and Carter, 2000). Thus in qualitative terms, *though not in quantitative ones*, the consequence for muscle lactate production on the new account remains much as the traditional concept would have predicted.

During exercise there is not a one-to-one relationship between lactate in muscle and lactate in blood, but lactate concentration ([lactate]) in the blood does give an indirect yet reproducible indication of the aerobic capacity of the working muscles. Lactate accumulation curves can therefore be used to assess changes in aerobic fitness and (by empirical rule of thumb) as guides to training intensity. The tests have obvious relevance for endurance athletes, but are also appropriate on the same sort of rule-of-thumb basis for players of multiple sprint games, who need to remove lactate quickly during recovery periods and even during support running.

GENERAL NOTES ON LACTATE METHODOLOGY

When assessing any individual there is a need for careful standardisation of repeat tests (exercise protocol and procedures for lactate sampling and assay). Capillary sampling from fingertips is generally utilised for safety and ease, but earlobe sampling is preferred by some laboratories, and is especially useful where the hands are in use, as on rowing, canoeing, skiing or arm crank ergometers. However, it should be noted that differences in blood [lactate] do occur according to the form of blood used (venous vs. arterial vs. capillary), the site of sampling and the post-sampling treatment and assay method. Likewise, there are substantial differences between plasma, whole blood and lysed blood. The most crucial practical implication of these differences is that any longitudinal study of a single athlete must adopt rigorously identical procedures for every sample. Other implications are in extrapolating any laboratory results to the field where different methods may be used, in the interpolation of performance at various blood lactate reference values, and in making comparisons between data from different laboratories and different individuals. In any case, calibration of lactate analysers and checking with standards of known concentration are essential and, when reporting results, details of the methodology used must be included.

TERMINOLOGY

Blood lactate concentrations are typically found to be significantly higher than resting values at work rates of 55–70% VO_{2max}. At higher work rates than these, blood concentrations are greater still. The work rate above which the blood [lactate] consistently exceeds the resting or baseline value (~1 $mmol \cdot l^{-1}$) has been given various names, including *Lactate Threshold* and *Anaerobic Threshold*. The latter term embodies an implicit assumption, namely the historical view that blood lactate accumulation reflected lactate production by the muscles resorting to anaerobic metabolism in conditions of hypoxia. As it is now recognised that fully aerobic muscles can produce lactate (see *Theoretical background*), the term Anaerobic Threshold is misleading. By contrast, the designation Lactate Threshold (LT) embodies no such assumptions, and simply represents what is observed; its use is therefore recommended. A blood [lactate] of 2 mmol·l^{-1} is accepted in some laboratories as a rough guide to the location of LT, but direct identification of the first 'break-point' on a blood lactate accumulation curve is greatly preferable (Figure 12.1).

A further frequently encountered term is *Onset of Blood Lactate Accumulation (OBLA)*. Though sometimes confused with LT, OBLA was initially approximated by the substantially higher reference value, 4 mmol·l^{-1}, and so was clearly not intended to represent the same functional intensity. Leading laboratories now note that, when blood [lactate] measured at the end of a 3–4 min exercise period is somewhere in the vicinity of 4 mmol·l^{-1}, it will not stay steady at that level but will *rise continually over time thereafter*. Thus it might reach 6 mmol·l^{-1} after 30 min *at the same work rate*. By contrast, a value of 2.5 mmol·l^{-1} at 4 min will typically fall to 2 or 1.5 mmol·l^{-1} if the same work rate is maintained for 30 min. 'OBLA' is thus intended to represent the minimum work rate at which a rise, rather than a fall, occurs over this kind of timescale. It will be evident that exactly defining this in practice is not easy.

The dynamics of blood lactate accumulation during constant-work-rate exercise are perhaps better embodied in the *Maximal Lactate Steady State (MLSS)* concept. The MLSS may be determined from 4 to 5 exercise bouts, each of up to 30 min duration and performed on separate days, with blood [lactate] determined at rest and after every subsequent 5 min of exercise (Figure 12.2). MLSS is defined as the highest work rate (or speed of running, etc.) at which blood [lactate] is elevated above baseline but is stable over time; in theoretical terms, therefore, it can be regarded as being infinitesimally lower than OBLA. In practice the MLSS is considered to have been exceeded when the increase in blood [lactate] between 10 and 30 min of exercise is greater than 1.0 mmol·l^{-1}. The assessment of MLSS is both time- and labour-intensive and, while remaining the 'gold standard', the MLSS is rarely directly measured but is often

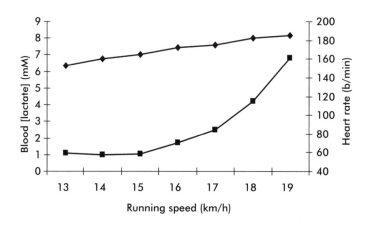

Figure 12.1 Typical blood lactate (closed squares) and heart rate (closed diamonds) responses to a multi-stage incremental treadmill test in an endurance athlete. In this example, the LT occurs at 15 km·h^{-1} while the LTP occurs at 17 km·h^{-1}

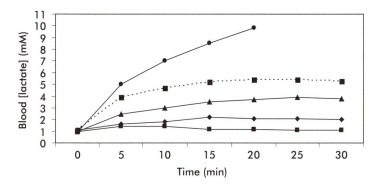

Figure 12.2 Determination of the running speed at the MLSS in an endurance athlete (the same as in Figure 12.1). This athlete completed five treadmill runs of up to 30 min duration at different running speeds (14, 15, 16, 17 and 18 km·h^{-1}) on different days. At 14 and 15 km·h^{-1}, blood [lactate] did not increase appreciably above that measured at rest; at 16 and 17 km·h^{-1}, blood [lactate] reached a delayed but elevated steady-state; and at 18 km·h^{-1}, blood [lactate] did not attain a steady state but increased inexorably until the athlete became exhausted. The running speed at the MLSS is therefore 17 km·h^{-1}

estimated using a variety of more practicable methods (Jones and Doust, 2001). Recent studies indicate that the work rate or speed at the *Lactate Turnpoint* (LTP), a second 'sudden and sustained' increase in blood [lactate] during incremental exercise (Figure 12.1), provides a good approximation of the work rate or speed at the MLSS (Smith and Jones, 2001; Pringle and Jones, 2002). Note, however, that the blood [lactate] measured at the LTP (~3 mmol·l^{-1}) will typically under-estimate the blood [lactate] at the MLSS (~4–6 mmol·l^{-1}).

A new term *Lactate Minimum Running Speed* (and hence by extrapolation to other sports, *Lactate Minimum Work Rate*), was introduced by Tegtbur *et al.* (1993). Though originally presented in a manner strongly suggestive of MLSS, the lactate minimum work rate has since been found to depend critically upon the test protocol and, as defined by certain interpretations of the original procedure, to relate more closely to LT than to MLSS. Since no approach to consensus has yet been reached, procedural details are not presented here.

DETERMINATION OF LT, LTP AND BLOOD LACTATE REFERENCE VALUES

The general procedures for an incremental protocol permitting the assessment of LT, LTP and various blood lactate reference values are described later. The reasons for selecting this protocol are three-fold: first, only 5–8 blood samples are usually required; second, it takes a minimum amount of time; and, finally, each exercise stage is sufficiently long for measurements of 'steady-state' oxygen uptake (VO_2) and heart rate (HR) to be made, so that exercise economy can be evaluated and training can be prescribed. This also enables the LT (for example) to be described as a metabolic rate (i.e. units of VO_2), which is technically correct. However, in situations where the relationship between blood [lactate] and work rate or exercise speed are not relevant, such as in patient populations, the LT can be conveniently determined (or estimated using non-invasive gas exchange procedures) to a close approximation with fast, non-steady-state, incremental protocols (Wasserman *et al.*, 1994). One other advantage to the protocol described later is that, if the test is continued to exhaustion, the measured peak VO_2 provides a close approximation of the VO_2 max (to within ~5%).

Protocol

The individual should report to the laboratory rested (i.e. having performed no strenuous exercise in the preceding 24–48 h), euhydrated and at least 3 h following the consumption of a light carbohydrate-based meal. No warm-up other than stretching is needed, as the initial exercise intensity should require no more than about 40% VO_{2max}. However, some individuals will prefer to complete 5–10 min of light exercise in preparation for the test, and this is strongly recommended for people with limited experience on the test ergometer. In this period, the test procedures along with safety considerations can be emphasised.

The protocol consists of exercising on an ergometer (treadmill, cycle, rowing, etc.), with the intensity (work rate, running speed, etc.) being increased every 4 min until the individual approaches or attains volitional exhaustion. Individual investigators may have a personal preference for the use of continuous or discontinuous protocols, although short breaks in exercise to facilitate 'clean' blood sampling do not seem to have a major influence on the derived blood lactate accumulation curve (Gullstrand *et al.*, 1994). [Lactate] is measured in blood samples obtained at the end of each 4-min stage for the subsequent determination of the work rate or exercise speed equivalent to LT, LTP and appropriate blood lactate reference concentrations (e.g. 2, 3, 4 mmol·l^{-1}). For children between 11 and 15 years and for well-trained athletes in whom a steady-state is reached more rapidly, exercise stages of 3-min duration are recommended. It is important that the exercise intensity selected for the first exercise stage is sufficiently low that it does not cause blood [lactate] to be appreciably elevated above the resting value; one should therefore 'err on the side of caution' in selecting this first intensity. The increment in exercise intensity between stages should be selected to allow the completion of a minimum of 5 and a maximum of 9 stages, with the number of stages being determined largely by the precision required in the determination of LT, LTP and the fixed blood lactate reference values. For example, in treadmill tests the increment will typically lie between 0.5 and 1.5 km·h^{-1} and in cycle or rowing ergometer tests the increment will typically lie between 20 and 50 W.

Expired air may be analysed continuously using an on-line system, or collected in Douglas bags between minutes 3–4, 7–8, 11–12, etc. for normal healthy adults, and between minutes 2–3, 5–6, 8–9 and 11–12 for children and trained athletes. The HR should be recorded during the final minute of each stage.

Blood sampling

Samples of capillary blood are obtained at rest – more than 2 h after previous exercise – and immediately after each 4-min stage; that is to say, after the cardio-respiratory measurements, but before the exercise intensity is increased. Blood sampling procedures should adhere to the appropriate health and safety guidelines (see Chapter 3).

Treatment of the blood lactate data

A graph of exercise intensity against blood [lactate] should be constructed. From this, the LT and LTP can be determined and/or the exercise intensities equivalent to appropriate blood lactate reference values can be interpolated. Plotting the HR response to the incremental test on a second y-axis enables the HR at the various 'thresholds' to be easily determined (Figure 12.1) and possibly used in the prescription of appropriate training intensities. Finally, it can also be useful to plot blood [lactate] against the relative exercise intensity (as $\%VO_{2max}$).

Reliability and sensitivity

Test-retest comparisons of ±0.2 mmol·l^{-1} (i.e. error of $< \pm10\%$) are generally considered to be acceptable in the assessment of blood [lactate] at a specific absolute work rate or exercise speed. The various lactate 'thresholds' are known to be very sensitive to improvements in aerobic capacity resulting, in particular, from endurance exercise training. A 'rightward shift' in the blood lactate curve when plotted against exercise intensity (along with a corresponding rightward shift in the HR response to exercise) is the expected outcome when an individual is re-tested following the commencement or continuation of an endurance training programme. In well-trained athletes, changes in the sub-maximal blood [lactate] profile can occur in the absence of any change in VO_{2max}. However, it is important to remember that the absolute blood [lactate] is also sensitive to factors such as muscle glycogen depletion and various dietary interventions, so that care should be taken to ensure that tests are carried out under standardised conditions (both laboratory- and individual-specific).

A NOTE ON THE ASSESSMENT OF BLOOD [LACTATE] FOLLOWING MAXIMAL-INTENSITY EXERCISE

Although this chapter has focused on the use of blood [lactate] measurement in the assessment of sub-maximal exercise performance, it should be noted that such measurements can also be valuable in other contexts. For example, the

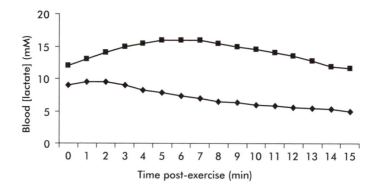

Figure 12.3 Schematic illustration of the response of blood [lactate] in the first 15 min of recovery from exhaustive high-intensity exercise in an endurance-trained athlete (closed diamonds) and a sprint-trained athlete (closed squares). Note that the peak blood [lactate] is higher and occurs later in recovery in the sprint-trained athlete

peak blood [lactate] measured in the recovery period following a bout or bouts of maximal-intensity exercise could be considered to provide a crude estimate of the extent to which energy has been supplied through anaerobic glycolysis. Strictly, it remains an upper bound figure, because a component of the lactate production remains aerobic as it is at work rates just above LT, but all the power output additional to that achieved at VO_{2max} has, of course, no alternative source of energy than anaerobic glycolysis, and this is likely to be the majority lactate source in a subject working at maximal intensity. In support of this general concept, sprint-trained athletes almost invariably demonstrate higher peak blood [lactate] values than their endurance-trained counterparts (Figure 12.3). Interestingly, the time at which the peak blood [lactate] is attained during the recovery period also differs between sprint and endurance athletes (being later in the sprint-trained) so that frequent blood samples are necessary if the peak value is to be accurately determined. Finally, whether one is considering maximal or sub-maximal exercise, it should be remembered that the blood [lactate] measured at any moment represents a conflation of the rate of lactate production in the active muscles, the rate of efflux of lactate from muscle to blood, and the rate at which lactate is cleared from the blood by muscle and other (chiefly visceral) organs (Brooks, 1991). Thus if subject A shows a lower blood [lactate] than subject B in a given test, A's muscles may be releasing less lactate in unit time than B's, *or* A's clearance mechanisms may be more active. In fact, both changes are likely consequences of aerobic training, but blood [lactate] measurements can provide no estimate of the balance between them.

REFERENCES

Brooks, G.A. (1991). Current concepts in lactate exchange. *Medicine and Science in Sports and Exercise*, 23: 895–906.

Connett, R.J., Honig, C.R., Gayeski, T.E.J. and Brooks, G.A. (1990). Defining hypoxia: a systems view of VO_2, glycolysis energetics and intracellular PO_2. *Journal of Applied Physiology*, 68: 833–842.

Gullstrand, L., Sjodin, B. and Svedenhag, J. (1994). Blood sampling during continuous running and 30-second intervals on a treadmill: effects on the lactate threshold results? *Scandinavian Journal of Medicine and Science in Sports*, 4: 239–242.

Holloszy, J.O. and Coyle, E.F. (1984). Adaptations of skeletal muscle to endurance exercise and their metabolic consequences. *Journal of Applied Physiology*, 56: 831–838.

Jones, A.M. and Carter, H. (2000). The effect of endurance training on parameters of aerobic fitness. *Sports Medicine*, 29: 373–386.

Jones, A.M. and Doust, J.H. (2001). Limitations to submaximal exercise performance. In R.G. Eston and T.P. Reilly (eds). *Exercise and Laboratory Test Manual*, 235–262, 2nd edn. E & FN Spon.

Pringle, J.S. and Jones, A.M. (2002). Maximal lactate steady state, critical power and EMG during cycling. *European Journal of Applied Physiology*, 88: 214–226.

Smith, C.G. and Jones, A.M. (2001). The relationship between critical velocity, maximal lactate steady-state velocity and lactate turnpoint velocity in runners. *European Journal of Applied Physiology*, 85: 19–26.

Spurway, N.C. (1992). Aerobic exercise, anaerobic exercise and the lactate threshold. *British Medical Bulletin*, 48: 569–591.

Tegtbur, U., Busse, M.W. and Braumann, K.M. (1993). Estimation of individual equilibrium between lactate production and catabolism during exercise. *Medicine and Science in Sports and Exercise*, 24: 620–627.

Wasserman, K., Hansen, J.E., Sue, D.Y., Whipp, B.J. and Casaburi, R. (1994). *Principles of Exercise Testing and Interpretation*, Philadelphia, PA: Lea & Febiger.

CHAPTER 13

RATINGS OF PERCEIVED EXERTION

John Buckley and Roger Eston

INTRODUCTION

Following two decades of research, Gunnar Borg's original rating of perceived exertion (RPE) scale was accepted in 1973 as a valid tool within the field of exercise science and sports medicine (Noble and Robertson, 1996). His seminal research provided the basic tool for numerous studies in which an individual's effort perception was of interest. It also provided the basis and incentive for the development of other scales, particularly those used with children. Borg's initial research validated the scale against heart rate and oxygen uptake. Later research focussed on the curvilinear growth of perceived exertion with lactate, ventilation and muscle pain responses, and led to the development of the category-ratio (CR-10) scale.

The general aim of using RPE is to quantify an individual's subjective perception of exertion as a means of determining the exercise intensity or regulating exercise intensity (Borg, 1998). In this way it acts as a surrogate or concurrent marker to key relative physiological responses including: percentage of maximal heart rate (%HRmax), percentage of maximal oxygen uptake (%VO_{2max}) and blood lactate. The strongest stimuli influencing an individual's RPE are breathing/ventilatory work and sensations of strain from the muscles (Cafarelli, 1977, 1982; Chen *et al.*, 2002). Other correlates include perceptions of limb speed, body temperature and joint strain (Robertson and Noble, 1997). A common misunderstanding is to assume that changes in heart rate, oxygen uptake and blood lactate, are factors, which influence RPE. However, one does not actually perceive heart rate, oxygen uptake or the accumulation of muscle and blood lactate. It is the sensations associated with increased ventilatory work and muscle and joint strains, which correspond with these physiological markers, that an individual perceives.

MODES OF USING THE RPE

Traditionally, RPE was developed as a dependent response variable to a given exercise intensity known as *estimation* mode (Noble and Robertson, 1996). Smutok *et al.* (1980) were one of the first to evaluate RPE as an independent exercise intensity regulator. They asked participants to adjust their treadmill running speeds to elicit a given RPE. This is known as *production mode*. This study also raised concern about the ability of some individuals to repeat the same heart rate and running speed for the same RPE, when RPE was used in production mode. From a practitioner's perspective, it is important to confirm that an individual can reliably elicit a given RPE for a given exercise intensity (heart rate or work rate) before being requested or directed to use RPE as a sole intensity regulator.

In this regard, one should not assume that exercise intensity measures derived from estimation–production paradigms are the same; they involve passive and active information processing procedures, respectively. In the *estimation–production* paradigm, objective intensity measures (*expected or derived*) from a previous estimation trial are compared to values produced during a subsequent exercise trial(s) in which the participant *actively* self-regulates exercise intensity levels using assigned RPEs. The memory of exercise experience is particularly relevant in the active paradigm. Following an exercise situation, memory will degrade and may impact upon future active productions. In comparison, the passive paradigm is based upon the interpretation of current stimulation. This information may then be used to compare responses between conditions after some form of intervention or to assist in the prescription of exercise intensity. These considerations are vital when RPE is used in clinical or older populations.

RPE AND RELATIVE MEASURES OF EXERCISE INTENSITY

Table 13.1 summarises the relationship between RPE scores and related physiological markers. During exercise testing or training, the robust relationship between RPE and objective physiological markers may allow the investigator to estimate the participant's relative exercise intensity. For example, once

Table 13.1 Summary of the relationship between the percentages of maximal aerobic power ($\%VO_{2max}$), maximal heart rate reserve ($\%HRR_{max}$), maximal heart rate ($\%HR_{max}$) and Borg's rating of perceived exertion (RPE)

$\%VO_{2max}$	<20	20–39	**40–59**	**60–84**	⩾85	100
$\%HRR_{max}$	<20	20–39	**40–59**	**60–84**	⩾85	100
$\%HR_{max}$	<35	**35–54**	**55–69**	**70–89**	⩾90	100
RPE	<10	10–11	**12–13**	**14–16**	17–19	19–20

Source: adapted from ACSM, 2005; Noble and Robertson, 1996; Pollock *et al.*, 1978

an individual has given an RPE greater than 16 on the RPE scale, or 6–7 on the CR-10 scale, it is highly probable that he/she has surpassed the level where lactate levels may lead to muscular fatigue and accelerated ventilation. This allows the investigator to prepare for test termination. Eston *et al.* (2005) have demonstrated the ability to predict VO_{2max} from a perceptually–guided submaximal exercise test using RPE as the independent variable (production mode). They observed that VO_{2max} values predicted from a series of submaximal RPE: VO_2 values in the range RPE 9–15 and RPE 9–17 were remarkably similar, and within (bias $\pm$ 1.96 $\times$ SDdiff) 2 $\pm$ 8 and $-1. \pm 6$ ml·kg^{-1}·min^{-1}, respectively.

Inspection of the values within Table 13.1 indicates that there is a range of up to 20% of any relative physiological measure for a given RPE. This can be largely attributed to two factors: (1) the subjective nature of rating perceived levels of exertion, and (2) inter-individual differences in training status (Berry *et al.*, 1989; Boutcher *et al.*, 1989; Brisswalter and Delignierè, 1997). Thus, physical training is characterised by a reduction in RPE for a given percentage of maximal HR or VO_{2max}.

FACTORS INFLUENCING RPE

During exercise testing, the inter-trial agreement of either RPE or a concurrent physiological response at a given RPE, increases with each use of the RPE scale (Buckley *et al.*, 2000, 2004; Eston *et al.*, 2000, 2005). Typically, the agreement is shown to be acceptable within three trials when the participant is exposed to a variety of exercise intensities.

Psychosocial factors can influence up to 30% of the variability in an RPE score (Dishman and Landy, 1988; Williams and Eston, 1989). Furthermore the literature has identified numerous modulators of RPE including: the mode of exercise, age, audio-visual distractions, circadian rhythms, gender, haematological and nutritional status, medication, muscle mechanics and biochemical status, the physical environment, and the psycho-social status or competitive milieu of the testing and training environment. These factors are exemplified in Borg's effort continua proposed in 1973 (Borg, 1998). Beta-blocking medication exerts an influence during extended periods of exercise and at intensities greater than 65% VO_{2max} (Eston and Connolly, 1996; Head *et al.*, 1997).

In healthy or clinical populations that may be fearful of the exercise-testing environment (e.g. cardiac patients), it is likely that they will inflate RPE (Morgan, 1973, 1994; Rejeski, 1981; Kohl and Shea, 1988; Biddle and Mutrie, 2001). Such inflation of RPE relates to individuals who either lack self-efficacy or who are unfamiliar or inhibited by the social situation of the exercise training or testing environment.

RPE AND STRENGTH/POWER TESTING AND TRAINING

Up until the late 1990s, most of the evidence in RPE focussed on application and research with aerobic type exercise. There is now a growing body of evidence in the use of monitoring somatic responses to local muscle sensations during resistive or strength training exercise (Borg, 1998; Gearhart *et al.*, 2002; Pincivero *et al.*, 2003; Lagally and Costigan, 2004). The important aspect to consider is that during short-term high-intensity exercise for a localised muscle group, where 8 to 15 repetitions are performed, RPE will grow by one point on the RPE or CR-10 for every 3 to 4 repetitions. For example if after 12 repetitions, one wishes to end his/her last repetition at an RPE of 15 or a CR-10 scale rating of 5 (hard, heavy), then the first or second repetition should elicit an RPE of ~12 or a CR-10 rating of 2 (between light and somewhat hard).

Which scale should I use?

In both Borg's RPE and CR-10 scale, the semantic verbal anchors and their corresponding numbers have been aligned to accommodate for the curvilinear nature (a power function between 1.6 and 2.0) of human physiological responses (Borg, 1998). The CR-10 scale, with its ratio or semi-ratio properties was specifically designed with this in mind. The RPE 6–20 scale was originally designed for whole body aerobic type activity where perceived responses are pooled to concur with the linear increments in heart rate and oxygen uptake, as exercise intensity is increased. The CR-10 scale is best suited when there is an overriding sensation arising either from a specific area of the body, for example, muscle pain, ache or fatigue in the quadriceps or from pulmonary responses. Examples of this individualised or differentiated response have been applied in patients with McArdle's disease (Buckley *et al.*, 2003) and chronic obstructive pulmonary disease (O'Donnell *et al.*, 2004).

PERCEIVED EXERTION IN CHILDREN

Simplified numerical and pictorial scales

There have been important advances in the study of effort perception in children in the last 15 years. The topic has been the subject of several critical reviews with the most recent being (Eston and Lamb, 2000; Eston and Parfitt, 2006). The idea for a simplified perceived exertion scale, which would be more suitable for use with children emanated from the study by Williams *et al.* (1991). They first proposed the idea for a 1–10 scale anchored with more developmentally appropriate expressions of effort. This led to a significant development in the measurement of children's effort perception in 1994 with the publication of two papers (Eston *et al.*, 1994; Williams *et al.*, 1994), which proposed and validated an alternative child-specific rating scale – the Children's Effort Rating Table (CERT, Figure 13.1, Williams *et al.*, 1994).

Compared to the Borg Scale, the CERT has five fewer possible responses, a range of numbers (1–10) more familiar to children than the Borg 6–20 Scale and verbal expressions chosen by children as descriptors of exercise effort. This type of scale facilitates the child's perceptual understanding and therefore the

1 Very, very easy
2 Very easy
3 Easy
4 Just feeling a strain
5 Starting to get hard
6 Getting quite hard
7 Hard
8 Very hard
9 Very, very hard
10 So hard I'm going to stop

Figure 13.1   The Children's Effort Rating Table (CERT, Williams *et al.*, 1994)

ability to use it in either a passive or active paradigm with greater reliability. The CERT initiative for a simplified scale containing more 'developmentally appropriate' numerical and verbal expressions, led to the development of scales which combined numerical and pictorial ratings of perceived exertion scales. All of these scales depict four to five animated figures, portraying increased states of physical exertion. Like the CERT, the scales have embraced a similar, condensed numerical range and words or expressions which are either identical to the CERT (PCERT, Yelling *et al.*, 2002), abridged from the CERT (CALER, Eston *et al.*, 2000; Eston *et al.*, 2001) or similar in context to the CERT (OMNI, Robertson, 1997; Robertson *et al.*, 2000).

The Pictorial CERT (PCERT), initially described by Eston and Lamb (2000), has been validated for both effort estimation and effort production tasks during stepping exercise in adolescents (Yelling *et al.*, 2002). The scale depicts a child running up a 45° stepped grade at five stages of exertion, corresponding to CERT ratings of 2, 4, 6, 8 and 10. All the verbal descriptors from the original CERT are included in the scale.

The OMNI Scale has various pictorial forms. It has been validated for cycling (Robertson *et al.*, 2000), walking/running (Utter *et al.*, 2002) and stepping (Robertson *et al.*, 2005). Robertson and colleagues have also proposed 'adult' versions of the OMNI Scale for resistance exercise and cycling, although we are doubtful of the need to develop such pictorial scales for normal adults, given the well-established validity of the Borg 6–20 RPE and CR-10 Scales. The original idea behind the development of pictorial scales was to simplify the cognitive demands placed on the child. This does not seem necessary in normal adults.

Roemmich *et al.* (2006) have recently validated the OMNI and PCERT scales for submaximal exercise in children aged 11–12 years. They observed no difference in the slopes of the PCERT and OMNI scores when regressed against heart rate or VO_2. There was also no difference in the percentage of maximal PCERT and OMNI at each exercise stage. In effect, the results showed that the two scales could be used with equal validity. Although pictorially different, their results are not that surprising since the scales utilise basically the same limited number range. It perhaps questions the need for pictorial scales for children of this age range.

All the pictorial scales developed so far to assess the relationship between perceived exertion and exercise intensity in children have used either a horizontal line or one that has a linear slope. Eston and Parfitt (2006) have proposed a pictorial 0–10 curvilinear scale which is founded on its inherently obvious face validity. As noted previously (Eston and Lamb, 2000), it is readily conceivable that a child will recognise from previous learning and experience that the steeper the hill, the harder it is to ascend.

FACTORS AFFECTING RPE IN CHILDREN

A discussion of the factors affecting RPE in children is provided in more detail by Eston and Parfitt (2006). The following identifies the key considerations in this group.

As indicated earlier, young children's ability to utilise traditional rating scales is affected by their numerical and verbal understanding. Pictorial scales with a narrower numerical range and fewer verbal references simplify the conceptual demands made on the child. An active paradigm places greater demands upon memory of exercise experience in order to generate a specific intensity in comparison to the passive paradigm that requires an instant response to the current exercise stimulation. Following an exercise situation, memory will degrade and be affected by a combination of factors associated with the three effort continua, particularly the interaction of perceptual and performance variables. This will impact upon future active productions and is an important consideration given the limited memory and range of experience in young children. As in adults, the accuracy of children's effort perception increases significantly with practice (Eston *et al.*, 2000).

The perceived effort response varies according to whether the exercise protocol is intermittent or continuous. The perceived exertion response appears to be higher in a continuous protocol. Intermittent protocols are therefore preferred with young children (Lamb *et al.*, 1997).

KEY POINTS FOR THE EFFECTIVE USE
OF RPE IN ADULTS AND CHILDREN

In considering the factors described throughout this chapter, the following points for instructing participants, patients and athletes have been recommended (adapted from Maresh and Noble, 1984; Borg, 1998, 2004).

1   Make sure the participant understands what an RPE is. Before using the scale see if they can grasp the concept of sensing the exercise responses (breathing, muscle movement/strain, joint movement/speed).
2   Anchoring the perceptual range, which includes relating to the fact that no exertion at all is sitting still and maximal exertion is a theoretical concept of pushing the body to its absolute physical limits. Participants should then be exposed to differing levels of exercise intensity (as in an incremental test or during an exercise session) so as to understand to what

the various levels on the scale feel like. Just giving them one or two points on the scale to aim for will probably result in a great deal of variability.

3   Use the earlier points to explain the nature of the scale and that the participant should consider both the verbal descriptor and the numerical value. The participant should first concentrate on the sensations arising from the activity, look at the scale to see which verbal descriptor relates to the effort he/she is experiencing and then linking this to the corresponding numerical value.

4   Unless specifically directed, ensure that the participant focuses on all the different sensations arising from the exercise being performed. For aerobic exercise, the participant should pool all sensations to give one rating. If there is an overriding sensation then additionally make note of this differentiated rating. Differentiated ratings can be used during muscular strength activity or where exercise is limited more by breathlessness or leg pain, as in the case of pulmonary or peripheral vascular disease, respectively.

5   Confirm that there is no right or wrong answer and it is what the participant perceives. There are three important cases where the participant may give an incorrect rating:

   (a)   When there is a preconceived idea about what exertion level is elicited by a specific activity (Borg, 1998).
   (b)   When participants are asked to recall the exercise and give a rating. As with heart rate, RPEs should be taken while the participant is actually engaged in the movements; not after they have finished an activity.
   (c)   When participants attempt to please the practitioner by stating what should be the appropriate level of RPE. This is typically the case when participants are advised ahead of time of the target RPE (e.g. in education sessions or during the warm-up). In the early stages of using RPE, the participant's exercise intensity should be set by heart rate or work rate (e.g. in METs) and participants need to reliably learn to match their RPE to this level in estimation mode. Once it has been established that the participant's rating concurs with the target heart rate or MET level reliably, then moving them on to production mode can be considered.

6   Keep RPE scales in full view at all times (e.g. on each machine or station or in fixed view in the exercise testing room) and keep reminding participants throughout their exercise session or test to think about what sort of sensations they have while making their judgement rating. Elite endurance athletes are known to be good perceivers, because in a race situation they work very hard mentally to concentrate (cognitively associate) on their sensations in order to regulate their pace effectively (Morgan, 2000).

REFERENCES

American College of Sports Medicine (ACSM) (2005). *Guidelines for Exercise Testing and Prescription*, 7th edn. Baltimore, MD: Lippincott, Williams and Wilkins.

Berry, M.J., Weyrich, A.S., Robergs, R.A., Krause, K.M. and Ingallis, C.P. (1989). Ratings of perceived exertion in individuals with varying fitness levels during walking and running. *European Journal of Applied Physiology and Occupational Physiology*, 58: 494–499.

Biddle, S.J.H. and Mutrie, N. (2001). *Psychology of Physical Activity; Determinants, Well-being and Interventions*. London: Routledge.

Borg, G. (1998). *Borg's Perceived Exertion and Pain Scales*. Champaign, IL: Human Kinetics.

Borg, G. (2004). *The Borg CR10 Folder. A Method for Measuring Intensity of Experience*. Stockholm, Sweden: Borg Perception.

Boutcher, S.H., Seip, R.L., Hetzler, R.K., Pierce, E.F., Snead, D. and Weltman, A. (1989). The effects of specificity of training on the rating of perceived exertion at the lactate threshold. *European Journal of Applied Physiology and Occupational Physiology*, 59: 365–369.

Brisswalter, J. and Deligniere, D. (1997). Influence of exercise duration on perceived exertion during controlled locomotion. *Perceptual and Motor Skills*, 85: 17–18.

Buckley, J.P., Eston, R.G. and Sim, J. (2000). Ratings of perceived exertion in Braille: validity and reliability in production mode. *British Journal of Sports Medicine*, 34: 297–302.

Buckley, J.P., Quinlivan, R.C.M., Sim, J. and Eston, R.G. (2003). Ratings of perceived pain and the second wind in McArdle's disease. *Medicine and Science in Sports and Exercise*, 35(Suppl. 5): 1604.

Buckley, J.P., Sim, J., Eston, R.G., Hession, R. and Fox, R. (2004). Reliability and validity of measures taken during the Chester step test to predict aerobic power and to prescribe aerobic exercise. *British Journal of Sports Medicine*, 38: 197–205.

Cafarelli, E. (1977). Peripheral and central inputs to the effort sense during cycling exercise. *European Journal of Applied Physiology and Occupational Physiology*, 37: 181–189.

Cafarelli, E. (1982). Peripheral contributions to the perception of effort. *Medicine and Science in Sports and Exercise*, 14: 382–389.

Chen, M.J., Fan, X. and Moe, S.T. (2002). Criterion-related validity of the Borg ratings of perceived exertion scale in healthy individuals: a meta-analysis. *Journal of Sports Sciences*, 20: 873–899.

Dishman, R.K. and Landy, F.J. (1988). Psychological factors and prolonged exercise. In D.R. Lamb and R. Murray (eds), *Perspectives in Exercise Science and Sports Medicine*, pp. 281–355. Indianapolis, IN: Benckmark Press.

Eston, R.G. and Connolly, D. (1996). The use of ratings of perceived exertion for exercise prescription in patients receiving β-blocker therapy. *Sports Medicine*, 21: 176–190.

Eston, R.G. and Lamb, K.L. (2000). Effort perception. In N. Armstrong and W. Van-Mechelen (eds), *Paediatric Exercise Science and Medicine*, pp. 85–91, Oxford, UK: Oxford University Press.

Eston, R.G. and Parfitt, G. (2006). Perceived exertion. In N. Armstrong (ed.), *Paediatric Exercise Physiology*. Elsevier, London (in Press).

Eston, R.G., Lamb, K.L., Bain, A., Williams, M. and Williams, J.G. (1994). Validity of a perceived exertion scale for children: a pilot study. *Perceptual and Motor Skills*, 78: 691–697.

Eston, R.G., Parfitt, G., Campbell, L. and Lamb, K.L. (2000). Reliability of effort perception for regulating exercise intensity in children using a Cart and Load Effort Rating (CALER) Scale. *Pediatric Exercise Science*, 12: 388–397.

Eston, R.G., Parfitt, G. and Shepherd, P. (2001). Effort perception in children: implications for validity and reliability. In A. Papaionnou, M. Goudas and Y. Theodorakis (eds), *Proceedings of 10th World Congress of Sport Psychology, Skiathos*, Greece, Volume 5, pp. 104–106.

Eston, R.G., Lamb, K.L., Parfitt, C.G. and King, N. (2005). The validity of predicting maximal oxygen uptake from a perceptually regulated graded exercise test. *European Journal of Applied Physiology*, 94: 221–227.

Gearhart, R.F. Jr, Goss, F.L., Lagally, K.M., Jakicic, J.M., Gallagher, J., Gallagher, K.I. and Robertson, R. (2002). Ratings of perceived exertion in active muscle during high-intensity and low-intensity resistance exercise. *Strength and Conditioning Research*, 16(1): 87–91.

Head, A., Maxwell, S. and Kendall, M.J. (1997). Exercise metabolism in healthy volunteers taking celiprolol, atenolol, and placebo. *British Journal of Sports Medicine*, 31: 120–125.

Kohl, R.M. and Shea, C.H. (1988). Perceived exertion: influences of locus of control and expected work intensity and duration. *Journal of Human Movement Studies*, 15: 225–272.

Lagally, K.M. and Costigan, E.M. (2004). Anchoring procedures in reliability of ratings of perceived exertion during resistance exercise. *Perceptual and Motor Skills*, 98(3 Pt 2): 1285–1295.

Lamb, K.L. and Eston, R.G. (1997). Effort perception in children. *Sports Medicine*, 23: 139–148.

Lamb, K.L., Trask, S. and Eston, R.G. (1997). Effort perception in children: a focus on testing methodology. In N. Armstrong, B.J. Kirby and J.R. Welsman (eds), *Children and Exercise XIX Promoting Health and Well Being*, pp. 258–266. London: E & FN Spon.

Maresh, C. and Noble, B.J. (1984). Utilization of perceived exertion ratings during exercise testing and training. In L.K. Hall, G.C. Meyer and H.K. Hellerstein (eds), *Cardiac Rehabilitation: Exercise Testing and Prescription*, pp. 155–173. Great Neck, NY: Spectrum.

Morgan, W.P. (1973). Psychological factors influencing perceived exertion. *Medicine and Science in Sports and Exercise*, 5: 97–103.

Morgan, W.P. (1994). Psychological components of effort sense. *Medicine and Science in Sports and Exercise*, 26: 1071–1077.

Morgan, W. (2000). Psychological factors associated with distance running and the marathon. In D. Tunstall-Pedoe (ed.), *Marathon Medicine*. London: Royal Society of Medicine Press.

Noble, B. and Robertson, R. (1996). *Perceived Exertion*. Champaign, IL: Human Kinetics.

O'Donnell, D.E., Fluge, T., Gerken, F., Hamilton, A., Webb, K., Make, B. and Magnussen, H. (2004). Effects of tiotropium on lung hyperinflation, dyspnoea and exercise tolerance in COPD. *European Respiratory Journal*, 23(6): 832–840.

Pincivero, D.M., Campy, R.M. and Coelho, A.J. (2003). Knee flexor torque and perceived exertion: a gender and reliability analysis. *Medicine and Science in Sports and Exercise*, 35(10): 1720–1726.

Pollock, M.L., Wilmore, J.H. and Fox, S.M. (1978). *Health and Fitness Through Physical Activity*. New York: Wiley: American College of Sports Medicine Series.

Rejeski, W.J. (1981). The perception of exertion: a social psychophysiological integration. *Journal of Sport Psychology*, 4: 305–320.

Robertson, R.J. (1997). Perceived exertion in young people: future directions of enquiry. In J. Welsman, N. Armstrong and B. Kirby (eds), *Children and Exercise XIX Volume II*, pp. 33–39. Exeter: Washington: Singer Press.

Robertson, R.J. and Noble, B.J. (1997). Perception of physical exertion: methods, mediators and applications. *Exercise and Sports Science Reviews*, 25: 407–452.

Robertson, R.J., Goss, F.L., Boer, N.F., Peoples J.A., Foreman A.J., Dabayebeh I.M., Millich N.B., Balasekaran G., Riechman S.E., Gallagher J.D. and Thompkins T. (2000). Children's OMNI Scale of perceived exertion: mixed gender and race validation. *Medicine and Science in Sports and Exercise*, 32: 452–458.

Robertson, R.J., Goss, J.L., Bell, F.A., Dixon, C.B., Gallagher K.I., Lagally K.M., Timmer J.M., Abt K.L., Gallagher J.D. and Thompkins T. (2002). Self-regulated cycling using the Children's OMNI Scale of Perceived Exertion. *Medicine and Science in Sports and Exercise*, 34: 1168–1175.

Robertson, R.J., Goss, J.L., Andreacci, J.L. Dube J.J., Rutkowski J.J., Snee B.M., Kowallis R.A., Crawford K., Aaron D.J. and Metz K.F. (2005). Validation of the Children's OMNI RPE Scale for stepping exercise. *Medicine and Science in Sports and Exercise*, 37: 290–298.

Roemmich, J.N., Barkley, J.E. and Epstein, L.H., Lobarinas, C.L., White T.M. and Foster J.H. Validity of PCERT and OMNI walk/run ratings of perceived exertion. (2006). Validity of the PCERT and OMNI-walk/run ratings of perceived exertion scales. *Medicine and Science in Sports and Exercise*, 38: 1014–1019.

Smutok, M.A., Skrinar, G.S. and Pandolf, K.B. (1980). Exercise intensity: subjective regulation by perceived exertion. *Archives of Physical Medicine and Rehabilitation*, 61: 569–574.

Utter, A.C., Robertson, R.J., Nieman, D.C. and Kang, J. (2002) Children's OMNI Scale of perceived exertion: walking/running evaluation. *Medicine and Science in Sports and Exercise*, 34: 139–144.

Williams, J.G. and Eston, R.G. (1989). Determination of the intensity dimension in vigorous exercise programmes with particular reference to the use of the rating of perceived exertion. *Sports Medicine*, 8: 177–189.

Williams, J.G., Eston, R.G. and Stretch, C. (1991). Use of rating of perceived exertion to control exercise intensity in children. *Pediatric Exercise Science*, 3: 21–27.

Williams, J.G., Eston, R.G. and Furlong, B. (1994). CERT: a perceived exertion scale for young children. *Perceptual and Motor Skills*, 79: 1451–1458.

Williams, J.G., Furlong, B., MacKintosh, C. and Hockley, T.J. (1993). Rating and regulation of exercise intensity in young children. *Medicine and Science in Sports and Exercise*, 1993, 25 (Suppl. S8) (Abstract).

Yelling, M., Lamb, K. and Swaine, I.L. (2002). Validity of a pictorial perceived exertion scale for effort estimation and effort production during stepping exercise in adolescent children. *European Physical Education Review*, 8: 157–175.

CHAPTER 14

STRENGTH TESTING

Anthony J. Blazevich and Dale Cannavan

INTRODUCTION

Maximum strength can be defined as the 'maximum force or torque a muscle or group of muscles can generate at a specified determined velocity' (Komi *et al.*, 1992: 90–102). Information regarding a person's strength is often sought in order to monitor longitudinal adaptations to training and rehabilitation, compare strength levels between individuals (or groups of individuals), determine the importance of strength to performance in other physical tasks, and to determine single limb or inter-limb strength inadequacies/imbalances.

The three main forms of strength testing are: isometric, isokinetic and isotonic (isoinertial). Importantly, each form of testing measures different qualities so the tests cannot be used interchangeably. This is largely due to the complex interaction of muscular, tendinous and neural factors impacting on strength expression. In order to determine the best possible battery of tests, it is important to consider issues of test specificity and reliability, the safety of subjects and the ease of test administration (and re-administration: for example, reproducibility of environment, subject motivation to re-perform, etc.). However, Abernethy and Wilson (2000: 149) ask five important questions to determine which form/s of strength assessment is/are most appropriate:

1   How reliable is the particular measurement procedure?
2   What is the correlation between the test score and either whole or part of the athletic performance under consideration? (If the performances are not related, are they specific enough to each other?)
3   Does the test item discriminate between the performances of members of heterogeneous and/or homogeneous groups?
4   Is the measurement procedure sensitive to the effects of training, rehabilitation and/or acute bouts of exercise?

5   Does the technique provide insights into the mechanisms underpinning strength and power performance and/or adaptations to training?

It is probably useful to examine each form of testing in relation to these questions.

ISOMETRIC TESTING

Isometric strength testing requires subjects to produce maximum force or torque against an immovable resistance. Force or torque can be measured by a force platform, cable tensiometer, strain gauge, or metal- or crystal-based load cell. Isometric tests can be easily standardised and have a high reproducibility (correlation $(r) = 0.85–0.99$; Abernethy et al., 1995), require minimal familiarisation, are generally easy to administer and safe to perform, can be used to assess strength over various ranges of motion, and can be conducted with relatively inexpensive equipment. Both maximum force (or torque) and the maximum rate of force development (RFD) can be quantified. RFD can be quantified as: (1) the time to reach a certain level of force, (2) the time to attain a relative force level (e.g. 30% of maximum), (3) the slope of the force–time curve over a given time interval or (4) the force or impulse (force $\times$ time) value reached in a specified time.

Despite the many benefits of isometric testing, the relationship between maximum isometric strength and athletic performance is generally poor (correlation coefficients <0.50). Also, isometric tests tend to be insensitive to changes in athletic performance or changes in isotonic or isokinetic strength (Baker et al., 1994; Fry et al., 1994). Thus, while isometric testing might provide information regarding isometric strength and RFD, its use in practical terms is questionable.

The lack of a strong relationship between isometric strength and various dynamic measures might be attributable to: (1) the significant differences in the neural activation of muscles during isometric versus dynamic movements (Nakazawa et al., 1993), or (2) the fact that many dynamic movements are performed with considerable extension and recoil of elastic structures in the muscle–tendon units. This is most notable in movements where the whole muscle–tendon unit is stretched rapidly before shortening, the so-called stretch–shorten cycle. Thus, isometric testing largely examines muscle function under a specific set of conditions, without accounting for the effects of the elastic elements.

Suggested protocol for isometric testing

1   Appropriate warm-up and the performance of several practice contractions should precede testing. While little practice is required for many isometric tests, appropriate muscle control strategies would be developed with consistent practice. Therefore, increases in force seen after several testing sessions might not be completely attributable to changes in the contractile component of the muscle, but indicate some 'learning' of the test.

2   Prolonged stretching performed prior to testing can reduce maximum force production (Fowles *et al.*, 2000; Behm *et al.*, 2001), so stretching during warm-up should be minimal, and repeated exactly in subsequent testing sessions.

3   Repeated testing should be conducted at the same time of day with the same environmental conditions (e.g. room temperature), and after the same pre-testing routine is performed (e.g. warm-up, food intake, training, stimulant use, sleep, etc.).

4   Participants should be highly motivated for every trial. Usually up to three trials should be performed with the best performance recorded.

ISOKINETIC TESTING

Isokinetic testing involves the measurement of force or torque during a movement in which the velocity is constant, and non-zero. Typically, isokinetic testing is performed on a specialised machine where a motor drives a lever or bar to move at a specified speed while force or torque is measured via a load cell or force platform. Isokinetic testing is commonly used to profile the force – velocity or torque – angle characteristics of a muscle group or limb, muscle fatigability and recovery, or joint range of motion. Both eccentric (muscle lengthening) and concentric (muscle shortening) strength can be tested, although it is not possible to test stretch–shorten actions with most standard isokinetic machines. Isokinetic testing has been largely thought to provide information about the capacity of muscle, rather than tendon. While this is largely true, since gravitational energy cannot be stored in elastic structures during isokinetic testing, recent research examining concentric actions has shown that there is significant tendon lengthening early in a movement with subsequent recoil of the tendon as the movement proceeds; this phenomenon is greater at higher movement speeds (Ichinose *et al.*, 2000).

The reproducibility of isokinetic force or torque depends on the type of movement and the speed at which it is performed. Subject force/torque reliability is generally good, or very good, provided the subject has had several familiarisation sessions and a strict testing protocol is followed (discussed later). A slight exaggeration of force is sometimes seen early in a movement – the so-called torque overshoot, which is small at slow speeds but larger at high speeds. This occurs as the limb gains momentum and impacts with the cuff or pad onto which the limb is moving. Modern systems use damping mechanisms or impose a controlled period of acceleration to reduce the overshoot, however it might be necessary to set the dynamometer to move through a larger range of motion than is required so that some data can be excluded from analysis. Isokinetic machines generally move within $1°\cdot s^{-1}$ of the set speed, with a larger discrepancy (up to $2°\cdot s^{-1}$) occurring in the period of overshoot, so speed measurements can be considered accurate and reliable. If necessary, the accuracy and reliability of movement speed can be assessed using motion analysis.

The validity of isokinetic testing is likely to decrease as the test movement pattern becomes less similar to the task movement pattern, particularly if the task involves a stretch–shorten cycle action or the complexity of the task increases above that of the isokinetic test. It is therefore necessary to examine the specificity of isokinetic testing before adopting it.

Protocol for testing on an isokinetic dynamometer

1   As per points 1–3 for isometric testing; although several familiarisation sessions may be required.
2   Participants should be highly motivated for every trial.
3   The subjects should be tightly secured to the seat or bench in order that extraneous movements do not significantly impact on force development (refer to manufacturer's guidelines). The limb to be tested should be tightly secured to the machine using the straps provided. Certain machines (e.g. Kin-Com; Chatanooga Inc., USA) require accurate recording of the positioning of the attachment so that torque (force × distance) can be reliably calculated.
4   The axis of rotation of the lever of the dynamometer should be adjacent to the centre of rotation of the joint being assessed. When large forces are produced, there will be unavoidable movement of the subject and flexion of the dynamometer mountings, so alignment will not be properly maintained. This is rarely problematic and is reasonably consistent between trials, however quantification of misalignment, and correction using mathematical means, can be performed by combining video analysis with force or torque measurement.
5   Gravity correction should be performed as per the manufacturer's guidelines. The error created when gravity correction is not performed is greater for fatigue tests than tests of maximum strength.
6   Ranges of motion and movement speeds need to be carefully considered.
7   Up to three trials should be performed with the best performance recorded, although it is often necessary to increase the number of repetitions when moving continuously at higher movement speeds (e.g. for knee extension, 4 and 5 trials should be performed at 180° and 300°·s^{-1}, respectively) in order for the subject to become accustomed to that speed. Fatigue tests may vary in their repetition number, however 30 and 50 repetitions are commonly used.
8   Rest intervals should be greater than 30 s, although up to 4 min may be required when testing at slower speeds where perceived exertion and contraction time are greater.
9   Test order generally progresses from slower speeds to faster and should be the same at consecutive testing sessions; this test order shows high reliability (Wilhite et al., 1992).

ISOTONIC (ISOINERTIAL) TESTING

Isotonic strength testing involves moving a fixed mass with constant acceleration and deceleration. Since acceleration changes with joint angle during a movement, these movements are probably more correctly described as isoinertial (Abernethy and Jürimäe, 1996). Common tests of isoinertial strength include the one-repetition maximum (1-RM) tests such as the maximal bench press or barbell squat tests, maximal concentric and eccentric strength tests, static and countermovement vertical and horizontal jumps (with and without additional load), throwing tests, cycle ergometer and sprint running/swimming tests. Performance can be measured via force platforms and load cells (with measures described as peak forces/torques, RFD, work/power, force decrements, etc.), by the maximum weight lifted, or by the distance/height thrown or jumped, etc.

Since many sports require the acceleration of a mass with constant inertia, such tests generally show higher task validity than isometric and isokinetic tests. Correlations with athletic task performance are generally high when the movement characteristics match those of the task, but are reduced as they differ more widely (see Table 14.1). Since the use of a variety of tests can provide a greater amount of information about the factors affecting strength (e.g. muscle recruitment potential, elastic energy storage and recovery, maximal muscle contraction force, work rates and fatigue indices, etc.), it might sometimes be necessary to use tests that do not correlate highly with task performance. Test performances are also very sensitive to change after periods of isoinertial strength, power or sprint training, although changes may not be seen after isometric or isokinetic training.

Reliability of isoinertial tests varies largely depending on the test performed and the experience of the subject. For traditional maximum strength tests such as the 1-RM bench press or squat, reliability is very high ($r = 0.92$–0.98 typically) when strict procedures are followed. When the 1-RM is predicted by mathematical equations from 3 to 10 RM lifting tests, reliability is slightly reduced ($r = 0.89$–0.96) and results may vary by up to 2.5 kg depending on which equation is used (see Table 14.2 for examples).

Table 14.1 Correlation between sprint running and selected test performances

Isoinertial test	Correlation coefficient	
	2.5 m sprint (ct ≈ 0.17–2.1 s)	Fastest 10 m time (ct ≈ 0.9–1.2 s)
F30 (t = 30 ms)	−0.46	−0.49
MAX RFD (t = 56 ms)	−0.62	−0.73
F100/WEIGHT (t = 100 ms)	−0.73	−0.80
MDS/WEIGHT (t = 121 ms)	−0.86	−0.69

Notes
The validity of isoinertial strength tests increases as the movement characteristics become more similar to the performance task. In the above example, correlations between sprint running performance (2.5 m time and 'fastest 10-m' times) and test performance are greatest when test duration is most similar to contact time (ct), for example, performance in the F100/weight test, which takes approximately 100 ms, correlates highest with 10 m time, where the foot-ground contact time is also approximately 100 ms

F30: force developed in the first 30 ms of a weighted (19 kg) jump squat from 120° knee angle; MAX RFD: maximum rate of force development in jump squat (which occurred [mean] 0.056 s after movement initiation); F100/WEIGHT: force applied 100 ms into jump squat; MDS/WEIGHT: maximum force developed during jump squat normalised for bodyweight (occurred [mean] 0.121 s after movement initiation). For more detail, see Young *et al.* (1995)

Table 14.2 Examples of equations used to calculate 1-RM lifting performances from multiple maximal repetitions

Test exercise	Equation	Correlation	Difference between achieved and predicted 1RM[d]	Cross validation reference
Bench press	100·repetition mass/ (52.2 + 41.9·exp[0.055·reps])[a]	$r = 0.992$	0.5 kg ± 3.6 kg	LeSuer *et al.*, 1997
Squat	100·repetition mass/ (48.8 + 53.8·exp[0.075·reps])[b]	$r = 0.969$	0.5 kg ± 3.5 kg	LeSuer *et al.*, 1997
Multiple exercises	100·repetition mass/ (102.78–2.78·reps)[c]	$r = 0.633$– 0.896	Not available	Knutzen *et al.*, 1999

Notes
a Formula from Mayhew *et al.* (1992)
b Formula from Wathan (1994)
c Formula from Brzycki (1993)
d Mean difference between predicted and actual 1RM from a cross validation of prediction equations (LeSuer *et al.*, 1997)

Also, while factors such as gender appear not to affect prediction accuracy, the reliability of estimates seems to be greater for some lifts (e.g. bench press and leg press) than others (e.g. barbell curl and knee extension) (Hoeger *et al.*, 1990). Reliability of weighted jumps measured on a force platform or contact mat are good, although it tends to be reduced as loads increase. Sprint running and cycling tests usually show very good reliability, although performance reliability of well-trained subjects is usually higher than that of lesser-trained subjects. Importantly, reliability of these tests should be determined in the test population before being used to assess performance.

Protocol for isotonic (isoinertial) testing

1   As per points 1–3 for isometric testing.
2   Subjects may require several familiarisation sessions before test reliability is acceptable; familiarisation of 1-RM lifts is usually more rapid than for higher-velocity tests.
3   Participants should be highly motivated for every trial.
4   Tests should be selected that are closely related to the athletic or rehabilitation task being trained for, and/or provide significant information about the functioning of a specific part of the neuromusculotendinous system.
5   Most tests can be performed with a wide variety of techniques. A technique should be chosen that has close specificity to the task being trained for, and should be performed identically on subsequent testing occasions (particular attention should be paid to the techniques adopted for weighted and drop jump tests).
6   For more detail regarding isoinertial strength testing, see Logan *et al.* (2000: 200–221).

OTHER CONSIDERATIONS IN TEST SELECTION

A range of tests should be selected so that as much information as possible is available to assess inter-individual differences, monitor training/rehabilitation progress, or examine a person's strengths and weaknesses. Consideration should be given to using a selection of isometric, isokinetic and isoinertial tests. For example, testing isokinetic knee extensor, knee flexor and ankle plantarflexor strength at a range of velocities will allow some determination of the capacity of the muscle groups to generate force over a range of contraction speeds and ranges of motion. Also, the ratio of squat (static) jump to countermovement jump height has been shown to provide a useful indication of the compliance of tendon structures in the lower limb (Kubo *et al.*, 1999). Thus, this test battery provides the examiner with information about both muscle and tendon function and allows appropriate training plans to be developed for a specific purpose. It is clear then, that the design of optimum test batteries requires a good scientific knowledge, and some creative design.

REFERENCES

Abernethy, P. and Jürimäe, J. (1996). Cross-sectional and longitudinal uses of isoinertial, isometric, and isokinetic dynamometry. *Medicine and Science in Sports and Exercise*, 28: 1180–1187.

Abernethy, P. and Wilson, G. (2000). Introduction to the assessment of strength and power. In C.J. Gore (ed.), *Physiological Tests for Elite Athletes*. Champaign, IL: Human Kinetics.

Abernethy, P., Wilson, G. and Logan, P. (1995). Strength and power assessment. Issues, controversies, and isokinetic dynamometry. *Sports Medicine*, 19: 401–417.

Baker, D., Wilson, G. and Carlyon, B. (1994). Generality versus specificity: a comparison of dynamic and isometric measures of strength and speed-strength. *European Journal of Applied Physiology*, 68: 350–355.

Behm, D.G., Button, D.C. and Butt, J.C. (2001). Factors affecting force loss with prolonged stretching. *Canadian Journal of Applied Physiology*, 26: 261–272.

Brzycki, M. (1993). Strength testing – predicting a one-rep max from reps-to fatigue. *Journal of Health Physical Education Recreation and Dance*, 64: 88–90.

Fowles, J.R., Sale, D.G. and MacDougall, J.D. (2000). Reduced strength after passive stretch of the human plantarflexors. *Journal of Applied Physiology*, 89: 1179–1188.

Fry, A.C., Kraemer, W.J., van Borselen, F., Lynch, J.M., Marsit, J.L., Roy, E.P., Triplett, N.T. and Knuttgen, H.G. (1994). Performance decrements with high intensity resistance exercise overtraining. *Medicine and Science in Sports and Exercise*, 26: 1165–1173.

Hoeger, W.W.K., Hopkins, D.R. and Barette, S.L. (1990). Relationship between repetitions and selected percentages of one repetition maximum: a comparison between trained and untrained males and females. *Journal of Applied Sports Science Research*, 4: 47–54.

Ichinose, Y., Kawakami, Y., Ito, M., Kanehisa, H. and Fukunaga, T. (2000). In vivo estimation of contraction velocity of human vastus lateralis muscle during 'isokinetic' action. *Journal of Applied Physiology*, 88: 851–856.

Komi, P.V., Suominen, H., Keikkinen, E., Karlsson, J. and Tesch, P. (1992). Effects of heavy resistance training and explosive type strength training methods on mechanical, functional and metabolic aspects of performance. In P.V. Komi (ed.), *Exercise and Sports Biology*, Champaign, IL: Human Kinetics.

Kubo, K., Kawakami, Y. and Fukunaga, T. (1999). Influence of elastic properties of tendon structures on jump performance in humans. *Journal of Applied Physiology*, 87: 2090–2096.

LeSuer, D.A., McCormick, J.H., Mayhew, J.L., Wasserstein, R.L. and Arnold, M.D. (1997). The accuracy of prediction equations for estimating 1-RM performance in the bench press, squat, and deadlift. *Journal of Strength and Conditioning Research*, 11: 211–213.

Logan, P., Fornasiero, D., Abernethy, P. and Lynch, K. (2000). Protocols for the assessment of isoinertial strength. In C.J. Gore (ed.), *Physiological Tests for Elite Athletes*. Champaign, IL: Human Kinetics.

Mayhew, J.L., Ball, T.E., Arnold, M.D. and Bowen, J.C. (1992). Relative muscular endurance performance as a predictor of bench press strength in college men and women. *Journal of Applied Sports Science Research*, 6: 200–206.

Nakazawa, K., Kawakami, Y., Fukunaga, T., Yano, H. and Miyashita, M. (1993). Differences in activation patterns in elbow flexors during isometric, concentric and eccentric contractions. *European Journal of Applied Physiology*, 66: 214–220.

Wathan, D. (1994). Load assignment. In T.R. Baechle (ed.), *Essentials of Strength Training and Conditioning*. Champaign, IL: Human Kinetics.

Wilhite, M.R., Cohen, E.R. and Wilhite, S.C. (1992). Reliability of concentric and eccentric measurements of quadriceps performance using the KIN-COM dynamometer: the effect of testing order for three different speeds. *Journal of Orthopaedic and Sports Physical Therapy*, 15: 175–182.

Young, W., McLean, B. and Ardagna, J. (1995). Relationship between strength qualities and sprinting performance. *Journal of Sports Medicine and Physical Fitness*, 35: 13–19.

CHAPTER 15

UPPER-BODY EXERCISE

Paul M. Smith and Mike J.Price

PREAMBLE

Upper-body exercise testing holds important practical applications for many populations including specifically trained competitors who pursue events such as canoeing and kayaking, and individuals who do not have the habitual use of their legs. Furthermore, this mode of exercise can be useful in clinical rehabilitation.

Several testing rigs such as kayak and wheelchair ergometers and swim benches have been designed to mimic the movement patterns and physiological demands of specific upper-body sports. However, arm crank ergometry (ACE) provides the sport and exercise scientist with a generic means by which physiological responses and adaptations of individuals to upper-body exercise can practically be examined. Work with ACE has concentrated principally on the development of protocols used to examine individual aerobic and anaerobic exercise capability. Nevertheless, few recommendations for exercise testing exist in this area. While electrically braked arm ergometers are now increasingly available, most laboratories use less expensive, suitably adapted friction-braked cycle ergometers to perform ACE tests. In some instances the use of a friction-braked ergometer will make the implementation of some of the testing protocols (e.g. ramp testing) problematic. Where such issues arise alternative protocol designs have been recommended.

AEROBIC TESTING

Previous studies have concentrated on methodological aspects including the use of continuous or discontinuous protocols, effects of crank rate selection and pattern by which exercise intensity changes. An important point to note is that the continuation of exercise during incremental ACE protocols is predominantly constrained by peripheral as opposed to centrally limiting factors. Consequently, in assessments of maximum oxygen uptake the term VO_{2peak} is preferred (refer to Chapter 5 for general information on methodological aspects of aerobic testing).

BODY POSITION

While some studies have required subjects to stand, the majority have adopted unrestrained, seated positions. The crankshaft of the ergometer is usually horizontally aligned with the centre of the glenohumeral joint. The subject is required to sit at a distance from the ergometer so that with their back vertical the arms are slightly bent at the furthest horizontal point of the duty cycle.

Variations in procedures used to brace either the legs and/or torso have been reported. To reflect what the athlete might experience in the field bracing is not necessary. However, it is recommended that a standard and consistent procedure be adopted where subjects should keep their back vertical with their feet flat on the floor and their knees at 90°.

Discontinuous protocols can be used in an attempt to postpone peripheral fatigue, though similar sub-maximal and peak physiological responses have been reported compared to continuous tests. The use of discontinuous protocols can be advantageous if a supplementary measurement such as blood pressure is required.

CRANK MODE AND RATE

The majority of studies adopt asynchronous cranking, though direct comparisons of physiological responses to exercise are available for synchronous and asynchronous modalities. It has consistently been demonstrated that influences in crank rate effect submaximal and peak physiological responses during ACE even when differences in the internal work needed simply to move the limbs is considered. At any given work rate during incremental exercise mechanical efficiency is lower using a faster crank rate resulting in greater energy expenditure. Previous editions of the BASES testing guidelines published in 1986 and 1988 recommended that a crank rate of 60 rev.min^{-1} be used with ACE. However, more recent work has shown that the use of a faster crank rate (70 and 80 rev.min^{-1}) elicits a higher and therefore, more valid peak physiological responses (Price and Campbell, 1997; Smith et al., 2001). The principal reason for this is that a faster crank rate will postpone the onset of peripheral muscular fatigue ensuring higher exercise intensities are achieved during incremental exercise (Smith et al., 2001). Faster crank rates also lead to lower differentiated ratings of perceived exertion (RPE) associated with perceptions of localised fatigue and strain in the active musculature. Conversely higher central ratings have being reported at the point of volitional exhaustion using a faster crank rate (please refer to Chapter 13 for further information on the use of RPE scales). It should be noted that if the crank rate employed is too slow (50 rev.min^{-1}) or too fast (90 rev.min^{-1}) premature fatigue can occur.

INCREMENTS

The initial exercise intensity and subsequent increases crucially influence the duration of a test designed to assess peak aerobic capacity. Step or ramp tests can elicit peak physiological responses, though it is important to note that they should not be used interchangeably (Smith *et al.*, 2004). Typically the VO$_{2peak}$ test should last between 8 and 15 min and a standard graded exercise test can be adopted, as illustrated in Figure 15.1. It is important to note that following the initial 3 min warm-up period, the total amount of work completed during each successive 2 min stage is equivalent between tests.

In any test to volitional exhaustion there is a trade-off between the duration and number of exercise stages that can be completed. Usually 2 min exercise stages during stepwise ACE protocols: (1) permit a valid measurement of peak physiological responses, and (2) evaluate the influence of changes in exercise intensity on the evolution of physiological responses to the point of volitional exhaustion.

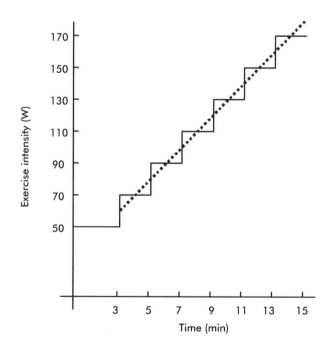

Figure 15.1   An illustration of step (solid lines) and ramp (broken line) patterns of increases in exercise intensity used to elicit peak physiological responses during a graded arm crank ergometry test

Table 15.1 Recommended starting and increments in exercise intensity (W) for graded arm crank ergometry tests

		Step test		Ramp test	
		Start (W)	Increment (W)	Start (W)	Increment (W)
Male	Trained	50	30 every 2 min	50	1 every 4 s
	Untrained	40	20 every 2 min	40	1 every 6 s
Female	Trained	30	15 every 2 min	30	1 every 8 s
	Untrained	20	10 every 2 min	20	1 every 10 s

PROTOCOL RECOMMENDATIONS

Table 15.1 summarises recommendations for trained and untrained, male and female subjects where step or ramp tests are used. Time permitting, we recommend an individualised approach to both starting and increments in exercise intensity. To this end, an initial incremental protocol should be conducted following the guidelines in Table 15.1 to establish the mean exercise intensity achieved during the final minute of the test or peak minute power (PMP). A test can then be designed with a starting (warm-up) intensity equivalent to 30% of PMP, with subsequent stepwise increments of 10% of PMP every 2 min. If a ramp test is used the same initial exercise intensity can be adopted for the warm-up with subsequent increments of 1% of PMP every 12 s.

TYPICAL VALUES AND REPRODUCIBILITY

Information relating to normative values and the reproducibility of the parameters associated with ACE testing is limited. In non-specifically trained groups of men, typical values of VO_{2peak} range from 2.5 to 3.2 $l \cdot min^{-1}$. In specifically trained men values in excess of 3.5 $l \cdot min^{-1}$ are frequently observed. Limited information is available for women however it is likely that non-specifically trained groups would typically achieve VO_{2peak} values from 1.5 to 2.0 $l.min^{-1}$. For trained women 1.7–2.6 $l \cdot min^{-1}$ could be anticipated. An acceptable typical error measurement for VO_{2peak} on a test–retest basis is ±5% for all groups.

The value of PMP achieved by a non-specifically trained group of men ranges from 110 to 160 W. Specifically trained men are able to achieve PMP values ranging from 170 to 270 W. For non-specifically trained women PMP values of 70–110 W can be achieved, while trained female groups may achieve values of 100–120 W. An acceptable typical error measurement for all PMP values for all groups on a test–retest basis is ±3%.

MAXIMAL INTENSITY EXERCISE

Considerable variations in methodological procedures associated with all-out exercise tests are also evident. Differences include equipment employed, body position adopted, resistive load used, and methods of power measurement and reporting.

Adapted friction-braked ergometers are most frequently used, with a standard resistive load usually determined according to a percentage of body weight. Studies tend to use resistive loads equivalent to 2.0–8.0% body weight. Commercially available computer software programs now permit the simultaneous measurement of uncorrected and corrected power outputs. Uncorrected data relate to, and can be used to assess, the force–velocity relationship of muscle. Corrected data is more concerned with the relationship between instantaneous torque production, which in turn, is calculated using algorithms that take into consideration information relating to the inertial characteristics of the heavy flywheel, the resistive (braking) load applied and the instantaneous rate of flywheel de-/acceleration. Most recent studies have recorded and reported values of corrected peak power output (PPO) over 1 s.

It is unlikely that subjects will be accustomed to the requirements placed upon them during an all-out upper-body sprint effort. Therefore, it is highly recommended that several practise tests are performed. For the purpose of standardisation, a similar seated position to that described earlier for the VO_{2peak} testing should be adopted though we recommend that for sprint tests the legs and ankles should be firmly braced.

PROTOCOL DESIGN

Although the Wingate Anaerobic Test (WAnT) was originally designed to last 30 s we have tended to use a 20 s test for the purpose of measuring values of PPO and mean power output (MPO) due to the rapid onset of fatigue. This is of particular relevance when high resistive loads are used. The test should be preceded by a standardised 3–5 min warm-up using a crank rate of 70 rev·min^{-1}, and a resistive load of 9.81 N. It is also recommended that during the latter stages of the warm-up several practise starts should be made with the resistive load to be used during the test. Thereafter, subjects are required to perform an all-out sprint effort for the full duration of the test. A rolling start of 70 rev·min^{-1} is recommended though it is important to start logging of data after the resistive load has been applied and before the subject begins to accelerate the flywheel so as to avoid a flying start. Once the subject has completed the test the resistive load should be reduced to 9.81 N and the subject should complete a warm-down using a crank rate of their choice. If repeated 20- or 30-s tests are performed it is recommended that at least 60 min of passive re covery be allowed between tests.

RESISTIVE LOAD

Recommendations associated with resistive loads are presented here as percentages of body weight. It is clear that the force–velocity relationship applies to the upper-body as it does for the legs. Corrected values of PPO are greater than uncorrected PPO values, and typically occur within the first 5 s of the test. However, few differences exist between uncorrected and corrected values of MPO.

When uncorrected PPO data are considered loading strategy is influential. Generally, uncorrected PPO increases as the resistive load increases (Figure 15.2). In contrast, a considerable inter-subject variation exists for corrected PPO and the resistive load used to elicit an optimal value (Figure 15.2). With this in mind, and time permitting, we recommend that an individual approach be used with respect to identifying the 'optimised' resistive load required to elicit corrected PPO. To achieve this, repeated 10 s tests are required employing the full range of resistive loads presented in Table 15.2. At least 20 min of passive recovery is required between tests and the order in which the resistive loads are used should be randomised. The test that elicits the highest value of corrected PPO represents the optimised resistive load. It is important to note that while an optimised value of corrected PPO will be achieved, this approach is unlikely to permit the concomitant assessment of an optimised value of uncorrected PPO or un-/corrected values of MPO. To achieve such measurements separate optimisation procedures would have to be conducted with respect to the power output measurement of interest.

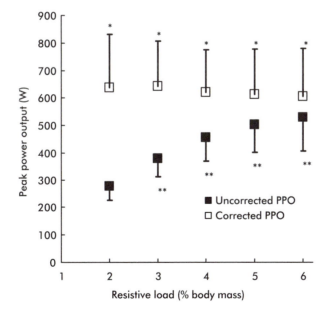

Figure 15.2 Mean (+/−SD) values of uncorrected and corrected PPO achieved using a range of resistive loads during an optimisation procedure conducted by 25 untrained men

Notes

* denotes corrected PPO greater ($P < 0.05$) than uncorrected PPO

** denotes the value of uncorrected PPO is greater ($P < 0.05$) compared to the value achieved using the previous resistive load

Table 15.2 Recommended ranges of resistive loads to be used in association with an optimisation procedure linked to corrected peak power output

Groups	Resistive load (% body weight)				
Trained males	4	5	**6**	7	8
Trained females	2	3	**4**	5	6
Untrained males	3	4	**5**	6	7
Untrained females	1	2	**3**	4	5

Note

Values in **bold and italicised** text represent the standard resistive loads that should be used by the respective subject populations if a single test is to be used

TYPICAL VALUES AND REPRODUCIBILITY

Typical values of uncorrected and corrected 1 s PPO for a non-specifically trained group of men range from 300 to 550 W and 400 to 700 W, respectively. For a specifically trained male group, respective values of 550–800 W and 600–1,100 W may be achieved. There is generally little difference between uncorrected and corrected values of MPO measured over 20 or 30 s. A non-specifically trained group of men would be able to achieve values ranging from 250 to 600 W, while a specifically trained group would achieve 400–700 W.

Typical values of uncorrected and corrected 1 s PPO for a group of non-specifically trained women range from 175 to 250 W and 200 to 300 W, respectively. For a specifically trained female group typical values would range from 200 to 300 W and 250 to 400 W, respectively. For 20 s MPO, a non-specifically trained group of women would be expected to achieve values from 175 to 250 W, while a specifically trained group can achieve values between 250 and 400 W. An acceptable level of typical error measurement for all power output values on a test–retest basis is ±5%.

REFERENCES

Price, M.J. and Campbell, I.G. (1997). Determination of peak oxygen uptake during upper body exercise. *Ergonomics*, 40, 491–499.

Smith, P.M., Price, M.J. and Doherty, M. (2001). The influence of crank rate on peak oxygen uptake during arm crank ergometry. *Journal of Sports Sciences*, 19, 955–960.

Smith, P.M., Doherty, M., Drake, D. and Price, M.J. (2004). The influence of step and ramp type protocols on the attainment of peak physiological responses during arm crank ergometry. *International Journal of Sports Medicine*, 25, 616–621.

PART 4

SPORT-SPECIFIC PROCEDURES

CHAPTER 16

MIDDLE- AND LONG-DISTANCE RUNNING

Andrew M. Jones

The physiological bases to success in endurance sports events, which include middle-distance (800 m and 1,500 m) and long-distance running (3,000 m steeple-chase to the marathon), have been extensively described (e.g. Sjodin and Svedenhag, 1985; Brandon, 1995; Coyle, 1995). In these events, oxidative phosphorylation represents the principal energy-producing metabolic pathway and, therefore, it is not surprising that the parameters of fitness which correlate most closely with performance are those related to oxygen uptake, that is, the maximal rate of oxygen uptake ($\dot{V}O_{2max}$), the oxygen uptake required to run at various sub-maximal speeds (running economy), and the oxygen uptake that can be sustained without appreciable accumulation of lactate in the blood (lactate threshold, maximal lactate steady state). However, although distance running events are principally aerobic in nature, there remains a small but significant contribution from substrate phosphorylation (phosphocreatine depletion and anaerobic glycolysis ending in lactate) in the middle-distance events that should not be neglected in the physiological evaluation of these athletes (Hill, 1999).

It can be estimated that, at the elite level: the 800 m event requires the energetic equivalent of ~120% $\dot{V}O_{2max}$; the 1,500 m, 110% $\dot{V}O_{2max}$; the 3,000 m steeplechase, ~100% $\dot{V}O_{2max}$, the 5,000 m, ~96% $\dot{V}O_{2max}$; the 10,000 m, ~92% $\dot{V}O_{2max}$; and, the marathon, ~85% $\dot{V}O_{2max}$ (Londeree, 1986). From this, it is clear that: anaerobic capacity becomes a progressively less important determinant of performance capability as the race distance increases; $\dot{V}O_{2max}$ is an important determinant of success in all events, but perhaps particularly so at 1,500–5,000 m; and 'sub-maximal' physiological parameters such as running economy and lactate threshold/turn-point become progressively more important determinants of success when the race distance exceeds 5,000 m.

The tests which follow are those that have been adopted for use in the physiological evaluation of middle-distance and long-distance runners by UK athletics. The 'main' treadmill test permits the measurement of running economy, lactate threshold and lactate turn-point, and $\dot{V}O_{2max}$, all within a period of 20–25 min. Other 'adjunct' tests which are useful in all middle- and long-distance runners are also described, along with recommended tests for the assessment of anaerobic capability in middle-distance specialists.

TREADMILL TEST

The treadmill test is administered in two parts. The first part of the test is a multi-stage incremental test for the determination of oxygen uptake, heart rate and blood lactate responses to a range of running speeds that might be used in training and competition. After a fingertip blood sample has been taken for the determination of resting blood [lactate], the athlete performs their individual warm-up regimen which might include 10–15 min jogging and some stretching. Athletes should be fitted with a telemetric heart rate monitor.

The treadmill grade should be set at 1% (Jones and Doust, 1996). The test starts by gradually increasing the treadmill belt speed to the speed required for the first stage. The speed for this first stage can be determined from Table 16.1.

Each stage is 3 min in duration and treadmill belt speed is increased by 1.0 km·h^{-1} at the end of each stage. Subjects should complete a minimum of five stages and a maximum of nine stages in this phase of the test. Half-way through each stage (i.e. after 90 s), the athlete should be handed the nose-clip and mouthpiece and asked to put them in position. If expired air is to be collected using Douglas bags, then the collection should be made over an accurately timed period within the last 50–60 s of each stage. The average heart rate in the last 30 s of each stage should be recorded. At the end of the stage, the athlete should place his or her hands on the guard-rails at the side of the treadmill, and lift their legs so that they are astride the moving treadmill belt. A fingertip blood sample should then be taken as quickly as possible using standardised procedures (i.e. alcohol swab, wipe with tissue, puncture, wipe away first drops of blood with tissue, collect into capillary tube). The blood collection procedure should take ~15–30 s. When the blood has been collected, the athlete should take their weight on their hands, match their cadence to treadmill belt speed and resume running. The treadmill belt speed is increased by 1.0 km·h^{-1} and the clocks are reset to time the next 3 min stage. This phase of the test should be terminated when the subject is close to, but has not

Table 16.1 Treadmill belt speed required for the first stage

World class performance male: 15.0 km·h^{-1}
World class performance female: 13.0 km·h^{-1}
World class potential male: 14.0 km·h^{-1}
World class potential female: 12.0 km·h^{-1}
Regional junior male: 13.0 km·h^{-1}
Regional junior female: 11.0 km·h^{-1}

reached, exhaustion. Generally, blood [lactate] will be >4 mM, heart rate will be within 5–10 b·min^{-1} of maximum, and the subjects would only be able to complete one more stage if they were required to continue. This phase of the test is used to determine running economy, lactate threshold and lactate turn-point (see later and also 'Lactate Testing' chapter), and to identify heart rate training zones. In 10,000 m and particularly marathon runners, in whom the accuracy of $\dot{V}O_{2max}$ determination is perhaps of relatively less importance, this multi-stage test can be continued until exhaustion; in this case, the highest $\dot{V}O_2$ attained during the last stage is typically within 5% of the $\dot{V}O_{2max}$ determined using more traditional tests. Also, in these longer-distance specialists, the increment in running speed might be reduced to 0.5 km·h^{-1} to permit a more accurate determination of the running speeds corresponding to lactate threshold and lactate turn-point.

The subject should be allowed ~15 min to recover (involving jogging and walking on the treadmill and around the laboratory). The exact time allowed for recovery is not especially important but the athlete should still feel warm, but recovered and ready to give a maximal effort. The second, and final, phase of the test is used to measure $\dot{V}O_{2max}$. The athlete should be instructed to put in a maximal effort and to run for as long as possible.

The treadmill belt speed to be used in this test can be calculated as follows:

final speed on the first part of the test – 2.0 km·h^{-1}

For example, an athlete whose speed on the final stage of the sub-maximal test was 21.0 km·h^{-1}, should complete the $\dot{V}O_{2max}$ test at a speed of 19.0 km·h^{-1}. The initial treadmill grade is set at 1%. The athlete should be taken up to the start speed of the test over a period of 1–2 min. When the start speed is reached, the clocks are re-set and the test begins. The treadmill belt speed remains the same throughout the test. Treadmill gradient should be increased by 1% at the end of each minute. Test duration is normally around 6–8 min in this phase of the test, but it can be as little as 5–6 min or as long as 9–10 min in some athletes. The athlete may run without respiratory apparatus for the first 3 min, but after this point the mouthpiece and nose-clip should stay in place until the end of the test. If Douglas bags are used, expired air should be collected over a timed period within the final 45 s of each minute. A fingertip blood sample should be taken immediately post-exercise and perhaps at 1, 3, 5, 7 and 10 min of recovery if peak blood [lactate] is of interest (see 'Lactate Testing' chapter).

DATA ANALYSES

Running economy

A common method for assessing an athlete's running economy is to look at the $\dot{V}O_2$ in ml·kg^{-1}·min^{-1} at 16.0 km·h^{-1} and 1% grade (i.e. 6:00 min·mile^{-1} pace). The average $\dot{V}O_2$ in well trained runners at this speed is ~52 ml·kg^{-1}·min^{-1}.

Therefore, a rough assessment of an athlete's economy can be made by comparing his or her $\dot{V}O_2$ at 16 km·h^{-1} with this value:

44–47 ml·kg^{-1}·min^{-1} is excellent
47–50 ml·kg^{-1}·min^{-1} is very good
50–54 ml·kg^{-1}·min^{-1} is average
55–58 ml·kg^{-1}·min^{-1} is poor.

However, because an individual athlete's economy can vary according to the speed used and because 16.0 km·h^{-1} is quite a high speed for younger athletes, it can be better to assess economy over a range of speeds or to choose a more appropriate speed such as 12.0 km·h^{-1} or 14.0 km·h^{-1} depending on the athlete's performance level. The expected $\dot{V}O_2$ for a certain speed can be calculated using the following equation:

$$\dot{V}O_2 = 13.5(\text{speed}) - 8.5$$

where speed is in metres per second. For example, the expected $\dot{V}O_2$ at 12. km·h^{-1} is $13.5 \times (3.33) - 8.5 = 36.5$ ml·kg^{-1}·min^{-1}.

Alternatively, running economy can be expressed in units of ml·kg^{-1}·km^{-1}. Irrespective of running speed, the typical value for running economy is 200 ml·kg^{-1}·km^{-1}. To convert the $\dot{V}O_2$ in ml·kg^{-1}·min^{-1} at a given speed into ml·kg^{-1}·km^{-1}, apply the following formula:

$$O_2 \text{ cost of running (ml·kg}^{-1}\text{·km}^{-1}) = \dot{V}O_2 \text{ ((ml·kg}^{-1}\text{·min}^{-1}) / \\ (\text{speed (km·h}^{-1}) / 60))$$

For example, for an athlete using 53 ml·kg^{-1}·min^{-1} of O_2 at 16 km·h^{-1}, 53 divided by 16/60 = 199 ml·kg^{-1}·km^{-1}.

Therefore:

170–180 ml·kg^{-1}·km^{-1} is excellent
180–190 ml·kg^{-1}·km^{-1} is very good
190–200 ml·kg^{-1}·km^{-1} is above average
200–210 ml·kg^{-1}·km^{-1} is below average
210–220 ml·kg^{-1}·km^{-1} is poor.

Lactate threshold

This is the first increase in blood [lactate] above baseline values. The speed at the lactate threshold is a strong predictor of the average speed that can be sustained in the marathon. The speed and heart rate at the lactate threshold are also useful in defining the transition between 'easy' and 'steady' running (discussed later). Using the protocol outlined earlier, the lactate threshold is easy to identify. For example, in Figure 16.1, the lactate threshold occurs at 16.0 km·h^{-1}.

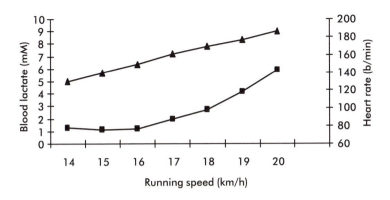

Figure 16.1 Typical blood lactate (filled squares) and heart rate (filled triangles) response to the multi-stage treadmill test in a well-trained distance runner. In this example, the athlete completed seven 3-min exercise stages, starting at 14 km·h^{-1} and finishing at 20 km·h^{-1}. The lactate threshold occurs at 16 km·h^{-1} and the lactate turn-point occurs at 18 km·h^{-1}

Lactate turn-point

The lactate turn-point is the running speed at which there is a distinct 'sudden and sustained' breakpoint in blood [lactate]. Typically, this occurs at 2.0–4.0 mM blood [lactate]. For example, in Figure 16.1, the lactate turn-point occurs at 18.0 km·h^{-1}. While the lactate turn-point is usually self-evident using the protocol detailed earlier, it can sometimes be difficult to discern due to the fact that it occurs on the curvilinear portion of the [lactate]–speed relationship. However, the lactate turn-point tends to occur at ~1–2 km·h^{-1} above the lactate threshold (the difference is smaller in longer-distance specialists and larger in middle-distance runners), and at ~90% maximal heart rate. The lactate turn-point speed can be held for a maximum of ~60 min and it therefore is useful in predicting performance over 10 miles to the half-marathon. For most athletes, it is the lactate turn-point that has a stronger relationship with performance than the lactate threshold. The lactate turn-point can also be used to define the transition between 'steady' running and 'tempo' running (discussed later).

Figure 16.2 provides an example of the way in which physiological data can be used by the coach to construct training programmes. Each training zone (defined by heart rate and speed) has a specific physiological purpose and the athlete and coach are better able to regulate training intensity using these guidelines.

Maximal oxygen uptake

This remains an important measure of performance capability in middle- and long-distance running. While factors such as economy and lactate threshold/turn-point can partially compensate for a relatively poor $\dot{V}O_{2max}$ in elite groups, entry to those elite groups is still limited by $\dot{V}O_{2max}$.

Table 16.2 provides information on the typical $\dot{V}O_{2max}$ values reported at various levels of performance in the United Kingdom.

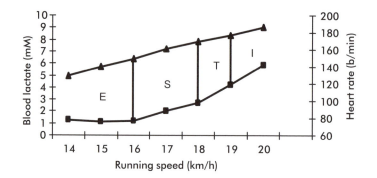

Figure 16.2 Illustration of the manner in which the blood [lactate] (filled squares) and heart rate (filled triangles) data garnered from the multi-stage test can be used in the prescription and regulation of exercise training, with each of the heart rate zones being employed for different purposes. E = easy running, S = steady running, T = tempo running and I = interval training. In this particular example, easy running would be performed at <16 km·h⁻¹ at a heart rate <149 b·min⁻¹, steady running would be performed at 16–18 km·h⁻¹ at a heart rate of 150–168 b·min⁻¹, tempo running would be performed at 18–19 km·h⁻¹ at a heart rate of 169–177 b·min⁻¹, and aerobic interval running would be performed at >19 km·h⁻¹ at a heart rate >177 b·min⁻¹

Table 16.2 Typical $\dot{V}O_{2max}$ values reported at various levels of performance in the United Kingdom

World class male: 80–90 ml·kg⁻¹·min⁻¹
World class female: 70–80 ml·kg⁻¹·min⁻¹
International male: 70–80 ml·kg⁻¹·min⁻¹
International female: 60–70 ml·kg⁻¹·min⁻¹
National male: 65–75 ml·kg⁻¹·min⁻¹
National female: 55–65 ml·kg⁻¹·min⁻¹
National junior male: 60–70 ml·kg⁻¹·min⁻¹
National junior female: 50–60 ml·kg⁻¹·min⁻¹

Speed at $\dot{V}O_{2max}$

The speed at $\dot{V}O_{2max}$ can be useful in predicting performance over 3,000 m (and also 1,500 and 5,000 m) and it has been argued that it is the ideal speed to train at to enhance $\dot{V}O_{2max}$ (Jones, 1998). To calculate the running speed at $\dot{V}O_{2max}$, simply multiply the $\dot{V}O_{2max}$ recorded during the second part of the treadmill test (in ml·kg⁻¹·min⁻¹) by 60 and divide by the running economy determined during the first part of the treadmill test (in ml·kg⁻¹·km⁻¹). To get an accurate economy measure, take the average of the economy in ml.kg.⁻¹km⁻¹ over the first 4–5 stages.

For example, an athlete with a $\dot{V}O_{2max}$ of 70 ml·kg⁻¹·min⁻¹ and a running economy of 190 ml·kg⁻¹·km⁻¹ would have an estimated running speed at $\dot{V}O_{2max}$ of (70 × 60)/190 = 22.1 km·h⁻¹.

OPTIONAL SUPPLEMENTARY MEASUREMENTS

Lung function measures (i.e. FVC, FEV_1, PEF, MVV) can be useful for general health screening and to rule out obvious pulmonary problems as a limitation to performance. The measurement of [Hb] can be invaluable especially in female runners as an 'early warning' of iron deficiency. However, the limitations to the analysis of [Hb] from capillary blood should be acknowledged. Athletes with low [Hb] measures should be encouraged to undertake a full blood test and to specifically request a measurement of serum ferritin. Flexibility can be an important fitness attribute in distance runners, especially middle-distance runners in whom poor flexibility might limit maximal speed. Therefore, a crude test such as the sit-and-reach test can be useful as a method of assessing changes in 'general' lower body flexibility over time. However, a large number of longer-distance runners perform at an extremely high level with what might be considered to be poor flexibility. The vertical jump test can be used to provide a crude measure of leg muscle power. One might expect reasonable (i.e. >40 cm) vertical jump test performances in middle-, but not long-distance, runners.

ADDITIONAL TESTS FOR MIDDLE-DISTANCE RUNNERS

The 800 m and 1,500 m running require a significant contribution from the anaerobic energy systems to the overall energy requirement. It is not easy to get a handle on the power and capacity of the anaerobic system during running and the tests given earlier only provide information on the parameters of aerobic fitness.

The peak blood [lactate] measured immediately after the $\dot{V}O_{2max}$ test provides a very indirect measure of anaerobic energy provision, that is, peak blood [lactate] tends to be higher in middle-distance runners (10–14 mM) than long-distance runners (6–10 mM) and it tends to increase when an athlete undergoes specific anaerobic-type training. The Wingate test provides information on power output during a 30 s cycle sprint test and *might* be useful in tracking changes with time but the mode of ergometry is not sport-specific. Estimation of anaerobic capacity from the maximal accumulated oxygen deficit test is yet another possibility but this method is fraught with the potential for error.

60 s RUN TEST

One reasonable approach with middle-distance runners would be to ask them to perform a 60 s run test at a supra-maximal intensity, an approach that is recommended by the Australian Sports Commission. Briefly, following a warm-up, the athlete is asked to straddle the treadmill belt. The treadmill is set to 4% grade and the speed is set at 20 km·h^{-1} for females and 22 km·h^{-1} for males.

A blood sample is collected before the test starts. When the athlete is ready, he or she lets go of the guardrails and begins running and the clock is started. At the completion of 60 s running, the athlete places his or her hands on the guardrails and straddles the treadmill belt. Another blood sample is taken. The idea is that as an athlete gets closer to competing, he or she should accumulate less lactate in response to this test.

A variation on this test is to require the athlete to continue to exhaustion. In this situation, as the athlete is closer to event-readiness, he or she might accumulate more lactate but should certainly be able to continue for longer. This test might well prove useful as a measure of 'anaerobic fitness'. However, it should be remembered that it is essentially a performance, not a physiological, test. Furthermore, performance on this test will not just reflect anaerobic capacity because the aerobic system will contribute significantly in an effort lasting longer than 60 s.

INTERIM TESTING

Full physiological assessments are generally scheduled every 3–4 months, typically around October, January, April and July to coincide with key transitions in an athlete's training and competition programme. However, athletes might prefer more regular monitoring. One approach would be for them to run at a moderately challenging speed (somewhere in the 'tempo' zone perhaps) on the treadmill for ~30 min every 4 weeks or so. Heart rate can be measured continuously and blood [lactate] and gas exchange can be measured periodically throughout the run. As fitness improves, heart rate and blood [lactate] should become lower. The respiratory exchange ratio should also become lower and this will be important in marathon runners as it will signify a greater use of fat (and reduced use of carbohydrate) while running at 'race pace'.

REFERENCES

Brandon, L.J. (1995). Physiological factors associated with middle distance running performance. *Sports Medicine*, 19: 268–277.

Coyle, E.F. (1995). Integration of the physiological factors determining endurance performance ability. *Exercise Sport Science Review*, 23: 25–63.

Hill, D.W. (1999). Energy system contributions in middle-distance running events. *Journal of Sports Science*, 17: 477–483.

Jones, A.M. (1998). A five year physiological case study of an Olympic runner. *British Journal of Sports Medicine*, 32: 39–43.

Jones, A.M. and Doust, J.H. (1996). A 1% treadmill grade most accurately reflects the energetic cost of outdoor running. *Journal of Sports Science*, 14: 321–327.

Londeree, B.R. (1986). The use of laboratory test results with long distance runners. *Sports Medicine*, 3: 201–213.

Sjodin, B. and Svedenhag, J. (1985). Applied physiology of marathon running. *Sports Medicine*, 2: 83–99.

CHAPTER 17

AMATEUR BOXING

Marcus S. Smith and Steve Draper

INTRODUCTION

The global appeal of amateur boxing is reflected in the 197 nations and territories affiliated to the world governing body (*Association Internationale de Boxe Amateur*, AIBA) (www.aiba.net, 2006). However, given: (1) its world-wide popularity, (2) the physiological demands that accompany a competitive bout and (3) the added complication of it being a weight classified sport, previous research is limited.

PHYSIOLOGICAL DEMANDS AND PROFILE OF ELITE BOXERS

The first weight classification system in boxing was introduced in 1867 and based upon the principle of 'equity' between competitors with reference to body weight (Prior, 1995). There are 11 Senior weight categories ranging from 48 to 91+ kg. Olympic boxers compete over four 2-min rounds, interspersed by a 1-min recovery, and may compete five times over 16 days (Smith, 2006). Success is determined by forcing an opponent to retire or via the accumulation of points. At the 2004 Olympic Games, 84% of contests were won on points with scores ranging from 17–11 to 54–27 (www.athens2004.com). Points are awarded using a computer system, which has lead to the development of punching force in preference to flair (Smith, 1998).

In general amateur boxers reduce body weight prior to competing. Smith (2006) reported a $7.0 \pm 0.8\%$ decrease in body weight over a 21-day period prior to competition amongst elite Senior England boxers. A consistent pattern of weight loss was observed that comprised a gradual [>7 days] ($1.7 \pm 0.7\%$)

and rapid [<7 days] (5.2 ± 0.4%) phase. Weight reduction in elite boxers is achieved by combining active and passive methods (Smith, 1998). Inappropriate weight making methods leading to dehydration and carbohydrate depletion have been associated with an earlier onset of fatigue, impaired boxing performance and increased risk to boxers' health (Smith, 1998).

The existence of a recovery period between the Official weigh-in and competition commencing provides a window of opportunity for the boxer to restore fluid balance and optimise carbohydrate stores. The duration of the recovery period varies between <2 h (domestic club competition) to 24 h (international match) (Smith, 1998). Smith (2006) reported a 4.4 ± 3.3% increase in body weight following a 24-h recovery period for Senior international boxers.

Urine osmolality has been shown to be a useful marker of hydratory status (Shirreffs and Maughan, 1998). Urine osmolality values >1,000 mOsm.kg^{-1} were recorded for England international boxers during domestic training camps, overseas training camps and multi-nation competitions (Smith, 2006).

Post-contest blood lactate concentrations >10 mmol/l emphasise the intense nature of competition and identify the critical role played by anaerobic glycolysis (Smith, 1998). The importance of boxers repeatedly experiencing high concentrations of blood lactate during training is well documented (Gosh et al., 1995; Smith, 1998).

The aerobic demands of competition are reflected in the maximum heart rate response observed in 'open sparring' (Smith, 1998). A well-developed aerobic capacity is considered as an important characteristic of the elite amateur boxer. A relative maximum oxygen uptake ($\dot{V}O_{2max}$) of between 57.5 ± 6.9 ml·kg^{-1}·min^{-1} to 63.8 ± 4.8 ml·kg^{-1}·min^{-1} for Italian, German and England Senior amateur boxers has been reported (Freidmann et al., 1999; Guidetti et al., 2002; Smith, 2006).

Aerobic and anaerobic threshold measurements can be used to set training intensities and monitor training adaptations. Smith (2006) reported a running speed for Senior England international boxers of 10.43 ± 1.48 km·hr^{-1} and 13.38 ± 1.13 km·hr^{-1} at 2 mmol·l^{-1} and 4 mmol·l^{-1}, respectively. The peak running speed of 16.7 ± 0.6 km·hr^{-1} was identical to the value reported by Freidman et al. (1999) for German national amateur boxers (16.7 ± 1.0 km·hr^{-1}).

The recent development of boxing-specific dynamometers is an important step towards ecological validity in analysing amateur boxing performance (Smith, 1998). Peak punching forces of 1,500–5,000 N have been recorded (Smith et al., 2000; Walilko et al., 2005) with differences related to the technical requirements of each type of punch (Fritsche, 1978; Hickey, 1980; Filimonov et al., 1983) and boxing ability (Smith et al., 2000).

PROTOCOLS TO ASSESS AND MONITOR PERFORMANCE

Laboratory-based assessment

At the beginning of each competitive season the boxer and coach will: (1) identify the boxers competition weight category, (2) identify major domestic and international tournaments and (3) construct a training programme aimed at peaking the boxer at the appropriate times. In discussion with the boxer and coach 3–4 laboratory visits should be time-tabled into the training programme to identify training intensities and training adaptations.

Peak punching force

Fundamental to successful boxing performance is optimum peak punching force. Any device claiming to measure punching force must have a robust method of calibration. Smith *et al.* (2000) describe the protocol for calibrating a tri-axial force measurement system by comparing the calculated and measured peak impact force (Table 17.1).

Above ≈500 N, in the common working punch impact range, the percentage error between the calculated and measured impact force was less than 3%.

Running – lactate profile, $\dot{V}O_{2max}$, economy, maximum running velocity

A high $\dot{V}O_{2max}$ has been recorded for elite boxers from different countries. This aerobic adaptation could be partly attributed to the volume and intensity of running undertaken by a boxer in their training programme. It is therefore appropriate that an incremental treadmill test is undertaken to determine $\dot{V}O_{2max}$ and establish sub-maximal training thresholds.

The exact protocol used to determine these parameters is dependent on the time available with each boxer. Ideally two tests are completed: an incremental sub-maximal test and a fast ramp test to exhaustion. However, if only a single test is possible these tests may be combined using shorter sub-maximal stages. The incremental test consists of 4–5 sub-maximal stages, 5–6 minutes duration, of increasing speed. The final stage must produce a blood lactate concentration >4 mmol·l^{-1} (lysed capillary sample). Mean heart rate and $\dot{V}O_2$ are also determined for the same period at the end of each collection. These data can then be used to evaluate the relationships between key variables ($\dot{V}O_2$, blood lactate concentration, heart rate and exercise intensity). Of particular importance is the determination of heart rate at 4 mmol·l^{-1}, as this is used to prescribe training intensities (Heck *et al.*, 1985). The ramp test should last ~8–12 min. A ramp test is preferable but an incremental test may be used (1-min stages) if the treadmill is not programmable. In either event a ramp rate of 1.2 km·hr^{-1} is used. The treadmill should be set to a moderate gradient for both tests, 5% is ideal. This is done to ensure that $\dot{V}O_{2max}$ is achieved, and the athlete is not cadence limited, during the ramp test and is also used for the incremental test since sub-maximal $\dot{V}O_2$ will be referenced to $\dot{V}O_{2max}$. If Douglas bags are used 30 s collections are taken continuously for the final minutes of the test and the highest single collection taken to be $\dot{V}O_{2max}$. If breath-by-breath equipment is available the data should be interpolated to second-to-second values and the highest 15 s moving average used.

Table 17.1 Comparison of calculated impact forces (mean ± sd) on a boxing manikin of simulated straight jab punches with recorded dynamometer forces

Mass (kg)	Acceleration $(m \cdot s^{-2})$	Calculated force (N)	Dynamomoter (punch units)	Mean difference	% difference
5.4	80.4 ± 2.5	434 ± 13.5	324 ± 2.1	109.9	33.8
9.8	53.2 ± 1.1	521 ± 11.6	513 ± 3.4	8.6	1.7
25.4	42.5 ± 1.1	1079 ± 29.8	1052 ± 5.5	26.7	2.5
50.4	37.3 ± 0.6	1882 ± 30.4	1857 ± 6.0	25.4	1.4
65.4	36.1 ± 2.0	2360 ± 133	2360 ± 5.0	0.3	0

Body fat

Central to the philosophy of manipulating body weight prior to boxing performance is the pursuit of minimal body fat. To date the majority of body fat assessments have used the equations generated by the 4-site upper body method described by Durnin and Womersley (1974). However, future assessments should use a 7-site model that incorporates upper and lower body sites, where changes in millimetres of body fat, rather than body fat%, should be monitored (Eston, 2003).

Feedback of results

It is imperative that verbal feedback is provided on the day of the test and within seven days in writing. Feedback should be delivered in a style that makes the boxer and coach feel comfortable and confident to ask questions. Failure to generate discussion indicates either a lack of interest or information being fed back at a level that is difficult to comprehend. At the end of the feedback session other stakeholders who could play a future role in supporting the boxer need to be identified and communication links established.

Field assessment

Experience has shown that it is vital to clarify roles and responsibilities before undertaking any applied work in the field (Collins *et al.*, 1999). This is particularly relevant to support work at competitions. Having gone through the process of role clarification the support work could take place at domestic or international training camps and competitions. The nature of field support work in boxing routinely falls into two categories: (1) monitoring of training intensity via heart rate telemetry, blood lactate measurement and breath-by-breath gas analysis; or (2) monitoring hydratory status using an osmometer.

REFERENCES

Collins, D., Moore, P., Mitchell, D. and Alpress, F. (1999). Role conflict and confidentiality in multidisciplinary athlete support programmes. *British Journal of Sports Medicine*, 3: 208–211.

Durnin, J.V.G.A. and Womersley, J. (1974). Body fat assessed from total body density and its estimation from skin fold thickness: measurements on 481 men and women from 16 to 72 years. *British Journal of Nutrition*, 32: 77–97.

Eston, R. (2003). Prediction of body fat from skinfolds: the importance of including sites from the lower limb. *Journal of Sports Sciences*, 21: 369–370.

Filiminov, V.I., Koptsev, K.N., Husyanov, Z.M. and Nazarov, S.S. (1983). Means of increasing strength of the punch. *National Strength and Conditioning Association Journal*, 7(6): 65–66.

Friedmann, B., Jost, J., Rating, T., Weller, E., Werle, E., Eckardt, K.-U., Bärtsch, P. and Mairbäurl, H. (1999). Effects of iron supplementation on total body haemoglobin during endurance training at moderate altitude. *International Journal of Sports Medicine*, 20(2): 78–85.

Fritsche, P. (1978). Ein dynamographisches informationssystem zur messung der schlagkraft beim boxen. *Leistungssport*, 2: 151–156.

Gosh, A.K., Goswami, A. and Ahuja, A. (1995). Heart rate and blood lactate response in amateur competitive boxing. *Indian Journal of Medicine*, 102: 179–183.

Guidetti, L., Musulin, A. and Baldari, C. (2002). Physiological factors in middleweight boxing performance. *Journal of Sports Medicine and Physical Fitness*, 42(3): 309–314.

Heck, H., Mader, A., Hess, G., Muccke, S., Muller, R. and Hollmann, W. (1985). Justification of the 4-mmol/l lactate threshold. *International Journal of Sports Medicine*, 6(3): 117–130.

Hickey, K. (1980). *Boxing – The Amateur Boxing Association Coaching Manual*. London, England: Kaye and Ward.

Prior, D. (1995). *Ringside with the Amateurs*. Milton Keynes, England: Stantonbury Parish Print.

Shirreffs, S.M. and Maughan, R.J. (1998). Urine osmolality and conductivity as indices of hydration status in athletes in the heat. *Medicine and Science in Sports and Exercise*, 30(11): 1598–1602.

Smith, M.S. (1998). Sport specific ergometry and the physiological demands of amateur boxing. Doctoral Thesis, University College Chichester, England.

Smith, M.S. (2006). Physiological profile of Senior and Junior England international amateur boxers. *European Journal of Sports Science and Medicine*, CSSI, 74–89.

Smith, M.S., Dyson, R.J., Hale, T. and Janaway, L. (2000). Development of a boxing dynamometer and its punch force discrimination efficacy. *Journal of Sports Sciences*, 18: 445–450.

URL:http://www.athens2004.com (2004). Boxing results.

URL:http://www.aiba.net (2006). National Federations: the knowledge of boxing.

Walilko, T.J., Viano, D.C. and Bir, C.A. (2005). Biomechanics of the head for Olympic boxer punches to the face. *British Journal of Sports Medicine*, January: 710–719.

CHAPTER 18

CYCLING

R.C. Richard Davison and Andrea L. Wooles

INTRODUCTION

Ergometry-based physiological testing has been the mainstay of many physiological investigations and thus there is a wealth of information on the physiological responses to such exercise. However, there is a much smaller volume of information where cyclists have used an ergometer which accommodates their own bicycle or with sufficient adjustability to match the appropriate cycling position. The development of mobile power measurement devices (i.e. SRM power meter) have also added greatly to our ability to assess the physiological demands of specific cycling events and affords the opportunity to develop a range of new specific tests in the future. Prior to the adoption of any new protocol it is important to establish that the test is reliable and sensitive to changes in performance. The tests described in this chapter are broadly similar to those recommended by British Cycling but despite their regular use there is limited information on some aspects of reliability and less information on the sensitivity to changes in performance.

In general terms the variable that is of most importance to the cyclist is maximal power output as it can be directly linked to performance and is more sensitive to changes in performance than VO_{2max}. There are a number of devices and ergometers that can measure power output accurately, however these devices must be calibrated regularly. This can be achieved statically by hanging calibrated weights from the chainring (Standard Error of the SRM power meter slope is $0.01\ Hz{\cdot}Nm^{-1}$, which equates to 1 W in 1000 W, Wooles *et al.*, 2005).

MAXIMAL AEROBIC TEST

As the majority of cycling events require a large aerobic capacity, maximal aerobic tests are the most common tests carried out on cyclists. In the majority of other aerobic sports the principal variable of interest is the VO_{2max} however maximal power output from an aerobic test is more relevant and a much better predictor of performance for cyclists (Balmer *et al.*, 2000a). Thus, in the literature there are a number of different tests which are designed to measure maximal aerobic power achieved at or near the end of a protocol of increasing intensity reaching volitional exhaustion after 8–60 min of exercise (Kuipers *et al.*, 1985; Hawley and Noakes, 1992; Padilla *et al.*, 1999, 2000). This variety of different protocols has also resulted in a range of different terms for maximal aerobic power (W_{max}, W_{peak}, MPO, MMP and RMP_{max}) being used to describe essentially the same physiological variable. However, while it is useful that the literature contains reference data from cyclists of wide ranging ability it is important to note that the type of protocol and method of determination does affect the maximal value (Smith, 2002) and data from different protocols cannot be used interchangeably.

The most widely used test is the ramp test where a starting intensity and ramp rate are chosen to reach volitional exhaustion in 8–15 min (see Table 18.1 for reference values). The maximal aerobic power being the highest power averaged over 60 s, with the term, maximal Ramp Minute Power (RMP_{max}) being the most accurate descriptor of this test. The reliability of this test has been established on a number of different types of ergometers, %CV for the Kingcycle was 2.0% (95% CI, 1.5–3.0%) and for the SRM Power Meter 1.3% (95% CI, 1.0–2.0%) (Balmer *et al.*, 2000b). The RMP_{max} obtained from this type of ramp test has also been shown to be a good predictor of performance over both 16.1 (SEE, 0.8%) and 40 (SEE, 4.7%) km time-trials (Balmer *et al.*, 2000a; Smith *et al.*, 2001). In addition this test is sensitive to changes in fitness across a training year with an equal ability to predict time-trial performance at different stages of the season (Smith, 2002).

Although absolute power is a reasonable predictor of flat time-trial performance this is somewhat influenced by body surface area. Similarly we need to consider body weight as cycling generally takes place on undulating terrain. Allometric scaling can be used to adjust for both these variables, however the complex interaction between body surface area, body weight, flat and uphill terrain has prevented the development of one universally acceptable exponent. Exponents have ranged from 0.32 (Swain, 1994) for level ground up to as much

Table 18.1 Reference values for ramp test

Weight (kg)	Male		Female	
	Start (W)	Ramp ($W \cdot min^{-1}$)	Start (W)	Ramp ($W \cdot min^{-1}$)
<45			100–120	13
45–49			110–130	14
50–54	140–160	17	120–140	15
55–59	150–170	18	130–150	16
60–64	160–180	19	140–160	16
65–69	170–190	20	150–170	18
70–74	180–200	21	160–180	19
75–79	190–210	22		
80+	200–220	23		

as 1.0 for steep gradients (Padilla *et al.*, 1999), however it is more common to use exponents in the range of 0.65–0.79. In most cycling events it is important to scale maximal aerobic power to weight probably using an exponent between 0.65 and 0.79, British Cycling use an exponent of 0.67 to create their fitness index.

MAXIMAL POWER TEST

A maximal power test is a method of assessing a cyclist's short-term muscular power. Ideally the test should be short (6–10 s) so that a rider's maximal power can be achieved in the first few seconds of accelerative effort. Martin *et al.* (1997) describe five different methods of assessing maximal power and concluded that the inertial-load method is the most valid and reliable. Many systems rely on a mathematical estimation of the inertial load of a flywheel as described by Lakomy (1986), but the optimal design would be an ergometer that directly measures the torque and angular velocity at a reasonably high frequency (i.e. 200 Hz). Unfortunately no such system currently exists and the best alternative is to use an SRM power meter or Ergometer, with torque option, to measure torque at 200 Hz and angular velocity once per revolution. Features of the protocol that clearly impact on maximal power are standing or rolling start, gear ratio and resistance load (MacIntosh *et al.*, 2003). However, it has been suggested that providing data have been corrected for the inertia of the flywheel and a standing start is used, there is no need to individualise optimal resistance (MacIntosh *et al.*, 2003).

Important variables from this type of test are both the maximal power and the optimal cadence for maximal power. As both of these variables are integrally linked by the force-velocity relationship of muscle they cannot be considered in isolation and play an important role in protocol design. The resistance chosen for the test must be large enough so that peak cadence does not limit the maximal power and not so large to limit cadence to unrealistic low values (Figure 18.1).

British Cycling determine maximal power using a 6-s test from a standing start on a SRM ergometer. The 'open ended' test mode provides a braking force which has a cubic relationship with speed, mimicking the effect of air resistance on a moving bicycle. The resistance is set by selecting an appropriate gear and the rider is asked to start from their strongest position on the bike and accelerate as fast and as hard as they can in a seated position. After 3 min of active recovery the test should be repeated and report the better of the two tests. All riders should perform the test in the same gear each visit. This test has a technical error of measurement (TEM) of 25 W(2.3%).

SUB-MAXIMAL TEST

For routine monitoring of improvements in aerobic fitness it is generally recommended to use a sub-maximal test in preference to repeated maximal

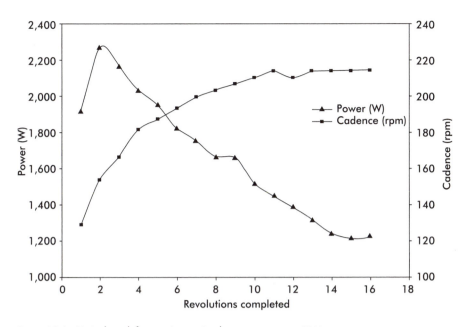

Figure 18.1 Typical result from a 6 s maximal power test on an SRM ergometer

testing. This test should be completed at 70% RMP_{max} as this represents a high intensity were 'steady state' is still attainable and represents an intensity approximately equating to 40 km time-trial pace (Davison *et al.*, 2000). During this test measures of HR, [La], VO_2, V_E and RER in repeated tests would indicate improvements in aerobic fitness.

It is important that the cyclist undergoes an identical preparation prior to each sub-maximal test as factors like over-training, fatigue, dietary preparation can have a major influence on the variables measured and thus not be a reflection of training adaptation. The protocol should consist of a 5 min ramp to reach 70% RMP_{max} which should be maintained for 10 min to attain an adequate 'steady state'.

When considering the results for the sub-maximal test it is important to consider the likely technical error and biological variation and that any changes between tests need to exceed these errors before being considered as a true change. Surprisingly there is no direct information on the reliability of this test but it is possible to estimate values using data reported for 40-km indoor time trials, which probably over estimate the variability. Smith (2002) reported that the coefficient of variation (95% CI) of heart rate, VO_2, V_E and [La] for repeated 40-km time trials were 3.2%(2.1–6.7), 3.0%(2.0–6.3), 5.4%(3.6–11.5) and 16.4%(8.4–31.6), respectively. This suggests that the worst case scenario HR would need to decrease by ~11 beats to be confident of a real change in fitness.

REFERENCES

Balmer, J., Davison, R.C. and Bird, S.R. (2000a). Peak power predicts performance power during an outdoor 16.1-km cycling time trial. *Medicine and Science in Sports and Exercise*, 32: 1485–1490.

Balmer, J., Davison, R.C. and Bird, S.R. (2000b). Reliability of an air-braked ergometer to record peak power during a maximal cycling test. *Medicine and Science in Sports and Exercise*, 32: 1790–1793.

Davison, R.C.R., Smith, M.F., Coleman, D.A., Balmer, J. and Bird, S.R. (2000). Variability of power output during 40-km outdoor time-trial cycling performances. *Medicine and Science in Sports and Exercise*, 32: S291-

Hawley, J.A. and Noakes, T.D. (1992). Peak power output predicts maximal oxygen uptake and performance time in trained cyclists. *European Journal of Applied Physiology*, 65: 79–83.

Kuipers, H., Verstappen, F.T.J., Keizer, H.A., Geurten, P. and van Kranenburg, G. (1985). Variability of aerobic performance in the laboratory and its physiologic correlates. *International Journal of Sports Medicine*, 6: 197–201.

Lakomy, H.L. (1986) Measurement of work and power output using friction loaded cycle ergometers, *Ergonomics*, 29: 509–519.

MacIntosh, B.R., Rishaug, P. and Svedahl, K. (2003). Assessment of peak power and short-term work capacity. *European Journal of Applied Physiology*, 88: 572–579.

Martin, J.C., Wagner, B.M. and Coyle, E.F. (1997). Inertial-load method determines maximal cycling power in a single exercise bout. *Medicine and Science in Sports and Exercise*, 29: 1505–1512.

Padilla, S., Mujika, I., Cuesta, G. and Goiriena, J.J. (1999). Level ground and uphill cycling ability in professional road cycling. *Medicine and Science in Sports and Exercise*, 31: 878–885.

Padilla, S., Mujika, I., Orbananos, J. and Angulo, F. (2000). Exercise intensity during competition time trials in professional road cycling. *Medicine and Science in Sports and Exercise*, 32: 850–856.

Smith, M.F. (2002). The physiological evaluation of 40-km time trial performance in cyclists. PhD Thesis, Canterbury Christ Church University College.

Smith, M.F., Davison, R.C., Balmer, J. and Bird, S.R. (2001). Reliability of mean power recorded during indoor and outdoor self-paced 40 km cycling time-trials. *International Journal of Sports Medicine*, 22: 270–274.

Smith, M.F., Balmer, J., Coleman, D.A., Bird, S.R. and Davison, R.C. (2002). Method of lactate elevation does not affect the determination of the lactate minimum. *Medicine and Science in Sports and Exercise*, 34: 1744–1749.

Swain, D.P. (1994).The influence of body mass in endurance bicycling. *Medicine and Science in Sports and Exercise*, 26: 58–63.

Wooles, A.L., Robinson, A.J. and Keen, P.S. (2005). A static method for obtaining a calibration factor for SRM bicycle power cranks. *Sports Engineering*, 8: 137–144.

CHAPTER 19

MODERN PENTATHLON

Gregory P. Whyte and David V.B. James

INTRODUCTION

Modern Pentathlon has featured in the modern Olympic Games since 1912 (Baker, 1983). The founder of the modern Olympic Games included the sport in an attempt to find the most complete athlete. For the first time in 2000, the Olympic programme included a competition for women. Great Britain has a strong tradition in the sport, particularly in the Olympic competition for women over recent years. Recently, the International Olympic Committee (IOC) has confirmed that Modern Pentathlon will remain in the Olympic programme post 2008. Given the sport's prominence in the Olympic programme since 1912, limited research exists examining the demands of the sport including the key determinants of performance, and athlete physiological characteristics.

PHYSICAL DEMANDS AND DETERMINANTS OF PERFORMANCE

Prior to the 1996 Olympic Games Modern Pentathlon changed to a one-day format (youth competitions remain over two day). The order of the events is not set, although the running event must take place last. Recently the distance of the running and swimming events changed to bring the men's and women's distances into line. The running event consists of a 3,000 m cross country time trial where the competitors start at intervals based on their point totals after four events. The swimming event is a 200 m sprint where competitors compete in heats determined by their previous season's best time. The shooting event includes 20 shots (40 s per shot) from a 4.5 mm air pistol at a distance of 10 m from the target. The fencing event is an epee one hit competition where all competitors fight each other. The show jumping requires the athlete to ride an unknown horse over a 450 m course of 12 obstacles, including a double and treble combination. A 'par' performance is awarded 1,000 points in each event, with points added or deducted for performances above or below the 1,000 point standard. The best athletes score in excess of 5,000 points. Indeed, the 2004 Olympic events were won with total point scores of 5,480 (men) and 5,448 (women).

Particularly with the one-day format, the accumulative fatigue over the five events is likely to be as important a determinant of performance as the determinants of performance in each event. The only published study examining accumulative fatigue during a Modern Pentathlon event was conducted during the 1966 world championships under the old five-day competition format (Hagerman, 1968). Despite the relatively prolonged duration between events, the study demonstrated that the best performers were those who fatigued the least over the five days. In addition, fatigue following the 65 bout fencing event was determined the single most decisive indicator of accumulative fatigue. The challenge, therefore, appears to be to emerge from the fencing event with a good performance and as little residual fatigue as possible. Some caution should be taken in generalising this finding to the one-day format, however, as the fencing event now involves fewer fencing bouts (e.g. 31 in the 2004 Olympic games), and each bout has a 1 min time limit in contrast to 3 min in the past.

Clearly Modern Pentathlon has numerous technical demands, particularly in the shooting, fencing and riding events. Identifying key physical determinants of performance in these events is difficult. It is acknowledged that the successful execution of several moves in epee fencing is contingent on explosive power in the legs (Nystrom et al., 1990). However, explosive power must be optimally applied through appropriate timing, balance and coordination. For example, Harmenberg et al. (1991) demonstrated that only an ecologically valid stimulus for a fencing movement allowed differentiation of the performance of the fencing movement between beginners and elite, and within an elite group. Furthermore, when an ecologically valid stimulus was used, reaction time, rather than movement time or overall response time for the movement was most strongly related to epee fencing performance. Although Nystrom et al. (1990) outline the importance of high aerobic power for epee fencing, the one-hit format (with a 1 min time limit) employed in pentathlon is unlikely to so extensively tax the aerobic energy system. Furthermore, Modern Pentathletes are likely to be sufficiently well aerobically trained through preparation for the other events (running event, discussed later). Therefore, we suggest that no key physiological determinants of performance in the epee fencing event in modern pentathlon are evident, and that fatigue resistance during the course of the fencing event is likely to be the most important determinant to focus on for athlete assessment.

It is possible that muscle endurance may become a limiting factor in the show jumping event, especially since Modern Pentathletes are known to spend relatively little time training for this event (Hagerman, 1968). Interestingly however, Westerling (1983) found no difference between trained riders and matched controls for isometric muscle strength for six muscle groups. Relative to athletes from other sports and untrained controls, Meyers and Sterling (2000) observed relatively low peak power in female collegiate riders during a 30 s sprint on a cycle ergometer. The aerobic demands of show jumping have been reported in only one published paper. Devienne and Guezennec (2000) reported that on average recreational riders attained 75% $\dot{V}O_{2max}$ during a 12 obstacle show jumping test, with some riders attaining $\dot{V}O_{2max}$. However, given the high aerobic power of well-trained Modern Pentathletes (discussed later), it is unlikely that aerobic power is itself a limiting factor. Therefore, we suggest that no key physiological determinants of performance in the show jumping event in Modern Pentathlon are evident from the literature, but it may be important that Modern Pentathletes spend time 'in the saddle' to ensure that local muscle endurance does not present a physical limiting factor.

Although the running and swimming events are more obviously limited by physiological determinants, technical ability remains important. The 200 m swimming event is completed in 2 min 30 s for a 1,000 point score, but the best athletes routinely score over 1,300 points (i.e. <2 min 5 s). Recent debates have questioned whether events of this duration, where $\dot{V}O_2$ may not have time to attain $\dot{V}O_{2max}$, should be considered as events within a separate 'extreme' exercise intensity domain (e.g. Hill *et al.*, 2002). Whether or not $\dot{V}O_{2max}$ is attained, an event of this duration is clearly dependent on high aerobic power in addition to high anaerobic capacity (e.g. Spencer and Gastin, 2001). However, good physiological capabilities must be optimally translated into propulsion, which requires considerable skill. Key physiological determinants of performance for the 200 m swimming event are therefore likely to be aerobic power, the potential to use that aerobic power during the short duration event (i.e. $\dot{V}O_2$ kinetics), and the anaerobic capacity.

The 3,000 m cross-country running event is completed in 10 min for a 1,000 point score, but the best athletes complete the distance in close to 9 min, which is an average speed of 20 $km \cdot h^{-1}$. In our experience, athletes of this standard achieve speeds in progressive treadmill exercise tests of ~22 $km \cdot h^{-1}$. Although empirical evidence is not available, it is likely that the 3,000 m cross-country run is completed in the severe exercise intensity domain. The assumption that athletes complete 3,000 m at severe exercise intensity appears safe, particularly if one considers the added air resistance during the cross-country run. In this exercise intensity domain, although the predicted $\dot{V}O_2$ from the sub-maximal $\dot{V}O_2$–speed relationship is less than $\dot{V}O_{2max}$, $\dot{V}O_2$ is known to increase with time until exercise is terminated or $\dot{V}O_{2max}$ is attained (Poole *et al.*, 1988). A high aerobic power and a good running economy ($\dot{V}O_2$ for a set speed) are key determinants of performance in the 3,000 m cross-country running event (Wood, 1999).

Only one study has been published examining the aerobic power of Modern Pentathletes compared with participants in other sports. Joussellin *et al.* (1984) examined 278 male and 133 female French athletes belonging to national teams, comprising 22 male and 6 female Modern Pentathletes, respectively. Aerobic power was determined using a discontinuous protocol on an inclined (3%) treadmill. The protocol consisted of 4 min stages with 1 min rest between each stage. Speed was increased by 2 $km \cdot h^{-1}$ per stage until 18 $km \cdot h^{-1}$

was reached whereby gradient increased from 3% in 3% increments. In absolute terms (mean ± SD) the Modern Pentathletes had a maximal aerobic power (maximal oxygen uptake; $\dot{V}O_{2max}$) of 4.94 ± 0.5 l· min^{-1} (male) and 3.13 ± 0.2 l· min^{-1} (female). The absolute aerobic power of the male Modern Pentathletes was only bettered by the rowers and equalled by the road cyclists and distance runners. In relative terms the Modern Pentathletes had a maximal aerobic power of 73.0 ± 11 ml·kg^{-1}·min^{-1} (male) and 55.5 ± 3.5 ml·kg^{-1}·min^{-1} (female). The relative aerobic power of the male Modern Pentathletes was only bettered by the distance runners. Classes *et al.* (1994) investigated the role of anthropometric characteristics in performance in 65 female Modern Pentathletes during the 1989 World Championships and found that fat mass as a proportion of body mass was inversely related to performance. Although the changed competition format has possibly altered the demands of the sport, as well as training approaches, these data provide a useful insight into the potential determinants of performance in Modern Pentathletes.

PROTOCOLS TO ASSESS KEY PHYSICAL DETERMINANTS OF PERFORMANCE

Laboratory-based assessment

Laboratory-based assessment should occur 3–4 times per year, based around the athletes competitive season. Information gained from laboratory assessments can be used in two ways. First, it can be used to define recommended training intensities for optimal physiological adaptation. Second, it can be integrated in the detection of both positive and negative training adaptations, allowing the individual's training programme to be retrospectively analysed and adjusted if necessary.

Aerobic power has been identified as a key determinant of performance for the running and swimming events, and may well be important in minimising fatigue in the more technical fencing and show jumping events. Furthermore, good aerobic fitness is likely to be useful in the shooting event, not least because heart rate at rest will be low.

RUNNING – LACTATE PROFILE, $\dot{V}O_{2max}$ ECONOMY, MAXIMUM RUNNING VELOCITY

For the running event, as well as a general measure of the aerobic fitness of the Modern Pentathlete, it is appropriate that a step incremental test is undertaken on a treadmill for the determination of maximal oxygen uptake ($\dot{V}O_{2max}$).

Protocol

Pentathletes complete 6×3 min incremental stages on a standard treadmill. At the end of each stage heart rate is recorded and a capillary blood sample is taken from the earlobe and analysed for blood lactate. The velocity of stages are determined from the pace for a recent 10 km run (preferably a race), setting this as the pace for the fourth stage. Higher stages are then calculated by successively adding 0.6 km·h^{-1} to the 10 km pace, lower stages by successive subtraction of 0.6 km·h^{-1}. Following these stages, the treadmill speed is increased by 1 km·h^{-1} every minute to maximum running velocity at which time treadmill grade is increased by 2%·min^{-1} until maximum volitional exhaustion. Expired air is measured throughout the entire test, for determination of running economy, fractional utilisation and $\dot{V}O_{2max}$.

SWIMMING – LACTATE PROFILE, MAXIMUM SWIMMING VELOCITY

Assessment of aerobic power for the swimming event is problematic. Although some researchers have successfully determined $\dot{V}O_2$ during swimming (see Demarie et al., 2001), a flume has been used as the 'ergometer' for progressive exercise. Such ergometry is not commonly available, so alternative approaches should be considered. The key consideration is that an ecologically valid mode of exercise is adopted. Use of arm cranking and/or swim benches is therefore considered inappropriate for accurately simulating swimming exercise. Our view is that due to the complexities associated with the determination of aerobic capacity testing should focus on those areas that can be accurately and reliably measured in an ecological valid environment (i.e. pool swimming).

Protocol

Using the 'Aqua-Pacer'© swim pacing devise, Pentathletes perform 6×300 m swims of incremental intensity (split times, discussed later). Following each swim, heart rate is recorded and a capillary blood sample is taken from the earlobe and analysed for blood lactate. Following the incremental swims and 10 min active recovery Pentathletes complete a 50 m sprint swim to determine maximum swimming velocity (Tables 19.1 and 19.2).

Protocol

400 m self-paced easy warm-up
200 m aquapacer habituation
6×300 m, 1 min. recovery (HR and blood sample collection)
400 m easy recovery
50 m sprint (push start)

Table 19.1   Split times vs. 200 m PB

200 m PB	Swim 1	Swim 2	Swim 3	Swim 4	Swim 5	Swim 6
2:50	28.4	27.3	26.2	25.3	24.4	23.5
2:45	27.3	26.2	25.3	24.4	23.5	22.6
2:40	26.2	25.3	24.4	23.5	22.6	21.7
2:35	25.3	24.4	23.5	22.6	21.7	20.8
2:30	24.4	23.5	22.6	21.7	20.8	19.9
2:25	23.5	22.6	21.7	20.8	19.9	19.0
2:20	22.6	21.7	20.8	19.9	19.0	18.1
2:15	21.7	20.8	19.9	19.0	18.1	17.2
2:10	20.8	19.9	19.0	18.1	17.2	16.3
2:05	19.9	19.0	18.1	17.2	16.3	15.4
2:00	19.0	18.1	17.2	16.3	15.4	14.5

Table 19.2   Split times (s)

25 m split	100 m split	300 m split
26.2	1:44.8	5:14.4
25.3	1:41.2	5:03.6
24.4	1:37.6	4:52.8
23.5	1:34.0	4:42.0
22.6	1:30.4	4:31.2
21.7	1:26.8	4:20.4
20.8	1:23.2	4:09.6
19.9	1:19.6	3:58.8
19.0	1:16.0	3:48.0
18.1	1:12.4	3:37.2
17.2	1:08.8	3:26.4

Feedback

Effective and appropriate feedback should be given verbally on the day of the test and in writing no later than 1 week after the test. The feedback should interpret the data in a way that the coach and athlete will be able to understand and allows them to use the information gained in order to further enhance the training programme. This process also facilitates a three-way education process between the sport scientist, athlete and coach and is one of the key features offered by a close and interactive support team.

Field assessment

Support should be offered in the field during training and competition in the United Kingdom and internationally. Advice should be offered for a variety of environmental conditions including warm-weather and altitude training and competition. Work performed in the field routinely includes the monitoring and verification of training intensities, and monitoring hydration status, markers of over-training and under-recovery.

REFERENCES

Baker, C.L. (1983). Comments on olympic sports Medicine: the Modern Pentathlon. *American Journal Sports Medicine*, 11(1): 42–45.

Claessens, A.L., Hlatky, S., Lefevre, J. and Holdhaus, H. (1994). The role of anthropometric characteristics in modern pentathlon performance in female athletes. *Journal Sports Sciences*, 12: 391–401.

Davis, J.A., Whipp, B.J., Lamrra, N., Huntsman, D.J., Frank, M.H. and Wasserman, K. (1982). Effect of ramp slope on determination of aerobic parameters from the ramp test. *Medicine and Science in Sports and Exercise*, 14(5): 339–343.

Demarie, S., Sardella, F., Billat, V., Magini, W. and Faina, M. (2001). The $\dot{V}O_2$ slow component in swimming. *European Journal of Applied Physiology*, 84: 95–99.

Devienne, M.-F. and Guezennec, C.-Y. (2000). Energy expenditure of horse riding. *European Journal of Applied Physiology*, 82: 499–503.

Hagerman, F.C. (1968). An investigation of accumulative acute fatigue in participants at the 1966 World Modern Pentathlon Championships, Melbourne (Victoria, Australia). *Journal of Sports Medicine*, 8: 158–170.

Harmenberg, J., Ceci, R., Barvestad, P., Hjerpe, K. and Nystrom, J. (1991). Comparison of different tests of fencing performance. *International Journal Sports Medicine*, 12(6): 573–576.

Hill, D.W., Poole, D.C. and Smith, J.C. (2002). The relationship between power and the time to achieve $\dot{V}O_{2max}$. *Medicine and Science in Sports and Exercise*, 34(4): 709–714.

Meyers, M.C. and Sterling, J.C. (2000). Physical, haematological, and exercise response of collegiate female equestrian athletes. *Journal of Sports Medicine Physical Fitness*, 40: 131–138.

Nystrom, J., Lindwall, O., Ceci, R., Harmenberg, J. Swedenhag, J. and Ekblom, B. (1990). Physiological and morphological characteristics of world class fencers. *International Journal of Sports Medicine*, 11(2): 136–139.

Poole, D.C., Ward, S.A., Gardner, G.W. and Whipp, B.J. (1988). Metabolic and respiratory profile of the upper limit for prolonged exercise in man. *Ergonomics*, 31(9): 1265–1279.

Spencer, M.R. and Gastin, P.B. (2001). Energy system contribution during 200- to 1500-m running in highly trained athletes. *Medicine and Science in Sports and Exercise*, 33: 157–162.

Westerling, D. (1983). A study of physical demands in riding. *European Journal of Applied Physiology*, 50: 373–382.

Wood, D.M. (1999). Physiological demands of middle-distance running. In J.L. Fallowfield and D.M Wilkinson (eds), *Improving Sports Performance in Middle and Long-distance Running*, 15–38. Chichester, UK: John Wiley & Sons Ltd.

CHAPTER 20

ROWING

Richard J. Godfrey and Craig A. Williams

Rowing is an Olympic sport in which athletes compete over a measured regatta course of 2,000 m. For the elite oarsman the competitive 2,000 m takes approximately 6 min. Two distinct disciplines exist within rowing; sweep rowing and sculling. In sweep rowing, boats require a crew of two ('pair'), four ('four') or eight rowers each using a single oar and row on either one side (e.g. bowside or starboard) or the other (e.g. strokeside or port) of the boat. A sculling boat can seat one ('single'), two ('double') or four ('quad') rowers each using two oars.

Physiologically, rowing is proposed to require a good balance between strength and endurance (Hagerman and Staron, 1983; Secher, 1983) so a contribution from both aerobic and anaerobic energy metabolism is required. The determinants of performance for elite heavyweight rowers have been assessed and five important physiological parameters identified (Ingham *et al.*, 2002). These are reported to be: power at $\dot{V}O_2$peak, maximum power, maximal force ($r = 0.95$; $P < 0.001$), $\dot{V}O_2$peak ($r = 88$; $P < 0.001$) and oxygen consumption at the blood lactate threshold ($r = 0.87$; $P < 0.001$). All of these findings are derived from the use of a Concept IIc rowing ergometer (Concept, Nottingham, UK) with a force transducer at the handle such that force profiles and power measurements all result from the rowing action. Two tests are used, the first being a seven stroke maximal power test in which the last five strokes are recorded and from which means for force and power values are derived.

The second is a discontinuous incremental test utilising five 4-min efforts, each one requiring a 25 W increase in power output and followed by a sixth and final 4-min effort at race pace (see Figure 20.1). Gas exchange, heart rate and blood lactate are monitored in the second test only.

An 'elite' rower is here defined as an individual who is currently part of the World Class Performance Plan. Accordingly, there are perhaps 100–150 elite rowers in the United Kingdom. In terms of participation rowing is, in some quarters, considered an elitist minor sport. Despite this, according to the

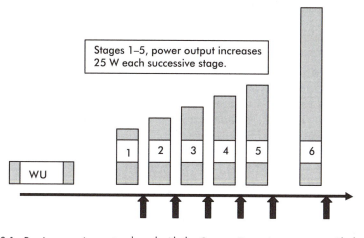

Figure 20.1 Rowing exercise protocol used with the Concept IIc rowing ergometer. Blocks 1–6 are work periods of 4 min. Blocks 1–5 have a 30-s rest interval during which a capillary (earlobe) blood sample (black arrow) is taken. A 2.5-min rest interval exists between stages 5 and 6. Stages 1–5 are sub-maximal, stage 6 is a maximal even-paced effort

Amateur Rowing Association, grassroots participation in rowing in the United Kingdom is estimated at more than 17,000 individuals. As a consequence of increasing media interest in rowing, resulting from long term and continuing Olympic success, there is greater awareness of the fact that elite rowers and their coaches view sport science as an essential component of their success. As a consequence more sports scientists, in a university setting, are being approached by club-level rowers seeking physiology support.

In prioritising which tests to implement in assessing the club level rower it could be argued that body composition, strength and power, and aerobic endurance are key. Body composition testing is important, particularly if the competitor being tested is a lightweight rower. DEXA (Dual Emission X-ray Absorpitometry) is currently considered the gold standard for assessing body composition and is currently being explored to characterise the body composition of elite rowers. Routinely, however, elite rowers continue to be tested using skinfold measurements. For almost 15 years the four-site skinfold method of Durnin and Womersley (1974) was used with GB elite rowing squads. Recently a seven-site method has been adopted at elite level in accordance with attempts to improve kinanthropometry methodology in accordance with standards laid down by the International Society for the Advancement of Kinanthropometry (ISAK). If, as a physiologist, one encounters a rower who has never been tested before and is not part of the World Class Plan, and has no such ambitions, then the Durnin and Womersley (1974) method may suffice.

Rowing is often referred to as a strength-endurance sport and, hence, being able to sustain a high power output over a prolonged period is key. Accordingly, in assessing changes in physiology the discontinuous incremental protocol (colloquially referred to as a 'step-test to max') presented in Figure 20.1 is tried and tested.

DATA COLLECTION

Ergometry

As mentioned earlier the preferred ergometer used in the United Kingdom is the Concept IIc (Concept, Nottingham, UK). When lab testing takes place with elite athletes a Concept IIc with force transducer at the handle is used, the Avicon system. This is an on-line system and force profiles for each stroke of the seven stroke power test and averages for each stage of the step test is recorded. This provides further feedback to the coach on technique. If testing sub-elite rowers a standard Concept IIc, without Avicon system, can be used and sufficient information provided to aid training of the individual.

Oxygen consumption

This is best accomplished using a breath-by-breath on-line gas analysis system (or 'metabolic cart') rather than Douglas bags. Since all on-line systems utilise a central processing unit that can integrate and provide a screen display in real time, there is the advantage of constantly being able to check the progress of the athlete towards maximal exercise.

Blood lactate

For a number of years now the EBIO plus (Eppendorf, Colone, Germany) laboratory-based analyser and the Biosen C_line (EKF Diagnostics, Germany) portable analyser have been used with elite rowers. These have superseded the long-term use of the Analox GM-7 (or portable PGM-7) (Analox Instruments Ltd, Hammersmith, UK). Generally lab-based systems have been preferred as their validity and reliability have been tested. Although it is possible to use new 'palm top' lactate analysers they are currently questioned with respect to validity and reliability but such research work is on-going. Hence, in the future, once their reliability and validity have been more fully demonstrated, they will prove to be the more practical equipment to use. With the EBIO plus and Biosen C_line it is possible to collect the capillary sample (earlobe is preferred as it limits pain of sampling, contamination of samples and is more consistent with health and safety). Blood is collected into a capillary tube which is immediately placed in an Eppendorf tube (containing a lysing agent) which can then be kept for 24 h without the need for refrigeration.

Heart rate

This is best measured using a telemetry system such as one of the monitors from the Polar range (Polar Electro Oy, Finland), with the preferred model being one which can record data to wrist watch monitor and which can be down-loaded after the event.

RESULTING DATA AND ITS USE

To reiterate, rowers should be tested using body composition (height, body mass and skinfold measures), seven stroke power test and step-test.

Seven stroke power test

The athlete carries out a warm-up on the ergometer, performs some light stretching and then a specific warm-up using hard efforts of 2, 3 and 4 strokes. The athlete then carries out the test with the first two strokes not recorded to allow the rower to overcome the inertia of the flywheel and to achieve a rating of 30 strokes per minute ($\pm$2 spm). From this test, work (J), mean force (N), mean power (W), stroke rate (spm) and stroke length (m) is reported from the five recorded strokes.

Step-test

The data collected and calculated from the step test includes, $\dot{V}O_2$peak, power at $\dot{V}O_2$peak, the percentage of maximum that can be sustained (i.e. $\dot{V}O_2$ at lactate threshold as a percentage of $\dot{V}O_2$peak), power at LT, maximum power, maximum force. Figure 20.2 shows the typical heart rate and blood lactate profile for the sub-maximal portion of a rowing step-test. The heart rate associated with LT can be used to determine a number of heart rate zones which can be used for training. In elite rowing heart rate training zones are no longer prescribed from the step-test and it is more commonly used to provide additional information on current status, for example illness.

Many coaches associated with the World Class Plan also use power at blood lactate reference values of 2 and 4 mM. Although, it is difficult to support the use of blood lactate reference values at all they can provide further information to the coach for comparison with future tests. However, if reference values are used it would be better to provide a range of values, for example power output at 2, 3, 4, 5 mM. Physiologists working with World

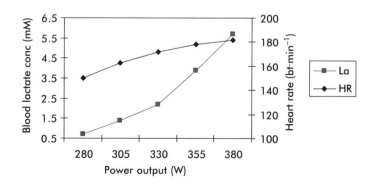

Figure 20.2 Heart rate and blood lactate response during a 'step-test' in an elite rower

Class rowing now place less emphasis on reference lactate values but some coaches still use them but in conjunction with all of the other data collected. Figure 20.3 demonstrates what happens to the blood lactate concentration – power output curve after a number of weeks training at the heart rate zones determined from the initial step-test. This rightward shift in the curve is interpreted as evidence for improvements in endurance.

Starting point for step-test design

For an individual who has not been tested before, a 2,000 m time trial on the Concept IIc should be performed first. This should be a maximal effort and the time for 2,000 m should be converted into a 500 m split time. For heavyweight men and women add 15 s to this time and you have the split for the third stage of the step-test. Subtract 25 W from this to get the power output (and split time) for stage 2 and subtract 50 W for stage 1. For stage 4 add 25 W and for stage 5 add 50 W. For lightweight men and women also add 15 s to the calculated 500 m split time to find the split for the third stage. However, it may be more appropriate to use a 15–20 W increment (rather than a 25 W increment) for lightweights.

Field-testing

Coaches in many sports are increasingly demanding that field-based testing replace laboratory-based testing. Generally, physiologists do not oppose a reduction in the frequency of lab-based testing and an increase in field-testing. However, it is very difficult to justify the elimination of lab-based testing altogether as it is simply impossible to collect data more objectively. Hence, only in the lab can conditions be appropriately standardised uninfluenced by the vagaries of changing gym, weather or water conditions. Indeed, GB elite rowers are still lab-tested two to three times per year with four to five field-based (step-test) sessions. In addition, the coach administers some tests such as an 18 km, 30 min, 2 km and 250 m rows on the water. On some occasions blood

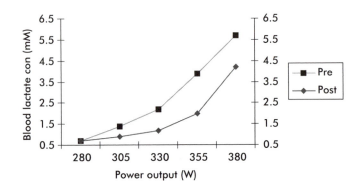

Figure 20.3 Changes in the blood lactate profile of an elite rower after months of endurance training

samples can be taken (by a physiologist) at the end of such rows or the 18 km row can be broken into 3 × 6 km rows with a 30–60 s rest interval for blood samples to be taken. In the 30 min row the athlete must complete the greatest possible distance at a fixed rate of 20 spm. This provides power output, heart rate and lactate data. The standard for elite women for this test is 7–8,000 m and for men 8.5–9,000 m.

At field camps overseas, early morning monitoring also occurs, prior to daily training. This involves the use of urine osmolality to monitor hydration status, blood urea as an index of the additional stress superimposed on training by an, often, extreme environment and morning body mass and supine resting heart rate. Data here are viewed in combination with a psychological inventory and if necessary discussion with the coach and athlete. The coach takes decisions on the necessity of modifying training with certain individuals as a consequence of all of this plus on-water and gym-based data. Further, decisions are taken 'weighing' the mass of data in the light of extensive coaching experience and the coach's personal knowledge of the individual.

REFERENCES

Durnin, J.V.G.A. and Womersley, J. (1974). Body fat assessed from total body density and its estimation from skinfold thickness: measurements on 481 men and women aged from 16 to 72 years. *British Journal of Nutrition*, 32(1): 77–97.

Hagerman, F.C. and Staron, R.S. (1983). Seasonal variations among physiological variables in elite oarsman. *Canadian Journal of Applied Sports Science*, 8: 143–148.

Ingham, S.A., Whyte, G.P., Jones, K. and Nevill, A.M. (2002). Determinants of 2000 m rowing performance in elite rowers. *European Journal of Applied Physiology*, 88: 243–246.

Secher, N. (1983). The physiology of rowing. *Journal of Sports Science*, 1: 23–53.

CHAPTER 21

SAILING

Neil Spurway

INTRODUCTION

The sport of sailing comprises many categories. The range of fitness challenges is at least as great as the range between say, downhill ski-racing and biathlon; since we are talking about the means of propulsion and load-bearing, one could even suggest that it compares with the range of all sports performed on foot! The physical demands depend strongly on the type of boat sailed, and the individual's position (and so function) in the boat. Fitness is most often taken seriously by:

1   '*Hikers*'. Those who hang out over the side of a light boat to balance it, with their feet hooked under straps near the centre line, and their thighs or calves against the edge of the hull. This is the posture of almost all single-handed dinghy sailors and the helmspersons in most double-handed boats.
2   *Trapezers*. These are the ones who 'stand' against the hull's edge, with body weight supported horizontally over the water by a wire from the mast. Theirs is more of a gymnastic challenge, but should be less of a fitness one.
3   *Boardsailors*. These stand on their ultra-light craft, controlling the sail by a rigid wishbone strut at chest height; under present rules they, too, have much of their weight supported by a harness when they lean back against the wind.
4   *Winch hands*. These are crew members of big yachts, whose specialist task is sporadically to crank in the sails as fast as possible. Between such bursts of effort they are negligibly active.

The UK governing body (Royal Yachting Association, RYA) has established a compromise group of simple fitness tests, which are applied in common to all these categories of sailor. Only the norms for performance, on the various tests within the group, reflect the contrasting requirements of the different specialists. Whatever one's judgement of this policy from the GB's standpoint, it may be assumed that a sailor consulting a sports scientist will want guidance more specific to his/her category. We confine these notes to activities (1) and (3) which are common and place high demands on fitness. Modifications appropriate to (2) and (4) should not be difficult for an experienced and suitably equipped sports physiologist.

BASIC TESTS, APPLICABLE TO BOTH CATEGORIES OF SAILOR

VO_{2max} by cycle or rowing ergometry

Cycling and rowing are more specific than running, since both hiking and boardsailing place more demands on quads than ankle extensors. It is also safer, as people with foot or knee problems, for whom running impacts are inappropriate, can be successful sailors; more severe knee injury, however, of course contraindicates even cycling or rowing.

Performance to be expected of the dinghy hikers is not high. A value of 50–60 ml·kg^{-1}·min^{-1} is typical of national squad members in some of the most demanding single-handed boats; in other sailors the figures will be lower. In the demanding boats in strong winds, heart rates of ~85% maximum may be sustained for long periods.

The limited available figures suggest, however, that top-level competitive boardsailors may be expected to have VO_{2max} values of 60–70 ml·kg^{-1}·min^{-1}, as theirs is a more aerobically active sport.

Sit and reach

Flexibility is at a considerable premium in both forms of sailing. The basic sit-and-reach test can appropriately be supplemented, for example by neck, shoulder and trunk-rotation tests (see Chapter 10) if work to improve these aspects of flexibility is contemplated for the individual. A goniometer and/or a Leighton flexiometer will be necessary for such extra tests.

After these two generalities, we come to the category-specific tests, discussed here.

SPECIFIC TESTS FOR HIKERS' FITNESS

Hiking itself

This places near-isometric endurance demands on tibialis anterior, quadriceps and abdominals; the ratios between the three vary, but in most people, sailing most shapes of boat, quadriceps are activated to the highest percentage of their maximum, and tibialis to the lowest.

Consider first what can be done in a *standard gymnasium*, with no sailing-specific apparatus:

Strength-endurance of quadriceps can be assessed by rhythmic half-squat against 30–50% body weight (selecting load according to subject fitness), timed by a metronome set to 25 min^{-1}. Seconds 0–50 of each minute follow this metronome, but during seconds 51–60 the separate presses are replaced by a static hold, with the knees at 120°.

For the abdominals, the maximum number of trunk curls per minute is often counted: this is far too rapidly dynamic, and should be replaced by a repetition endurance test exactly analogous to that for the quadriceps. Thus the curls are performed to the 25 min^{-1} metronome, but during seconds 51–60 of each minute they are replaced by a static hold, with the back just clear of the floor.

Tibialis anterior is not normally tested. However, if your sailor complains of fatigue-pain there during hard-weather sailing, a third test to the same rhythm as the others can be performed with the subject lying supine and flexing his/her ankles against a loaded strap.

All these functions (and more – discussed later) can be tested at once with a single sailing-specific item, the 'hiking bench'. This is a simple frame, with a seat at side-deck height and foot-straps in a location equivalent to the position of the straps in the individual sailor's boat. If seat height, distance from it to foot-straps, and degree of slack in these straps are adjustable, the same bench can be used for sailors of all likely dinghies (Figure 21.1).

'Hiking out' on such a bench directly loads all leg-muscle groups and abdominals in an essentially isometric way – the way they are loaded in the boat. Basically, the test consists simply of timing the period for which the hiking position can be held. However, realism is enhanced (and the risk of boredom reduced) by adding trunk rotations and curls. Have your subject do one curl (to trunk-vertical) every 30 s, and two rotations of shoulders about hips in each direction during the middle part of the time between the curls (Figure 21.2). (The sailor may well describe the curls as 'sitting in' – in to the boat, as against out from it.)

Each movement should take 2 s – that is 1 s up (or twist), 1 s down (or straighten).

(Note: Dynamic sailing simulators of different designs, but all much more complex than the adjustable hiking bench, are used in most laboratories which explore the physiology of dinghy sailing. For basic fitness testing, however, they are unnecessary and, indeed, of questionable advantage; their use is as research tools.)

Back strength-endurance

This may be less expected than the first group of tests, but is highly desirable nonetheless; dinghy sailors are more than averagely prone to lower-back pain, and the best precautionary measure is to encourage back-strengthening exercises. Risk-free tests, however, are elusive.

With everyone who is highly active in their sport and not suffering back trouble at the time, it is recommended to use the prone lift – the subject, face down

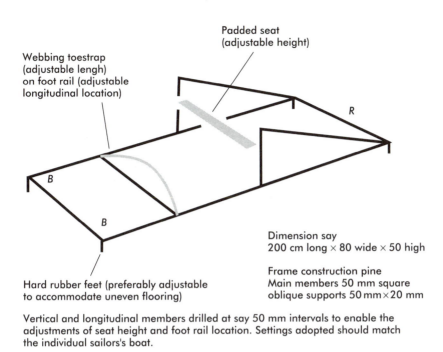

Padded seat
(adjustable height)

Webbing toestrap
(adjustable lengh)
on foot rail (adjustable
longitudinal location)

R

B

B

Dimension say
200 cm long × 80 wide × 50 high

Frame construction pine
Main members 50 mm square
oblique supports 50 mm × 20 mm

Hard rubber feet (preferably adjustable
to accommodate uneven flooring)

Vertical and longitudinal members drilled at say 50 mm intervals to enable the
adjustments of seat height and foot rail location. Settings adopted should match
the individual sailors's boat.

For sheeting simulations, fix multipart rubber shock cord on inner face of frame
at R, and pulley blocks at BB. Tie midpoint of 'mainsheet' to free end of shock cord
[roughly under the seat] and lead working ends through blocks to sailor's hands.
Work can be standardised by markings on ropes, where they pass through blocks.

Figure 21.1   Hiking bench

Curl_____Curl

Twist right, twist left, twist right, twist left

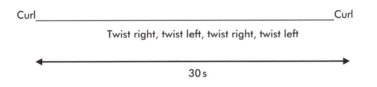

30 s

Figure 21.2   Sequence of trunk movements during each 30-sec period on hiking bench

on floor or table, raises all four limbs and upper trunk together, and holds this
position for 5 s, rests 5 s and repeats. Anyone who can hold form fully through 6
repetitions is in reasonable shape; 12 repetitions are excellent, and need not be
exceeded. This is not intended to be a test to exhaustion, just a check that the back
flexors' strength-endurance is adequate to meet the likely demands of the sport.

More and less severe tests can clearly be devised – less severe ones being
essential for people whose backs are already suspect. An example is to adopt
an 'all fours' position, and from this raise one leg and the contralateral
arm together from the floor for 5 s, then the opposite pair; rest in 'all fours'
position 5 s, then repeat the lifts. Provided a given sailor reports to the same

scientist for follow-ups, changes over time will be satisfactorily documented from these or any other reasonable tests.

Arm strength-endurance

Sailors control the sails by pulling on ropes, which are typically 10–12 mm in diameter. Grip dynamometers do not test this kind of grip-strength; nor do the handles of multi-gym machines offer anywhere near a comparable challenge. So test arm-flexion performance by having your sailors actually pull on a rope: ideally, an old 'mainsheet' (rope controlling the main sail) from a dinghy.

In the gym, this can be attached to a weight-machine, so that the line of pull leads under a pulley at floor level. On the hiking bench, lead two ropes (one for each hand) around blocks on the front cross-member (Figure 21.1), and back to springs or shock-cord beneath the subject; load/extension calibration marks for either of these can be made on the floor or light wooden strips parallel and close to the extensible component. Alternatively, but at greater cost, spring balances can be attached direct to the bench cross-member, and the loads applied by the respective arms directly read from their dials.

Loads should be selected to suit the subject: 10% body weight for males, 8% for females, are guideline starting values. The same rhythm as for gym-testing of the hiking muscles is suitable, that is, pulls at 25 min^{-1}, interrupted by 10 s static holds at the end of every minute. Time to volitional exhaustion, as before.

SPECIFIC TESTS FOR BOARDSAILORS' FITNESS

Even within boardsailing ('windsurfing'), there are a variety of approaches. However, the acrobats who somersault off waves in Honolulu, or even Tiree, are unlikely to present themselves for Sports Science assessment. Racing boardsailors are much more likely to, and are the people considered here. Though actually at the less dynamic end of the range within boardsailing, their activity is more mobile than most dinghy sailing. (This is especially the case, though the non-sailor does not expect this, in light winds, because the boardsailor then 'pumps' the board along by working the sail with muscle power against slow-moving air.)

Consequently, general fitness tests are more valid than for dinghy sailors. However, they must concentrate on the upper body; rowing ergometry and/or arm-cranking are worth taking to VO_{2max} if your lab has these facilities, but running and cycling have negligible relevance. More specific tests follow:

Arm extension strength-endurance

Traditional press-ups are all that are needed here. Place feet on a bench to make them harder, and simply count how many can be done to a 25 or 30 min^{-1} metronome.

Arm flexion strength-endurance

Chinning the bar is a rough approximation to the sailing action, but it is much better simulated if the arms work straight out from the chest, at right angles to the spine. So lie the sailor supine, feet on one bench and trunk on another. Arrange a strong bar, say 4 cm diameter, to be reached by arms extended straight up from shoulders. When the weight is taken on this bar, remove the trunk bench, quickly replacing it by a thick mat, and start the metronome. Count how many chinnings of *this* bar can be done before the pace is lost.

Grip strength-endurance

Because a wishbone is substantially thicker than a rope, grip dynamometry is much more applicable to the boardsailor than the dinghy person. For each hand in turn, first establish maximum grip strength (best of three trials). Now you want your subjects to do a series of repeats at a fixed percentage of max. If you anticipate testing a lot of boardsailors, construct your own dynamic instrument, measuring grip force on a load cell with digital readout: a 12 min^{-1} rhythm, of grips measuring 40% max ($\pm$ 5% for any one effort) discriminates well. For one-off tests stick with the grip dynamometer, and work out a rhythm which allows the instrument to be re-cocked between contractions. 6 min^{-1} should be comfortably feasible; because of the longer intervals the appropriate grip-target in this case is about 55% max strength.

Quads strength-endurance

This is the only test which can fruitfully be shared between board and dinghy sailors, but for the boardsailor the severity should be at the top of the dinghy-sailors' range, or a little higher. Rhythmic half-squats against ~50% body weight are timed by a metronome set to 25 or 30 min^{-1}. As before, seconds 0–50 of each minute follow this metronome, but during seconds 51–60 the subject holds static, with knees at 120–130°.

CHAPTER 22

SWIMMING

Kevin G. Thompson and Suzan R. Taylor

INTRODUCTION

For the aerobic assessment of competitive swimmers a number of tests have been used which include:

- 10×100-m test (Touretski, 1993);
- 5×300-m freestyle incremental test (Pyne, 1994);
- 5×200-m incremental test (backstroke, breaststroke, butterfly) (Pyne, 1994);
- 10×100-m test (Goldsmith, 2001);
- 7×200-m test (Lakomy and Peyrebrune, 2003).

The 7×200-m incremental protocol devised by the Australian Swimming Inc. Sports Science Advisory Group (1997) has been widely adopted around the world. In 2003, Lakomy and Peyrebrune issued an adapted version of the protocol on behalf of British Swimming. Improvements in fitness are proposed from shifts in the heart rate–speed and blood lactate–speed relationships. Studies suggest that training increases the swimming speed at the lactate threshold and at absolute lactate concentrations around 2 mmol·l^{-1} to 6 mmol·l^{-1} from well-trained to elite swimmers, although these changes may or may not be associated with improvements in competition performance (Touretski, 1993; Olbrecht, 2000, p. 81; Pyne *et al.*, 2001; Thompson *et al.*, 2003). It is widely accepted that a highly developed aerobic metabolism and a highly efficient stroke technique are required to be successful in most of the competitive swimming events.

7 × 200-m step test protocol (Lakomy and Peyrebrune, 2003)

A standardised warm-up (e.g. up to 1,000 m) is advised, to ameliorate differences in physiological responses between morning and evening swimming at least at speeds approximating to the maximum lactate steady state (Martin and Thompson, 2000). Following the warm-up, a measurement of blood lactate concentration of less than 1.5 mmol·l^{-1} is recommended to ensure a good progression through the test protocol (Table 22.1).

Each repetition begins 5 min after the previous one started for the freestyle and backstroke, and 6 min after the previous one started for the butterfly and breaststroke. The swimmer is required to pace evenly, so as to ensure there is no more than 2 s between each 100-m split time. Stroke count (strokes per length) and stroke rate (strokes per minute, Base 3 function stopwatch) measurements can be used to identify poor efficiency, stroke adjustments and developing fatigue. Immediately after each repetition the swimmer's heart rate is recorded, as is the RPE (the swimmer can self-record these measurements with a waterproof pencil on an acetate sheet after each repetition). Within 30–45 s of the swimmer completing the repetition a capillary blood sample is taken for the determination of lactate concentration. Following the final repetition, further blood sampling takes place every 2 min until a peak lactate value is determined. Heart rate recovery may also be determined by taking a measurement every 30 s.

Sports scientists generally produce simple line plots or 2nd or 3rd order polynomial plots of the data. A recent book by Olbrecht (2000), suggests that the interpretation of data, to prescribe changes in training, is more complicated than previously thought. Arguably, in addition to the 7 × 200-m step-test the scientist should also measure the previously mentioned variables following

Table 22.1   7 × 200-m step test protocol

Step	Approximate swimming intensity (%)	Approximate heart rate below maximum (b·min^{-1})	Seconds slower than personal best time – males (s)	Seconds slower than personal best time – females (s)
1	70	60	30	24
2	75	50	25	20
3	80	40	20	16
4	85	30	15	12
5	90	20	10	8
6	95	10	5	4
7	100	0	0	0

Source: Reproduced with permission from H.K.A. Lakomy and M.C. Peyrebrune (2003)

Note
The swimmer should be habituated to swimming on a tether prior to the assessment.

maximal efforts over shorter distances (50-m, 100-m), to provide greater insight into the kinetics and development of the aerobic and anaerobic pathways.

For a specialist freestyle 50-m swimmer, an alternative protocol (Touretski, 1993; discussed later) may be more appropriate for determining variable vs. swimming speed plots.

> $3 \times$ 100-m, 30-s recovery @ aerobic (slow) pace;
> $3 \times$ 100-m, 45-s recovery @ threshold (moderate) pace;
> $3 \times$ 100-m, 60-s recovery @ high speed, but not maximal pace;
> Maximal effort 100 m.

Goldsmith (2001) has also produced several similar protocols which vary depending on the swimming stroke, for the monitoring of age-group swimmers.

It is suggested that for all of the protocols described, turning times (the time taken for the swimmer to travel 7.5 m, or alternatively 5 m, into and out of a turn) are measured, particularly during swimming in a 25 m pool, where they make up a significant proportion of the distance covered. It is possible for a shorter swimming time to occur, as a result of the swimmer turning faster rather than swimming faster, with the physiological response remaining largely unchanged (Thompson et al., 2004). Turning times, at least during breaststroke swimming, appear to be able to differentiate between better swimmers (Thompson et al., 2003) and are acutely sensitive to fatigue (Thompson et al., 2000, 2002, 2004).

There are other tests that can be used to identify key performance factors such as the determination of: maximal swimming speed (a maximal 25 m sprint), start time (dive and sprint to 15 m) and an analysis of stroke effectiveness (10 $\times$ 100-m repetitions, with the speed of each 100-m repetition progressively increasing from slow to fast speeds). Finally, a major problem with swimming-based protocols has been the accuracy of pacing. Various pacing methods have been tried over the last 30 years including the use of controllable measurement–equipment platforms to tow or pace, swimming flumes, light sequencing and audible signals. A recent innovation has been the Aquapacer™ (Challenge and Response, Inverurie, Scotland) which consists of a programmable coin-size audible pacing device placed in the swimmer's cap close to the ear. There is some evidence that swimmers can quickly habituate to this device and elicit pacing which is more accurate than self-pacing (Martin and Thompson, 2000; Thompson et al., 2002).

NOTES

Please note that some elite swimmers, with a highly trained aerobic metabolism, might not increase their blood lactate concentration above 2 mmol·l^{-1} until after repetitions 4 or 5. In this situation it would be better to reduce the time for step 1, perhaps by beginning at step 2 or even 3 and to also reduce the target time by 3 s thereafter in the six consecutive stages that follow. Conversely specialist 50-m swimmers may find the protocol too demanding, in which case another protocol (discussed later) may be a more acceptable alternative.

REFERENCES

Australian Swimming Inc, Sports Science Advisory Group (1997). *National Testing Protocols for Physiology*. Brisbane.

Goldsmith, W. (2001). Ten × 100-m swim test protocol: a simple and effective test for swimmers of all ages. *Swimming in Australia*, March/April; 9–15.

Lakomy, H.K.A. and Peyrebrune, M.C. (2003). *Aerobic Fitness Assessment: 7 × 200-m Step test*. British Swimming.

Martin, L. and Thompson, K.G. (2000). Reproducibility of diurnal variation in swimming. *International Journal of Sports Medicine*, 21(6): 387–392.

Olbrecht, J. (2000). *The Science of Winning. Planning, Periodizing and Optimizing Swim Training*. Luton, England: Swimshop.

Pyne, D.B. (1994). *Sports Science for Age Groupers: Exercise Physiology*. Australian Swimming Coaches and Teachers convention 5th–8th May, Gold Coast, Australia: ASCTA.

Pyne, D.B., Lee, H. and Swanwick, K.M. (2001). Monitoring the lactate threshold in world-ranked swimmers. *Medicine and Science in Sports and Exercise*, 33(2): 291–297.

Thompson, K.G., Haljand, R. and MacLaren, D. (2000). An analysis of selected kinematic variables in national and elite male and female 100-m and 200-m breaststroke swimmers. *Journal of Sports Sciences*, 18: 421–431.

Thompson, K.G., MacLaren, D., Lees, A. and Atkinson, G. (2002). Accuracy of pacing during breaststroke swimming using a novel pacing device, the Aquapacer™. *Journal of Sports Sciences*, 7: 537–546.

Thompson, K.G., MacLaren, D., Lees, A. and Atkinson, G. (2003). The effect of even, negative and positive pacing on metabolic, kinematic and temporal variables during breaststroke swimming. *European Journal of Applied Physiology*, 88(4–5): 438–443.

Thompson, K.G., MacLaren, D., Lees, A. and Atkinson, G. (2004). The effects of changing pace on metabolism and stroke characteristics during high speed breaststroke swimming. *Journal of Sports Sciences*, 22: 149–157.

Touretski, G. (1993). Footprints to success. An analysis of the training for an Olympic sprint champion. Alexandre Popov. *Australian Swim Coach*, X(V): 5–11.

ANAEROBIC TESTING IN SWIMMERS

Introduction

Compared to the large number of tests used to monitor and evaluate aerobic capacity few tests have been developed to assess changes in anaerobic performance. Data gathered using a cycle ergometer or non-motorised treadmill cannot simply be transferred to swimming due to the differences in body position, motor recruitment patterns and the fact that the sport takes place in water. The systems currently available in swimming research can be divided into two main categories, that is, dry-land and water-based. The dry-land method has mainly involved the measurement of power using a biokinetic swim bench, which simulates swimming using an ergometer, but does not replicate swimming completely as only the power of the arms can be assessed. Water-based tests can be sub-divided into indirect and direct methods. Indirect methods have been criticised as they oversimplify a complex action and are not favoured by physiologists.

Direct methods involve the use of a tethering device. Semi-tethered swimming involves swimming against a weighted resistance for a distance of ~10 m but this only stresses the ATP-PCr energy system. However recently, British Swimming have attempted to improve upon this and built an oversized semi-tethered system allowing a 50-m swim to be accommodated, which would stress the ATP-PCr and glycolytic systems. The semi-accommodating device works on a similar principle to the semi-tethered device, except it only allows the swimmer to travel a distance of 13 m. In contrast the fully tethered swimming does not allow any forward movement, which prevents the calculation of power per se, but propulsive force can be recorded and used as an indication of power (Lakomy, 1994). Unlike other systems, fully tethered swimming allows the duration of the swim to be manipulated, and therefore offers greater flexibility in the assessment of anaerobic performance.

TESTING PROTOCOL FOR FULLY TETHERED SWIMMING

The tethered swimming system

Currently there are no fully tethered swimming systems commercially available. The tethered swimming system used by Taylor (2003) consisted of a laptop computer (Toshiba Satellite 230CX, Tokyo, Japan) and software (PowerLab™ System, Chart for Windows®, ADI Instruments, Castle Hill, Australia), a starting block (used to anchor the force transducer), an amplifier (FE 359 TA 12v conversion, Fylde, Electronic Labs, Preston, UK), a PowerLab™/400 system (ADI Instruments, Castle Hill, Australia), a 100 kgf force transducer (V4000, Maywood Instruments, Hampshire, UK), 3 karabiners (1000 kN, EB Viper, Bangor), 6 m of pre-stretched rope (diameter 0.5 cm) and a climbing belt (Trat, Arizona, USA). A diagrammatic representation of the tethered swimming system is given in Figure 22.1.

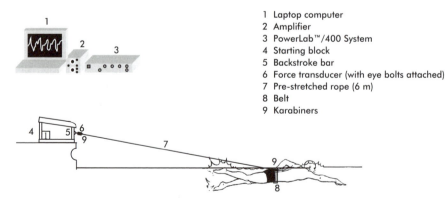

1 Laptop computer
2 Amplifier
3 PowerLab™/400 System
4 Starting block
5 Backstroke bar
6 Force transducer (with eye bolts attached)
7 Pre-stretched rope (6 m)
8 Belt
9 Karabiners

Figure 22.1 A schematic diagram of the fully tethered swimming system (Taylor, 2003)

Table 22.2 Normative data for age group swimmers of county to national level standard

	Age in years					
	10	11	12	13	14	15
Peak force (N)						
Boys	64.0 ± 7.4	68.6 ± 11.3	76.4 ± 18.2	86.2 ± 18.8	110.3 ± 15.3	118.6 ± 13.4
Girls	69.4 ± 8.4	70.5 ± 12.8	72.1 ± 12.4	74.6 ± 13.7	85.8 ± 12.2	92.1 ± 9.8
Mean force (N)						
Boys	45.0 ± 5.4	49.0 ± 6.8	54.0 ± 10.8	62.6 ± 11.3	79.0 ± 9.5	83.6 ± 8.4
Girls	48.1 ± 7.1	50.2 ± 9.2	52.0 ± 8.0	55.4 ± 9.0	63.8 ± 9.0	68.8 ± 9.1

Warm-up

See example in aerobic section.

The test

The swimmer should always habituated to swimming on a tether prior to the assessment. The swimmer commences the test from a rolling start, which involves taking up the slack in the rope and swimming sub-maximally until a whistle is blown. To standardise the procedure the starting whistle should be blown on the sixth or seventh stroke, when the fingertips of the right hand enter the water (on freestyle). Swimmers should be verbally encouraged throughout the 30 s test, instructed to avoid pacing, and to maintain maximal effort for the duration of the test. The swimmer should try to stay in the middle of the lane whilst swimming. After the 30 s period is complete the whistle is blown again to signify the end of the test.

The results

The raw data should be expressed as a mean for each second of the 30 s period; from this peak and mean force should be calculated. Peak force represents the greatest force generated, whereas mean force is the average (mean) force generated over the 30 s period.

Normative data

Only limited normative data are available in the literature as tethered force is dependent on the age and the standard of the swimmer (Table 22.2).

REFERENCES

Lakomy, H.K.A. (1994). Assessment of anaerobic power. In M. Harries, C. Williams, W.D. Stanish and L.J. Micheli (eds), *Oxford Textbook of Sports Medicine*, pp. 180–187. New York: Oxford University Press.

Taylor, S.R. (2003). The analysis of anaerobic performance in age group swimmers. Unpublished PhD Thesis, Liverpool John Moores University, UK.

CHAPTER 23

TRIATHLON

Les Ansley

INTRODUCTION

Triathlons are performed over a variety of distances. The most common (in ascending order of distance (km)) are sprint (0.75/20/5), Olympic (1.5/40/10), long course (2.5/80/20) and Ironman (3.8/180/42) triathlons. Because of the entirely different demands imposed by the different types of triathlon it is not realistic to discuss them together and therefore this chapter will deal exclusively with the two short-distance triathlons.

Due to the multidisciplinary nature of a triathlon, competitors are well-trained in all the events but not to the same extent as athletes who only compete in one event (Kohrt *et al.*, 1987). However, it seems that the relationships between the physiological variables (such as maximal aerobic power) and performance variables (such as 40 km time trials) of the individual disciplines both within the context of a triathlon (Bentley *et al.*, 1998, 2003) and in isolation are still valid. Consequently, most research into triathlon has adopted a reductionist approach by breaking the sport down into its element components. However, competing in more than one event within a single race does result in changes in physiological and biomechanical parameters in subsequent events that are absent from single event disciplines (Guezennec *et al.*, 1996; Hausswirth *et al.*, 1997; Hue *et al.*, 1999, 2000).

The chapters on the individual sports that make up triathlon have detailed the testing protocols for their particular sport. Should you wish to apply specific swimming, cycling or running tests please refer to the relevant chapter. In order to avoid repetition, this chapter will deal mainly with the idiosyncrasies of triathlon.

TRANSITIONS

An almost unique feature of triathlon is the transition between disciplines. The transition from one event to another can have profound implications on physiological and kinematic measures (Hue *et al.*, 1999; Heiden and Burnett, 2003) and can affect both the perceived effort (Delextrat *et al.*, 2003) and performance in the remaining events (Margaritis, 1996). Both cycling and running economy are compromised when preceded by swimming (Delextrat *et al.*, 2005) and swimming and cycling (Guezennec *et al.*, 1996; Hausswirth *et al.*, 1997), respectively. Although the extent of these changes in economy appear to be related to the ability level and training status of the athlete (O'Toole *et al.*, 1987; Millet *et al.*, 2000) and highlights the need for specific training in the transition between events, which develops skill and improves physiological responses during the transition phase (Hue *et al.*, 2002; Heiden and Burnett, 2003).

Despite the importance of the transition between events, it is not an area that many athletes and coaches objectively assess and there is void in the scientific literature regarding validation of specific physiological tests. One methodology that has been used in the past by the British Triathlon Association is to compare heart rate, stride frequency and blood lactate concentration between an incremental running test performed in isolation and an incremental running test that is performed immediately after an incremental cycle test (Figure 23.1). A change in stride frequency, due to the changes in knee and hip angle kinematics, is considered an important marker in assessing running fatigue (Hanon *et al.*, 2005). Stride frequency can be a useful variable, along with heart rate, as it is relatively easy to measure in the field.

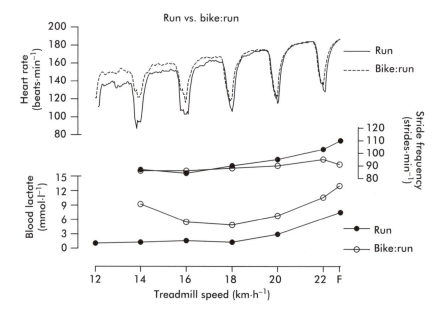

Figure 23.1 Results for an elite triathlete comparing an isolated incremental treadmill test with an incremental treadmill test following an incremental bike test

DRAFTING

An important feature of elite short-distance triathlons nowadays is the sanction to draft on the bicycle. Drafting on the swim leg of a triathlon has always been permitted for all competitors and has been shown to decrease the perception of effort on the subsequent cycle (Delextrat *et al.*, 2003). Similarly, the energy cost of cycling as a drafter during the bike leg is between 26% and 39% (depending on how many cyclists are in the bunch) lower than cycling alone (McCole *et al.*, 1990). This is reflected in improved running speed following a sheltered ride (Hausswirth *et al.*, 1999). Moreover, the benefits derived from drafting on the bike are greater for elite triathletes (Millet *et al.*, 2000) and stronger runners (Hausswirth *et al.*, 1999) since at higher cycling velocities, the proportion of effort required to overcome air resistance is progressively greater (di Prampero *et al.*, 1979). In addition to the impact of prior events, these differences between performances in draft and non-draft races should be considered when applying testing protocols and interpreting the results.

PREDICTING RACE PERFORMANCE

Total race time is significantly related to time spent running ($r^2 = 0.94$) and cycling ($r^2 = 0.66$) but not swimming ($r^2 = 0.09$) (Dengel *et al.*, 1989), which is not entirely surprising since swimming, cycling and running contribute roughly 15%, 55% and 30%, respectively, to the total duration of an Olympic distance triathlon. This discrepancy in relative contributions has lead some to argue that a fairer competition would be an equilateral triathlon, in which each leg would take an approximately equal time (2.7 km swim, 22.4 km cycle and 10 km run) (Wainer and De Veaux, 1994), although this idea is not widely supported.

The average percentage VO_2 maintained during a sub-maximal laboratory-simulated triathlon, in which each discipline is performed at a pace equivalent to 60% maximal aerobic capacity, has been identified as a good predictor of overall triathlon performance (Miura *et al.*, 1997). Also the fractional VO_2 utilisation during sub-maximal exercise (1 m·s^{-1} for swimming; 200 W for cycling; and 201.2 m·min^{-1} for running) is an excellent predictor of specific event times (Dengel *et al.*, 1989).

Several of researchers have also devised formulae to predict race times from maximal and sub-maximal physiological variables. Hue (Hue, 2003) found that triathlon performance time for drafting Olympic distance races in elite athletes was best predicted from variables measured during a cycle-run test (30 and 20 min, respectively). Blood lactate concentrations at the end of the 30 min cycle leg ($[La]_b$) and distance covered on the 20 min run leg (m) were the best predictors of race time ($r^2 = 0.93$):

Predicted race time (s) = (1.128)*m) + 38.8*$[La]_b$ + 13.338

Schabort *et al.* 2000) have predicted race times with a similar level of accuracy ($r^2 = 0.81$) from the blood lactate concentration during sub-maximal cycling at 4 W·kg body mass^{-1} (La) and the peak treadmill speed (km·h^{-1}) (PTS) attained during a maximal incremental running test in which treadmill speed was increased 1 km·h^{-1} every 60 s.

$$\text{Predicted race time (s)} = -129)^*(\text{PTS} + 122^*[\text{La}]_b + 9456$$

CONCLUSION

Most often the physiological tests performed on triathletes are those that are designed for testing athletes in the individual sports that make up a triathlon. Although the correlates between the outcomes in these tests and competitive outcomes in the field are very good, it is clear that there are unique aspects of a triathlon that cannot be measured by these generic protocols. Unfortunately, there are few physiological tests that have been designed specifically for triathlon. Therefore, it is important to carefully consider exactly what aspect of performance needs to be assessed before designing or implementing a testing regime.

REFERENCES

Bentley, D.J., Wilson, G.J., Davie, A.J. and Zhou, S. (1998). Correlations between peak power output, muscular strength and cycle time trial performance in triathletes. *Journal of Sports Medicine and Physical Fitness*, 38: 201–207.

Bentley, D.J., McNaughton, L.R., Lamyman, R. and Roberts, S.P. (2003). The effects of prior incremental cycle exercise on the physiological responses during incremental running to exhaustion: relevance for sprint triathlon performance. *Journal of Sports Science*, 21: 29–38.

Delextrat, A., Tricot, V., Hausswirth, C., Bernard, T., Vercruyssen, F. and Brisswalter, J. (2003). Influence of drafting during swimming on ratings of perceived exertion during a swim-to-cycle transition in well-trained triathletes. *Perception and Motor Skills*, 96: 664–666.

Delextrat, A., Brisswalter, J., Hausswirth, C., Bernard, T. and Vallier, J.M. (2005). Does prior 1500-m swimming affect cycling energy expenditure in well-trained triathletes? *Canadian Journal of Applied Physiology*, 30: 392–403.

Dengel, D.R., Flynn, M.G., Costill, D.L. and Kirwan, J.P. (1989). Determinants of success during triathlon competition. *Research Quarterly in Exercise and Sport*, 60: 234–238.

di Prampero, P.E., Cortili, G., Mognoni, P. and Saibene, F. (1979). Equation of motion of a cyclist. *Journal of Applied Physiology*, 47: 201–206.

Guezennec, C.Y., Vallier, J.M., Bigard, A.X. and Durey, A. (1996). Increase in energy cost of running at the end of a triathlon. *European Journal of Applied Physiology*, 73: 440–445.

Hanon, C., Thepaut-Mathieu, C. and Vandewalle, H. (2005). Determination of muscular fatigue in elite runners. *European Journal of Applied Physiology*, 94: 118–125.

Hausswirth, C., Bigard, A.X. and Guezennec, C.Y. (1997). Relationships between running mechanics and energy cost of running at the end of a triathlon and a marathon. *International Journal of Sports Medicine*, 18: 330–339.

Hausswirth, C., Lehenaff, D., Dreano, P. and Savonen, K. (1999). Effects of cycling alone or in a sheltered position on subsequent running performance during a triathlon. *Medicine and Science in Sports and Exercise*, 31: 599–604.

Heiden, T. and Burnett, A. (2003). The effect of cycling on muscle activation in the running leg of an Olympic distance triathlon. *Sports Biomechanics*, 2: 35–49.

Hue, O. (2003). Prediction of drafted-triathlon race time from submaximal laboratory testing in elite triathletes. *Canadian Journal of Applied Physiology*, 28: 547–560.

Hue, O., Le, G.D., Boussana, A., Chollet, D. and Prefaut, C. (1999). Ventilatory responses during experimental cycle-run transition in triathletes. *Medicine and Science in Sports and Exercise*, 31: 1422–1428.

Hue, O., Le, G.D., Boussana, A., Chollet, D. and Prefaut, C. (2000). Performance level and cardiopulmonary responses during a cycle-run trial. *International Journal of Sports Medicine*, 21: 250–255.

Hue, O., Valluet, A., Blonc, S. and Hertogh, C. (2002). Effects of multicycle-run training on triathlete performance. *Research Quarterly in Exercise and Sport*, 73: 289–295.

Kohrt, W.M., Morgan, D.W., Bates, B. and Skinner, J.S. (1987). Physiological responses of triathletes to maximal swimming, cycling, and running. *Medicine and Science in Sports and Exercise*, 19: 51–55.

McCole, S.D., Claney, K., Conte, J.C., Anderson, R. and Hagberg, J.M. (1990). Energy expenditure during bicycling. *Journal of Applied Physiology*, 68: 748–753.

Margaritis, I. (1996). Factors limiting performance in the triathlon. *Canadian Journal of Applied Physiology*, 21: 1–15.

Millet, G.P., Millet, G.Y., Hofmann, M.D. and Candau, R.B. (2000). Alterations in running economy and mechanics after maximal cycling in triathletes: influence of performance level. *International Journal of Sports Medicine*, 21: 127–132.

Miura, H., Kitagawa, K. and Ishiko, T. (1997). Economy during a simulated laboratory test triathlon is highly related to Olympic distance triathlon. *International Journal of Sports Medicine*, 18: 276–280.

O'Toole, M.L., Hiller, D.B., Crosby, L.O. and Douglas, P.S. (1987). The ultraendurance triathlete: a physiological profile. *Medicine and Science in Sports and Exercise*, 19: 45–50.

Schabort, E.J., Killian, S.C., St Clair, G.A., Hawley, J.A. and Noakes, T.D. (2000). Prediction of triathlon race time from laboratory testing in national triathletes. *Medicine and Science in Sports and Exercise*, 32: 844–849.

Wainer, H. and De Veaux, R.D. (1994). Resizing triathlons for fairness. *Chance*, 7: 20–25.

CHAPTER 24

TENNIS

Polly Davey

An essential component of playing tennis has always been skill, but physiological fitness is increasingly recognised as being of great importance. There is a paucity of research reflecting a direct relationship between skilled on court performance and performance in the laboratory (Roetert *et al.*, 1992: 225; Perry *et al.*, 2004: 136). Therefore, in the light of current knowledge the following protocol has been developed, focusing on the key aspects of the physiological parameters, important to tennis.

Players should ideally be tested three times a year for those tests outlined below including lung function (once annually – normal healthy individuals) and body composition (stature (m), body mass (kg) and percent body fat). Typical values for the following tests are provided in Table 24.1.

AEROBIC CAPACITY: TREADMILL TEST

The use of both the treadmill test and field test designed by Smekal *et al.* (2000: 243) is recommended. Initial workload on a motorised treadmill of 8 km·h^{-1} with an increase of 2 km·h^{-1} every 3 min at a constant grade of 1.5%. The test ends when players are volitionally exhausted.

AEROBIC CAPACITY: FIELD TEST

A tennis ball machine fires tennis balls at a frequency of 12 shots per minute with an increase of 12 shots per minute, every 3 min (Smekal *et al.*, 2000: 243). Ball feeds alternate to both the forehand and backhand sides (landing point of ball feed 2 m in front of the baseline at the intersection of baseline with sideline) requiring players to return balls towards a target measuring 1 m² placed in the rear corner of the singles court (intersect of baseline and side line). Strokes must be hit no higher than 1.5 m above a net band (the use of a ribbon stretched above the net band enables this distinction). The test ends when a player is no longer able to perform their strokes with acceptable precision or technique. Stroke ratings are described further in Smekal *et al.* (2000: 243).

Physiological measurements in both tests consist of the following: Capillary blood samples obtained from the earlobe during a 30 s break between each stage and analysed for blood lactate concentration. Online breath-by-breath respiratory gases should be measured using a portable system with averaged values taken over the last 30 s of each stage. Heart rate should be recorded throughout with averaged values taken for the last 30 s of each stage. Maximal oxygen uptake is determined by an increase in oxygen uptake of less than 2 ml·kg^{-1}·min^{-1} or 3% with an increase in exercise intensity (BASES, 1997: 64). Individual lactate threshold is determined by plotting workload against blood lactate concentration and through the use of polynomial interpolation using cubic splines.

For on-court tennis drilling, players should be prescribed heart rates corresponding to their individual anaerobic threshold in the field test to prevent stroke deterioration associated with high lactate levels (Ferrauti *et al.*, 2001: 235; Davey *et al.*, 2002: 311). For running training, players should be prescribed heart rates corresponding to their individual anaerobic threshold obtained from the treadmill test.

POWER AND IMPULSE: VERTICAL JUMP

Players are asked to perform three attempts at a vertical jump (highest score recorded), bare footed using a counter movement with the arms (Harman *et al.*, 1991: 116). A contact mat, is used for this purpose which records details of the height jumped (cm) and ground time (s). *Peak power* (W, W·kg) and *Mean power* (W) estimations are calculated based on the following equations derived by Harman *et al.* (1991):

$$\text{Peak power (W)} = (61.9 * \text{best jump height (cm)}) + (36 * \text{body mass (kg)}) - 1822$$
$$\text{Peak power (W·kg)} = \text{Peak power (W)} / \text{body mass (kg)}$$
$$\text{Mean power (W)} = (21.2 * \text{best jump height (cm)}) + (23 * \text{body mass (kg)} - 1393$$

POWER AND IMPULSE: OVERHEAD MEDICINE BALL THROW

A 2 kg medicine ball should be used. The player is required to start in the service stance position behind a start line and propel the medicine ball maximally from an elevated flexed position behind the head; ball held in two hands

(Buckeridge *et al.*, 2000: 391). Leg flexion/extension and trunk/shoulder rotation are permitted; however players are not allowed to step forwards. A release angle of approximately 45° is advised to achieve optimal distance of throw. A total of three throws are performed with the greatest distance recorded (cm) as the test score.

POWER AND IMPULSE: SIDEARM MEDICINE BALL THROW

A 2 kg medicine ball should be used. The player is required to stand sideways behind the start line and propel the medicine ball maximally from a nearly straightened arm position in front of the body in a near horizontal plane (Buckeridge *et al.*, 2000: 391). The medicine ball is held in two hands, with the right hand placed at the back of the ball and left hand under the ball for the right side throw and vice versa for the left side throw, respectively. Players are allowed to take a backswing and use trunk/shoulder rotation, however players are not allowed to bend their arms or step. The arms should move in a horizontal plane throughout the movement. A total of three throws are performed on each side, with the greatest distance recorded (cm) as the test score. The reader is directed to McClellan and Bugg (1999: 22), for a pictorial explanation of the forehand medicine ball throw.

POWER: WINGATE MULTIPLE ANAEROBIC TEST

Players undergo a 5 min warm-up on a Monark cycle ergometer against a resistance equivalent to 1% body weight. A 2–3 min rest is permitted. This test has been adapted from the original 30 s Wingate test (Bar-Or *et al.*, 1977: 326) (and has been used in the past by the Lawn Tennis Association as a component of their testing battery) to simulate the intermittent demands placed upon players in actual game conditions based upon match analyses outlined in Davey *et al.* (2003). Players start with a stationary start and are instructed to pedal maximally for a period of 5 s (representing the mean time in a point) and rest for 20 s (representing the time allowed between points). This is repeated 10 times (five, 5 s sprints (representing 5 rallies per game) interspersed with 20 s rest followed by 1 min 30 s rest (representing maximum time allowed to rest in a changeover between games) and a further five 5 s sprints with 20 s rest interspersed: Figure 24.1). Recommended resistances are 6% and 8% body weight for males and females, respectively. Wingate software provides the following outputs: *Peak power* (W, W·kg) is taken as the highest mechanical power generated over any of the ten 5 s sprints; *time to peak power(s)* is the time taken to reach peak power within any given sprint; *total amount of work (kJ)* total work performed per sprint; *Percent fatigue index* is produced for each of the 5 s sprints. Further indices of *fatigue* and *recovery* can be calculated

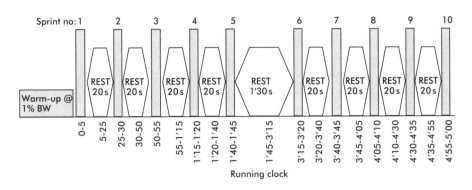

Figure 24.1   Wingate multiple anaerobic test

over the 10 sprints: [*Fatigue index* (($\bar{x}$ of sprints 8 − 10/$\bar{x}$ of sprints 1 − 3) * 100 − 100)], recovery index [(sprint 6/sprint 5) * 100 − 100].

SPEED: SPRINT TESTS (5 m, 10 m)

Timing gates set at 0, 5 and 10 m intervals are set up to assess the players sprint time and split time for 5 m and 10 m. Players initiate the start from behind a start line. The player completes three trials and the best time for both 5 m and 10 m are recorded.

AGILITY: HEXAGON TEST

This test has been shown by the United States Tennis Association to be a valid predictor of tennis performance (Roetert *et al.*, 1992: 226). Create a hexagon on a non-slippery surface, measuring 61 cm (24 in) per side meeting to form 120° angles. Players must be warmed-up and familiarised with the test. Players face forwards and begin and end in the centre of the hexagon. On the command 'Go' players perform a double-footed hop in and out of the hexagon in a clockwise direction until three revolutions around the hexagon are complete. Time for three revolutions is recorded (s). Players are permitted three attempts with the fastest time being recorded. Players must remain facing the same direction throughout and if a line is touched or an incorrect sequence is performed the test must be restarted. Players are only permitted one restart.

STRENGTH AND ENDURANCE: GRIP STRENGTH/ENDURANCE TEST

Players perform seven sets of five maximal isometric grips with 20 s rest between each set followed by a 1 min 30 s rest and a subsequent eighth set on the dominant and non-dominant sides (Davey, 2002: 1115). Starting with the arm in full extension above the shoulder, players maximally grip the dynamometer on each effort, whilst rotating the arm about an anterior-posterior axis to the diametrically opposite position. No involvement of the non-exercising arm is permitted. Record each grip strength (kgf). *Peak grip strength* (kgf) (highest grip strength achieved), *fatigue index* [($\bar{x}$ of last 3 grips in set 7/$\bar{x}$ of first 3 grips in set 1) * 100 − 100], *recovery index* [$\bar{x}$ of first 3 grips in set 8/$\bar{x}$ of last 3 grips in set 7)* 100 − 100] are calculated.

Table 24.1 Typical range of values for elite adult male and female players

Test parameter	Unit	Males	Females
Maximal oxygen uptake	ml·kg^{-1}·min^{-1}	62–66	53–57
Vertical jump			
Peak power	Watts	3,500–4,000	2,500–3,000
Mean power	Watts	850–1,100	700–1,000
Overhead medicine ball throw	cm	11–14	8–12
Sidearm medicine ball throw			
Right	cm	14–18	11–16
Left	cm	13–18	10–14
Wingate multiple anaerobic test			
Peak power	Watts	950–1,100	550–650
Mean time to peak power	s	3–4	2–3
Total work	kJ	2.5–3.5	1.85–1.95
Sprint fatigue index	%	2.0–3.0	4.5–5.5
Fatigue index (10 sprints)	%	−0.05–0.5	−0.05–0.5
Recovery index	%	>10%	>10%
Sprint tests			
5 m	s	1.0–1.2	1.0–1.3
10 m	s	1.7–2.0	2.0–2.2
Hexagon test	s	12.0–12.5	13.0–13.5
Grip strength			
Peak grip strength	kgf	50–52	38–40
Fatigue index	%	<10	<10
Recovery index	%	>10	>10
Sit and reach	cm	50–55	55–60

Source: Based on personal data and adapted from Baechle and Earle (2000), Buckeridge *et al.* (2000), Davey (2002) and Roetert *et al.* (1992)

STRENGTH AND ENDURANCE: PRESS UPS

Athlete lies prone with the lower body weight on the toes and the hands shoulder width apart. The player extends their arms whilst maintaining a straight body. The number of press ups performed in 60 s or to failure if within 60 s is recorded. The arms must be completely extended with the upper reaching parallel to the floor and the body kept in alignment to be counted as a complete press up (Roetert *et al.*, 1996: 140).

STRENGTH: ISOKINETIC

Ideally tennis players should be assessed isokinetically for knee flexion and extension, shoulder external and internal rotation and shoulder diagonal power, respectively (Perry *et al.*, 2004: 138). A warm-up of five repetitions at 50% perceived exertion should be performed followed by a 1 min recovery, followed by three maximal efforts, at 3.14 rad·s^{-1} (medium speed). Peak torque and mean power (W) is recorded.

FLEXIBILITY: SIT-AND-REACH TEST

Players warm-up the hamstrings and lower back with non-ballistic exercises. The shoeless player sits with their legs extended in front of them with their feet 30 cm apart (toes pointed upwards). The point at which the plantar surface of the feet abutt the front of the box is exactly at 38.1 cm. The player slowly extends forwards with both hands as far as possible, holding their position briefly and pushing for a further time whilst exhaling and dropping their head between their arms when stretching (Baechle and Earle, 2000: 301). A score of less than 38 cm (15 in.) indicates that the player was not flexible enough to reach further than the base of their feet. Ideally in addition to performing the sit-and-reach test, players should undergo assessment of flexibility in the areas of shoulder elevation, trunk rotation (to determine ankle, hip, trunk and shoulder mobility) and shoulder rotation (the reader is referred to the general procedures section), respectively.

REFERENCES

Baechle, T.R. and Earle, R.W. (2000). *Essentials of Strength Training and Conditioning*, 2nd edn, National Strength and Conditioning Association. Champaign, IL: Human Kinetics.

Bar-Or, O., Dotan, R. and Inbar, O. (1977). A 30-second all out ergometric test – its reliability and validity for anaerobic capacity. *Israel Journal of Medical Sciences*, 13: 326.

British Association of Sport and Exercise Sciences (BASES) (1997). The assessment of maximal oxygen uptake. In S. Bird and R. Davison (eds), *Physiological Testing Guidelines*, 3rd edn. Leeds: BASES.

Buckeridge, A., Farrow, D., Gastin, P., McGrath, M., Morrow, P., Quinn, A. and Young, W. (2000). Protocols for the physiological assessment of high performance tennis players. In C.J. Gore (ed.), *Physiological Tests for Elite Athletes*. Australian Sports Commission. Champaign, IL: Human Kinetics.

Davey, P.R. (2002). Grip strength/endurance testing of elite tennis players versus non-racket players. Paper presented at 7th Annual Congress of the European College of Sports Science, Athens: Paskalis Medical Publishers.

Davey, P.R., Thorpe, R.D. and Williams, C. (2002). Fatigue decreases skilled tennis performance. *Journal of Sports Sciences*, 20: 311–318.

Davey, P.R., Thorpe, R.D. and Williams, C. (2003). Simulated tennis matchplay in a controlled environment. *Journal of Sports Sciences*, 21: 459–467.

Ferrauti, A., Pluim, B.M. and Weber, K. (2001). The effect of recovery duration on running speed and stroke quality during intermittent training drills in elite tennis players. *Journal of Sports Sciences*, 19: 235–242.

Harman, E.A., Rosenstein, M.T., Frykman, P.N., Rosenstein, R.M. and Kraemer, W.J. (1991). Estimation of human power output from vertical jump. *Journal of Applied Sports Science Research*, 5: 116–120.

McClellan, T. and Bugg, B.S. (1999). Lunge variations to enhance specificity in tennis. *Strength and Conditioning Journal*, 21: 18–24.

Perry, A.C., Wang, X., Feldman, B.B., Ruth, T. and Signorile, J. (2004). Can laboratory – based tennis profiles predict field tests of tennis performance? *Journal of Strength and Conditioning Research*, 18: 136–143.

Roetert, E.P., Brown, S.W., Piorkowski, P.A. and Woods, R.B. (1996). Fitness comparisons among three different levels of elite tennis players. *Journal of Strength and Conditioning Research*, 10: 139–143.

Roetert, E.P., Garrett, G.E., Bran, S.W. and Camaione, D.N. (1992). Performance profiles of nationally ranked junior tennis players. *Journal of Applied Sports Science Research*, 6: 225–230.

Smekal, G., Pokan, R., von Duvillard, S.P., Baron, R., Tschan, H. and Bachl, N. (2000). Comparison of laboratory and 'on-court' endurance testing in tennis. *International Journal of Sports Medicine*, 21: 242–249.

CHAPTER 25

TABLE TENNIS

Matt Cosgrove and Sarah L. Hardman

Table tennis is an explosive sport involving dynamic rapid movement patterns such as lunges and jumps, with simultaneous multidirectional body turns and delivery of strokes with great force and rotational velocity on the ball (Drianovski and Otcheva, 2002). Players are required to perform complex movement patterns in fractions of a second throughout the duration of a match (Drianovski and Otcheva, 2002). Adaptability of movement patterns in spilt second intervals during repetitive bouts of intermittent vigorous activity is therefore necessary for performance at a high level.

Generally, the aerobic demands of table tennis are relatively light. However, players need to be able to cope with the demands of training, both technical and physical, and endure numerous matches in a single day throughout successive days in competition. The table tennis calendar allows little recovery between competitions. A reasonable level of aerobic conditioning is therefore required to aid recovery, reduce the risk of injury and to enable players to sustain a high level of performance throughout training and competition.

Given the dynamic, explosive nature of table tennis play, a well-developed anaerobic energy system is vitally important. With matches lasting between 20 and 50 min, and the average rally lasting 3 s with a mean of 9 s recovery between rallies (elite male singles), glycolytic pathways are utilised (Drianovski and Otcheva, 2002). These typical exercise to rest ratios mean blood lactate concentrations rarely exceed 3 mmol·l^{-1}. The ball is in play for 51–83% of the total playing time (Drianovski and Otcheva, 2002).

Therefore excellent speed, power and agility, combined with a well-developed aerobic and anaerobic capacity are essential requirements for elite table tennis players. The physiological assessment of elite table tennis players must therefore assess these many requirements. The following sections outline the tests utilised by the Welsh Institute of Sport.

LABORATORY-BASED TESTS FOR ELITE TABLE TENNIS PLAYERS

Tests are designed to assess the specific physiological requirements of competing in international table tennis. Tests of anthropometry (stature, body mass and sum of seven skinfolds), lung function, flexibility and aerobic capacity are performed. The tests for anthropometry (stature, body mass and sum of seven skinfolds), lung function and flexibility are all performed in accordance with the relevant BASES guidelines.

AEROBIC CAPACITY (MAXIMAL AEROBIC POWER)

The principles described earlier in this text apply. A continuous incremental protocol is utilised to elicit exhaustion within 8–12 min. This is typically an initial treadmill speed of 5 km·hr^{-1} and an initial gradient of 5%. The treadmill speed is increased by 1 km·hr^{-1} every minute at 5% gradient until volitional exhaustion. Expired air analysis is performed throughout the test. Heart rate is recorded throughout the test and peak blood lactate concentration is determined.

This protocol has been selected due to the greater running speeds utilised with more standard treadmill protocols becoming too great for a group of individuals who do not perform large volumes of running as part of their training. We have demonstrated that table tennis players achieve higher VO_{2max} scores with this protocol than standard protocols that utilise higher running speeds.

FIELD-BASED TESTS

Lower body strength – standing vertical jump

Athletes are tested for both double and single leg power. Players start in an upright position, shoulders level and feet shoulder width apart. Players jump as high as they can in a vertical direction (arm swings and counter-movement are permitted) and the height of the jump is measured by an electronic jump mat. Players have three attempts at each jump and the best of the three attempts is recorded to the nearest centimeter.

Drop jump

Athletes are tested for double leg power only. Players start in an upright position, shoulders level, standing on a 40 cm bench. Players drop down onto an electronic jump mat and immediately jump as high as they can in a vertical direction. The best of three trials is recorded to nearest centimetre.

Speed and agility

A table-tennis-specific speed and agility test has been developed to assess these parameters in the players' own environment. The test is based around a table tennis table and is best illustrated by the use of a diagram (Figure 25.1). An electronic timing system is utilised to assess the players' movement speed. The player follows movement patterns 1–4 in numerical order as illustrated in the diagram. Light beams at positions 1–4 must be broken with the table tennis paddle to give intermediate times for each movement pattern. The touch pad, placed in the middle of the start box, must be touched at the start and finish of the test as well as in-between each movement pattern. The player has three attempts and the best score is recorded.

Reaction time

This table-tennis-specific test is designed to assess reaction time. The equipment and layout of the test is the same as for the speed and agility test. The touch pad, placed in the middle of the start box, must again be touched at the start and finish of the test. The player moves left or right to break a light beam depending on the direction indicated by the light reaction system located on the

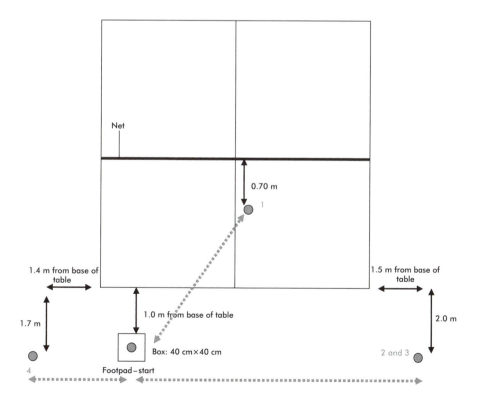

Figure 25.1   Layout of test equipment around the table for speed and agility test

opposite side of the net. The player has a minimum of three attempts each way and the best score is recorded (Figure 25.2).

Anaerobic capacity

This test is performed on a badminton court. The players run from the service line to the rear of the court and back twice and then to the net and back twice for a total of eight shuttles in the shortest time possible. Players should face the side of the court and run sideways as they would around a table tennis table. At each point the players must place one foot on or past the marker or line.

Each of the eight shuttles begins at 30 s intervals. If the player takes less than 30 s the remaining time is given as rest. Each shuttle time is recorded and a drop-off in performance is expected over the course of the eight shuttles. The total time and fastest and slowest shuttle are recorded. This information is then used to calculate the fatigue index or percentage drop off in performance over the eight shuttles. The following formula is used to calculate fatigue index:

$$\frac{(\text{slowest shuttle time} - \text{fastest shuttle time}) \times 100}{\text{fastest shuttle time}}$$

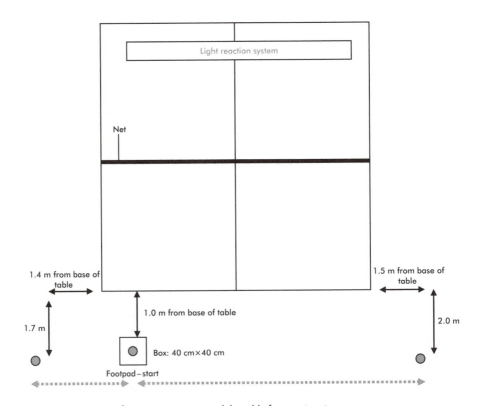

Figure 25.2 Layout of test equipment around the table for reaction time test

Table 25.1 Normative data for international table tennis players (unpublished data)

	Females	Males
Standing vertical jump height (cm)		
Single leg	24.8	28.7
Both legs	37.8	47.2
Drop jump (cm)		49.1
Agility test (s)	5.64	5.35
Reaction test (s)		
Right	1.74	1.71
Left	1.70	0.98
Aerobic capacity (ml$\cdot$kg$^{-1}\cdot$min^{-1})	46.10	54.50

SUMMARY

The tests outlined assess the major attributes required to be an elite level table tennis player. The laboratory-based tests give an indication of standard physiological variables that impact on sports performance. The field-based tests measure specific attributes required to be an elite table tennis player and are performed in the training and competition environment to make the tests and results gained as specific to performance as possible.

REFERENCE

Drianovski, Y. and Otcheva, G. (2002). Survey of the game styles of some of the best Asian players at the 12th World University Table Tennis Championships (Sofia, 1998). In N. Yuza, S. Hiruta, Y. Iimoto, Y. Shibata, Y. Tsuji and J.R. Harrison (eds), *Table Tennis Sciences, Number 4 and 5*, pp. 3–9. Lausanne: International Table Tennis Federation.

CHAPTER 26

SQUASH

Michael Wilkinson and Edward M. Winter

INTRODUCTION

Squash imposes diverse physiological demands on cardio-pulmonary endurance, muscle endurance, explosive strength, speed and flexibility (Sharp, 1998). At élite level, squash has been classed as a high-intensity intermittent activity with mean rally lengths of 16–21 s and recovery times of 10–16 s between rallies (Montpetit, 1990; Hughes and Robertson, 1998). Professional matches can last for 3 h during which players are active for up to 67% of the time (Montpetit, 1990). This makes marked challenges to energy supply.

Due to the short, explosive nature of rallies, squash is often mistaken for an anaerobic sport. However, heart rate quickly reaches a steady state equivalent to 80–90% of predicted maximum (Blanksby *et al.*, 1980; Docherty, 1982; Mercier *et al.*, 1987; Brown and Winter, 1995). Oxygen uptake reaches mean values of ~42 ml·kg^{-1}·min^{-1} ($\approx$74%$\dot{V}O_{2max}$) (Gillam *et al.*, 1990; Todd *et al.*, 1998), and mean lactate concentrations of between 2 and 4 mmol·l^{-1} have been reported (Beauchamp and Montpetit, 1980; Noakes *et al.*, 1982; Mercier *et al.*, 1987). These low concentrations together with high and stable heart rate and $\dot{V}O_2$ suggest that energy for the short duration, high-intensity rallies is provided largely by intramuscular phosphates and O_2 stores that are replenished by oxidative metabolism during the short recovery periods. The $\dot{V}O_{2max}$ values of 62–66 ml·kg^{-1}·min^{-1} and Wingate peak powers of 12.5–13.5 W·kg^{-1} in élite male players confirm the importance of high anaerobic and aerobic power for successful performance (Chin *et al.*, 1995; Brown *et al.*, 1998).

In common with multiple-sprint activities such as soccer, basketball and other racket sports, the specific movement patterns and demands of squash provide a unique challenge to physiologists attempting to produce valid and reliable assessments of the physiological factors relevant to squash performance. The challenge is to combine the control of laboratory procedures with the ecological validity of tests carried out in the specific movement patterns of the sport. Only a squash-specific yet controlled test can provide truly useful data from which to assess players' strengths and weaknesses for training purposes, and to track sport-specific training adaptations that might otherwise go undetected by conventional non-specific laboratory procedures.

TEST PROCEDURES FOR AEROBIC FITNESS

Recently, there have been attempts to produce controlled tests to replicate squash-specific physiological demands (Todd *et al.*, 1998; Sherman *et al.*, 2004). However, these tests were designed to simulate match play demands rather than assess squash-specific fitness. Although several groups are working in this area, only two previous papers describing on-court squash protocols developed for assessment purposes have been published (Steininger and Wodick, 1987; Girard *et al.*, 2005). While both protocols attempt to replicate the demands of squash movement, the procedures of Steininger and Wodick (1987) are more suited to physiological assessment. This section will outline the procedure and comment briefly on its applicability.

Steininger and Wodick's (1987) squash-specific field test

Background

This test was devised to mimic the physiological demands and techniques specific to squash movement but in clearly defined increments to allow the assessment of squash-specific endurance.

Procedure

Six lamps are suspended to hang just below the height of the 'out' line on the side walls of a squash court (3 on each side). The first pair is positioned near to the front corners of the court (numbers 1 and 2), the second pair (lamps 3 and 4) level with mid-court, and the third pair (5 and 6), are located mid-way between lamps 3 and 4 and the rear wall. Squash balls are suspended on string under the pairs of lamps, at knee-height for 1 and 2, eye-height for lamps 3 and 4, and at hip-height for lamps 5 and 6.

   The lamps are connected to a sequencing device off-court which alters the frequency of light flashes. When a lamp lights, the player must move from the T position to strike the ball below the lamp in a technically appropriate manner, and return to the T before the next lamp is lit. The first intensity level begins with 12 light pulses per minute, equivalent to 36 dashes over the 3 min stage duration. Intensity is increased by 6 pulses (or moves) per 3 min level (2 moves per minute) until volitional exhaustion is reached.

In the original protocol, 45 s rest intervals between stages were used to collect blood samples for lactate determination and for measurement of final-stage heart rate by 3-lead ECG. Using modern portable gas analysers, it would also be feasible to measure $\dot{V}O_2$ during the test (Table 26.1).

Strengths

Ranked performance data from the test correlated with ranked playing fitness coefficients estimated from competitive results and a coaches subjective estimates of match fitness ($r = 0.9$, $p < 0.05$). There was a modest correlation ($r = 0.5$, $p < 0.05$) between ranked playing fitness coefficients and ranked performance on a laboratory incremental treadmill test (Steininger and Wodick, 1987). This finding highlights the ability of the test to assess physiological capacities in squash-specific movements.

Limitations

Steininger and Wodick (1987) used a fixed blood lactate of 4 mmol·l^{-1} to determine anaerobic threshold, and reported that most players (who were national

Table 26.1 Physiological responses to the squash test for individual players

Test person	Max. heart rate	Heart rate 5 min after load end	Max. blood lactate (mmol·l^{-1})	Max performance Last load step	Max performance Time (min)	Performance at the anaerobic threshold (light pulses per min)
A	176	107	8.6	6	0.5	54
B	190	110	9.7	5	2.25	56.1
C	185	115	5.5	5	1.25	50.1
D	182	110	6.1	5	1.0	55.5
E	195	119	10.1	4	3.0	33.6
F	184	117	8.5	5	1.0	44.4
G	175	116	11.4	5	0.5	45.3
H	180	120	8.4	5	1.5	45.3
I	190	122	9.0	5	1.5	45.3
J	190	121	8.3	5	0.75	49.2
K	178	120	4.8	4	2.75	48.9
L	195	122	6.1	4	2.0	34.5
M	190	124	7.8	3	3.0	32.1
X̄	185.3	117.1	8.0			45.7
SD	6.5	5.1	1.8			7.7

Source: Adapted from Steininger and Wodick, 1987

standard) had surpassed this value by intensity level 2 of the test. Chin *et al.* (1995) used the test with the Hong Kong men's national squad, and showed similarly high lactate responses early in the test. This suggests that either the initial intensity is too high, that the step increase in intensity is too great, or both. It brings into question the value of the test for assessing players of a lesser standard and fitness, as it is likely that the test will be too difficult to gain any useful data.

Whilst the movement patterns created by the test clearly replicate those of squash play, it appears that the element of randomness of true squash movement has not been accounted for. The predictable sequence of light flashes (i.e. 1–6) means the player must only move *through* the T area *en-route* to the next court position. In contrast, match play often requires players to stop in the T area before determining the next movement direction. The muscular demands of accelerating from a static start are likely to be much greater and different from those required to simply re-direct existing momentum. The muscular ability to accomplish this rapid acceleration is a crucial performance characteristic in squash (Behm, 1992) but is likely to go undetected by a test using predictable movement sequences.

TESTS PROCEDURES FOR ANAEROBIC POWER, SPEED AND AGILITY

Currently there appear to be no published squash-specific tests for anaerobic capabilities, speed or agility. However, as with endurance capability, several groups are investigating squash-specific on-court sprint manoeuvres but until these have been validated and published, non-specific field- and laboratory-based procedures could be used. Careful study of movement patterns and game demands should allow the selection of a test that at least comes close to assessing the performance characteristic of interest.

While sport-specific procedures are preferable, they might not always be necessary. For example, the explosive effort following the 'split step' that characterises the first movement from the T in most squash rallies can be assessed using a standard counter movement jump test from a force platform or simply measuring maximum jump height as in the Sargent jump (Brookes and Winter, 1985). On face value, an explosive jump test is likely to be more ecologically valid than laboratory cycle tests of peak power output due to the similarity of mechanics with the squash movement.

Appropriate general field tests for speed and agility assessment need to consider the multi-directional, short distance nature of squash movements. Vučković *et al.* (2004) reported that more than 40% of movements in a rally occurred within 1 m of the T position, and the maximum single movement distance is unlikely to be more than 3.5 m (the diagonal distance from T to one racquet length from the court corner). As such a straight sprint of >3.5 m is unlikely to be of value as a measure of squash speed for example. In fact most squash movements are not in a straight line, so a truly specific speed test would also need to encompass an element of agility or the ability to control multiple changes of direction at speed.

FUTURE DIRECTIONS

There are clearly many gaps which need to be filled in the development of controlled squash-specific physiological assessments. Until this is achieved, the rationale for training interventions based on non-specific assessment is questionable, as is the ability of physiologists to accurately track adaptations resulting from squash training and play.

Work is currently underway to validate a newly developed on-court protocol for the replication of match physiological demands, and assessment of aerobic and anaerobic capacities using controlled intensities of randomised squash movements. Other ongoing research is concerned with the validation of a combined squash-specific speed and agility test. Publication of these protocols will be a significant move towards the provision of valid and reliable tools for the assessment of competitive squash players.

REFERENCES

Beauchamp, L. and Montpetit, R.R. (1980). The oxygen consumption during racquet-ball and squash matches. *Canadian Journal of Applied Sports Sciences*, 5: 273A.

Behm, D.G. (1992). Plyometric training for squash. *National Strength and Conditioning Association Journal*, 14(6): 26–28.

Blanksby, B.A., Elliot, B.C., Davis, K.H. and Mercer, M.D. (1980). Blood pressure and rectal temperature responses of middle-aged active 'A' grade competitive male squash players. *British Journal of Sports Medicine*, 16: 133–138.

Brookes, F.B.C. and Winter, E.M. (1985). A comparison of 3 measures of short duration, maximal performance in trained squash players. *Journal of Human Movement Studies*, 11: 105–112.

Brown, D. and Winter, E.M. (1995). Heart rate responses in squash during competitive matchplay. *Journal of Sports Sciences*, 14: 68–69.

Brown, D., Weigland, D.A. and Winter, E.M. (1998). Maximum oxygen uptake in junior and senior elite squash players. In A. Lees, I. Maynard, M. Hughes and T. Reilly (eds), *Science and Racket Sports II*, pp. 15–19. London: E & FN Spon.

Chin, M., Steininger, K., So, R.C.H., Clark, C.R. and Wong, A.S.K. (1995). Physiological profiles and sport specific fitness of Asian élite squash players. *British Journal of Sports Medicine*, 29(3): 158–164.

Docherty, D.A. (1982). A comparison of heart rate responses in racket games. *British Journal of Sports Medicine*, 16: 96–100.

Gillam, I., Siviour, C., Ellis, L. and Brown, P. (1990). The on-court energy demands of squash on elite level players. In J. Draper (ed.), *Third Report on National Sport Research Program*, pp. 1–22. Australian Sports Commission, Canberra.

Girard, O., Sciberras, P., Habrard, M., Hot, P., Chevalier, R. and Millet, G.P. (2005). Specific incremental test in elite squash players. *British Journal of Sports Medicine*, 39: 921–926.

Hughes, M. and Robertson, C. (1998). Using computerised notational analysis to create a template for elite squash and its subsequent use in designing hand notation systems for player development. In A. Lees, I. Maynard, M. Hughes and T. Reilly (eds), *Science and Racket Sports II*, pp. 227–234. London: E & FN Spon.

Mercier, M., Beillot, J., Gratas, A., Rochcongar, P. and Lessard, Y. (1987). Adaptation to work load in squash players: laboratory tests and on-court recordings. *Journal of Sports Medicine*, 27: 98–104.

Montpetit, R.R. (1990). Applied physiology of squash. *Sports Medicine*, 10(1): 31–41.

Noakes, T.D., Cowling, J.R., Gevers, W. and Van Niekerk, J.P. de V. (1982). The metabolic response to squash including the influence of pre-exercise carbohydrate ingestion. *South African Medical Journal*, 62: 712–723.

Sharp, N.C.C. (1998). Physiological demands and fitness for squash. In A. Lees, I. Maynard, M. Hughes and T. Reilly (eds), *Science and Racket Sports II*, pp. 1–13. London: E & FN Spon.

Sherman, R.A., Creasey, T.J. and Batterham, A.M. (2004). An on-court, ghosting protocol to replicate physiological demands of a competitive squash match. In A. Lees and Jean-Francois Kahn (eds), *Science and Racket Sports III*, pp. 3–8. Oxon: Routledge.

Steininger, K. and Wodick, R.E. (1987). Sports-specific fitness testing in squash. *British Journal of Sports Medicine*, 21(2): 23–26.

Todd, M.K., Mahoney, C.A. and Wallace, W.F.M. (1998). The efficacy of training routines as a preparation for competitive squash. In A. Lees, I. Maynard, M. Hughes and T. Reilly (eds), *Science and Racket Sports II*, pp. 91–96. London: E & FN Spon.

Vučković, G., Dežman, B., Erčulj, F. and Perš, J. (2004). Differences between the winning and the losing players in a squash game in terms of distance covered. In A. Lees and Jean-Francois Kahn (eds), *Science and Racket Sports III*, pp. 202–207. Oxon: Routledge.

CHAPTER 27

BADMINTON

Michael G. Hughes and Matt Cosgrove

PHYSIOLOGICAL DEMANDS OF BADMINTON

Badminton has many physiological similarities with other repeated sprint sports: short duration, high-intensity activity is followed by recovery periods throughout a match. Specifically, matches at the elite level vary in length between around 20 min to over 1 h. Average rally length is usually less than 10 s with rest periods typically lasting around 15 s. However, these figures vary with many factors including style of play, tactics and the level of competition. Additionally, the characteristics of match play are quite different between the sub-disciplines of the game (singles or doubles for both men and ladies).

Considering only the physical elements, Badminton is a highly explosive sport where strength, power, flexibility, movement speed and agility are essential for competitive success. Movement technique is highly specialised and is of vital importance in determining success. Physiological measures also demonstrate that Badminton presents significant levels of aerobic stress during match play. Typically, elite players maintain heart rates above 75% of maximum in the doubles disciplines and over 80% in the singles disciplines. Blood lactate levels rarely exceed 3–4 mmol·l^{-1} during simulated match play.

The combination of these varied physiological and technical requirements demonstrates the need for fitness assessment in elite Badminton players to incorporate many factors. The following tests currently are used in the assessment of fitness for elite competitors within this sport.

FITNESS TEST PROCEDURES

The following tests have been devised for the Badminton England and the Welsh Institute of Sport. The field tests have been developed to be highly sport-specific and easily administered by players and coaches irrespective of their access to sport science support staff. Specific procedures for some of these tests such as court markings, timings of endurance test, etc. can be obtained from the English governing body (www.badmintonengland.co.uk, 01908 268400). The 'norm' values quoted are the result of many tests performed by physiology staff exclusively on national squad players between 1999 and the time of publication.

Preliminary procedures and anthropometry

Tests of anthropometry (stature, body mass and sum of seven skinfolds), lung function, flexibility and anaerobic and aerobic capacity are performed in accordance with the relevant BASES guidelines. Players are instructed to prepare for fitness testing as they would do for a competition; specifically, only light training in the day prior to testing and to ensure they are well hydrated and have eaten well prior to the tests.

Jump tests

Jump tests are performed using vertical jump and standing long jump procedures. Vertical jump is measured using Vertec (JumpUSA, CA, USA) apparatus. Counter-movement is allowed and the dominant hand only is raised in execution of the test jump. The standing long jump is measured from the front of the feet (taken as zero point) to the back of the foot which travelled the least distance upon landing. The best result from at least four attempts is recorded as the result.

Movement speed

Speed tests were designed to assess (1) general movement speed and (2) specific Badminton speed. The 'general' speed test requires players to make 10 lateral movements across the width of a court. The Badminton-specific test requires a total of eight movements in an ordered sequence in all directions around the court (see Figure 27.1; from points 1 to 4, twice in succession). This test requires Badminton-specific 'shadow play' movements (i.e. players make shots at each position but no shuttlecock is used). The best time from at least two attempts is taken as the test result. Our experience has shown that the correlation between jump test and the general speed test results is highly significant ($P < 0.01$ for males and females). In contrast to this, the Badminton-specific speed test results appear fairly independent of jump test and general speed test performance. This demonstrates the importance of differentiating between general and specific characteristics where the highly technical movements of Badminton are employed (Hughes and Bopf, 2005).

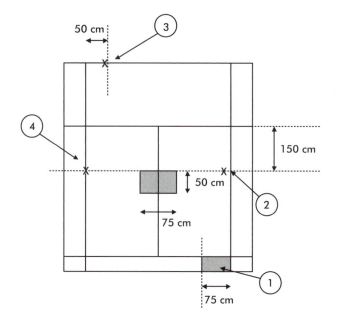

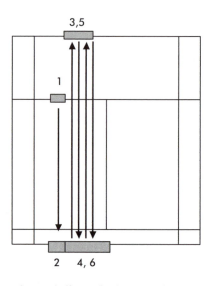

Figure 27.1 Layout of the Badminton half-court for the specific speed test. Numbers indicate order of movement (set-up shown for right-handed player only)

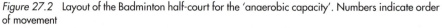

Figure 27.2 Layout of the Badminton half-court for the 'anaerobic capacity'. Numbers indicate order of movement

Anaerobic capacity

This test is performed on a badminton half-court. The players run from the service line (point 1, Figure 27.2) to the rear of the court (point 2 etc.) and back and then to the net and back twice in the shortest time possible (finishing at point 6). This is repeated 8 times with each bout beginning at 30 s. Players should always return to the baseline by running backwards, after reaching the markers by putting one foot on or past the marker or line.

Each shuttle time is recorded, and a drop-off in performance is expected over the course of the eight shuttles. The information presented in Table 27.1 includes average time (mean of all 8 shuttles) and a fatigue index:

$$FI = \frac{(\text{slowest shuttle time} - \text{fastest shuttle time})}{\text{fastest shuttle time} \times 100}$$

Aerobic performance test

The high aerobic demands of Badminton, coupled with the specific requirements of movement technique and economy dictate that a specific aerobic test is performed. An incremental, intermittent on-court aerobic test has been developed whereby Badminton movements are performed at movement speeds controlled by a series of audio tones. The test is made up of series of ~20-s movement bouts followed by around 10-s recoveries. Movements are made, in a standard order, to specific points on the court using 'shadow play' technique (see Figure 27.3). Results from the aerobic test are highly correlated to laboratory VO_{2max} in senior elite players ($P < 0.01$). Indeed, mean maximal heart rate was significantly higher (195 vs. 190 beats·min^{-1}) in this field test than in a laboratory treadmill maximal test in a group of 17 senior elite players (Hughes et al., 2002).

Table 27.1 Normative data for senior elite players

	Females		Males	
	Singles	Doubles	Singles	Doubles
Vertical jump height (cm)	50.0	47.6	61.9	62.5
Standing long jump (cm)	208	204	244	247
General movement speed test (s)	15.4	16.4	14.2	14.4
Specific movement speed test (s)	13.3	13.8	12.3	12.0
Anaerobic capacity test				
Mean test time (s)	13.6*		12.2*	
Fatigue index (%)	7.5*		7.3*	
Aerobic performance test results				
Reached (Level, stage)	14,1	13,2	17,1	16,1
Peak HR (b·min^{-1})	193	196	197	190
Aerobic capacity (ml·kg^{-1}·min^{-1})	55.4*		60.8*	
Body fat (%)	24.0	26.1	11.4	13.3

* denotes pooled data for singles and doubles

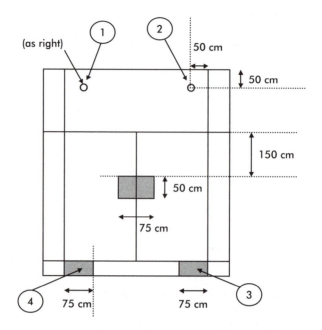

Figure 27.3 Layout of the Badminton half-court for the on-court aerobic test. Numbers indicate order of movement (set-up shown for right handed player only)

Aerobic capacity

The general principles for this laboratory treadmill test are described earlier in this text. A continuous incremental protocol is utilised to elicit exhaustion within 8–12 min. This is typically an initial treadmill speed of 5 km·h^{-1} and an initial gradient of 5%. The treadmill speed is increased by 1 km·h^{-1}·min^{-1} at 5% gradient until volitional exhaustion. Expired air analysis is performed throughout the test. Heart rate is recorded throughout the test and peak blood lactate is determined.

This protocol has been selected due to the greater running speeds utilised with more standard treadmill protocols becoming too great for a group of individuals who do not perform large volumes of running as part of their training. We have demonstrated that badminton players achieve higher VO$_{2max}$ scores with this protocol than standard protocols that utilise higher running speeds.

Additional assessments

In addition to these performance-based assessments, players also receive regular assessments of flexibility and strength performed by physiotherapists and strength and conditioning experts, respectively.

SUMMARY

The predominance of field-based tests here emphasises the perceived importance of sport-specificity in testing for competitors in this sport. The exception to this is the test of aerobic capacity which is considered sufficiently important as a general capacity that is likely to be a correlate of competitive success. The use of both general tests (i.e. aerobic capacity, 'general movement speed test') in conjunction with highly specific tests ('specific movement speed test', 'anaerobic capacity test', aerobic performance test) is recommended as this helps to identify players whose performance is restricted by movement technique or general fitness.

REFERENCES

Hughes, M.G. and Bopf, G. (2005). Relationships between performance in jump tests and speed tests in elite Badminton players. *Journal of Sport Sciences*, 23(2): 194–195.

Hughes, M.G., Andrew, M. and Ramsay, R. (2002). A sport-specific, endurance performance test for elite Badminton players. *Journal of Sport Sciences*, 21(4): 237–238.

CHAPTER 28

ARTISTIC GYMNASTICS

Simon L. Breivik

INTRODUCTION

Female artistic gymnasts compete on four 'apparatus': the vault, the uneven bars, the balance beam and the floor while their male counterparts compete on six: pommel horse, the vault, the floor, the high bar, the parallel bars and the rings. To all intents and purposes, an apparatus can be regarded as a sport in its own right, each characterised by the intensity, duration and nature of the effort required to perform on or in it.

For example, success in vault, an event lasting just 5 s, is highly dependant on a gymnast's ability to accelerate down the vault run and carry their momentum into a high and long post-flight (Sands and McNeal, 1995). A floor routine, on the other hand, can take over a minute to complete, during which time blood lactate concentrations can reach in excess of 7 mmol·l^{-1} (Goswami and Gupta, 1998).

On the whole, performance in artistic gymnastics relies on a combination of a gymnast's short-term, and to a larger extent, long-term anaerobic energy systems with a small contribution from aerobic mechanisms.

ASSESSMENTS

Movement patterns in gymnastics are difficult to replicate in the laboratory and so force the physiologist to be innovative and produce valid yet sensitive field-based fitness tests. Most physiological testing in gymnastics is carried out in the field with tests ranging from sprints to multiple jumps.

Here follows a brief description of a battery of tests currently suited to assessing the fitness profiles of elite artistic gymnasts.

BODY COMPOSITION

Elite female gymnasts have a body mass of approximately 50 kg and reach peak performance in their mid- to late-teens while equivalent values for males are ~65 kg and late teens to early twenties, respectively. Due to the rotational element of gymnastics, participants benefit from being small in stature and keeping their centre of gravity close to their axis of rotation with a low moment of inertia. Mean stature for male and male gymnasts' have been reported to be 161 cm and 168 cm, respectively (Bale and Goodway, 1990).

A gymnast's power-to-weight ratio influences their ability repeatedly to move their body mass against gravity during complex moves. Fat mass is detrimental to acrobatic performance, adding to a performer's weight without contributing to their power output (Salmela, 1979).

In addition to technical performance, gymnasts are judged on aesthetic appeal, which adds to the pressure on performers to be lean, an image considered more appealing by judges. Therefore, gymnasts possess low amounts of body fat.

Skinfold thickness

Measurements of skinfold thickness can be used to assess body composition but care should be taken in attempts to determine percentage body fat in gymnastic populations.

Where such attempts have been made, estimates of 13–16% in females and 3–6% in males have been recorded (Breivik, unpublished data).

MAXIMAL INTENSITY EXERCISE

Anaerobic energy systems contribute predominantly to performance in gymnastics with energy derived from a combination of phosphagen stores (short-term) and anaerobic glycolysis (long-term).

During the vault and elements of routines on other apparatus, such as floor, a gymnast's ability to produce vertical and horizontal force quickly is crucial to their success. Measurements of immediate power or impulse should be assessed and in ways that replicate gymnastic movement.

Leg power and impulse

Countermovement jump test

To replicate the stretch-shortening cycle activity that predominates in the vertical element of gymnastics (jumping and landing), the *countermovement jump* should be used to assess short-term explosive performance. A switch mat is a convenient substitute to the *Sargent Jump* as it automatically calculates jump height to the nearest 0.1 cm based on the flight time of the athlete.

Participants should be instructed to flex rapidly at the joints in the legs, and so pre-load the stretch-shortening cycle and then proceed to jump vertically as high as possible. Participants should be prohibited from falsely increasing their jump time by flexing at the knees mid-flight. The higher or highest score achieved from two or three attempts, respectively should be recorded as the final score.

Female artistic gymnasts generally score between 45 and 50 cm while their male counterparts score between 55 and 60 cm in the countermovement jump (Breivik, unpublished observations).

Speed

20-m sprint test

A gymnast's ability to accelerate down the vault-run affects their performance on vault (Sands and Cheetham, 1896). A 20-m sprint should be used to assess a gymnast's speed. In the interests of validity, the test should be performed, where possible, in the gymnasium on the vault-run carpet, provided that there is sufficient space for the test to be carried out safely.

Electronic timing gates should be placed at 0 m and 20 m. Participants should set off from a stationary position 1 m before the 0 m timing gates and complete the 20-m course as quickly as possible. Timing will start once the first beam is broken and stop when the participant passes through the 20-m timing gates. Speed is measured to the nearest 0.01 s with the quicker result from two trials recorded as the score.

Results for elite male and female gymnasts on the 20-m sprint range from 2.8 to 3.0 s and 3.0 to 3.2 s, respectively (Breivik, unpublished observations).

LONG-TERM ANAEROBIC PERFORMANCE

'Long-term' performance in artistic gymnastics should be considered in the context of routines that last from between 5 and 90 s. This duration would comprise medium-term in most other sports. Consequently, the major source of energy is provided by anaerobic glycolysis; the vault being the exception (Jemni *et al.*, 2000).

The Wingate test

The 30-s Wingate test has been used to characterise maximal intensity exercise in athletes and provide the physiologist with measurements of peak power, mean power and a fatigue index (Bar-Or *et al.*, 1977).

Limitations associated with the Wingate test is that it does not sufficiently replicate gymnastic movement patterns that are plyometric in nature and muscle force–velocity relationships might not be satisfied by the resistive loads used (Sargeant *et al.*, 1981). In light of this, the Bosco test (Bosco *et al.*, 1983) was developed to provide a more valid, sport-specific protocol to assess the long-term anaerobic profiles of gymnasts.

The Bosco test

This comprises 60 s of repeated jumping.

A switch mat with multiple jump function allows the physiologist to set a 60-s protocol that can be used to measure both the flight and ground contact time of consecutive vertical jumps. Gymnasts should be instructed to jump as high and as rapidly as possible for the duration of the test with knee flexion reaching 120° during each landing. Horizontal and lateral displacement should be minimised to avoid unaccounted work output. Participants should be instructed to keep their hands on the hips throughout each jump to minimise contribution from the upper body. Mean power output should be suitably expressed relative to body mass. Typical results for female gymnasts have been reported to be 23–28 $W \cdot kg^{-1}$ (Sands, 2000; Sands *et al.*, 2001) (Table 28.1).

FLEXIBILITY

Gymnasts are among the most flexible of athletes as a result of the extreme positions they must adopt during their performance of skills. Flexibility is joint-specific so the physiologist must measure the range of movement about the joints where flexibility is most important to a gymnast's performance – specifically the hip, groin, back, hamstrings and shoulders.

Flexibility is measured using a goniometer while the gymnast performs sport-specific movements, including the 'splits', 'box splits', 'Y scale' and 'front leg-lift'. Methods whereby gymnasts are graded out of 5 points for each movement are used in gymnastics.

Table 28.1 Typical test results for female and male artistic gymnasts

Test	Female	Male
Body fat (%)	13–16	3–6
Countermovement jump (cm)	45–50	55–60
20-m sprint (s)	3.0–3.2	2.8–3.0
Wingate test ($W \cdot kg^{-1}$)	10–12	12–14
Bosco test ($W \cdot kg^{-1}$)	23–28	—

Source: Breivik, unpublished work; Sands, 2000; Jemni *et al.*, 2001; Sands *et al.*, 2001

REFERENCES

Bale, P. and Goodway, J. (1990). Performance variables associated with the competitive gymnast. *Sports Medicine*, 10(3): 139–145.

Bar-Or, O., Dotan, R. and Inbar, O. (1977). A 30 second all-out ergometer test – its reliability and validity for anaerobic capacity. *Journal of Medical Science*, 13: 126–130.

Bosco, C., Luthanen, P. and Komi, P.V. (1983). A simple method for measurement of mechanical power in jumping. *European Journal of Applied Physiology*, 50: 273–282.

Breivik, S. (2002–2005). Unpublished work with British Gymnastics squads.

Goswami, A. and Gupta, S. (1998). Cardiovascular stress and lactate formation during gymnastics routines. *Journal of Sports Medicine and Physical Fitness*, 38: 317–322.

Jemni, M., Friemel, F., Lechevalier, J. and Origas, M. (2000). Heart rate and blood lactate concentration analysis during a high-level men's gymnastics competition. *Journal of Strength and Conditioning Research*, 14(4): 389–394.

Jemni, M., Friemel, F., Sands, W. and Mikesky, A. (2001). Evaluation of the physiological profile of gymnasts over the past 40 years. A review of the literature. *Canadian Journal of Applied Physiology*, 26(5): 442–456.

Salmela, J.H. (1979). Growth patterns of elite French Canadian female gymnasts. *Canadian Journal of Applied Sports Sciences*, 4: 219–222.

Sands, W.A. (2000). Olympic preparation camps 2000 physical abilities testing. *Technique*, 20(10): 6–19.

Sands, W.A. and Cheetham P.J. (1986). Velocity of the vault run: junior elite female gymnasts. *Technique*, 6: 10–14.

Sands, W.A. and McNeal, J.R. (1995). The relationship of vault run speeds and flight duration to score. *Technique*, 15(5): 8–10.

Sands, W.A., McNeal, J.R. and Jemni, M. (2001). Anaerobic Power Profile. *USA Gymnastics Online: Technique*. 5.

Sargeant, A.J., Hoinville, E. and Young, A. (1981). Maximum leg force and power output during short term dynamic exercise. *Journal of Applied Physiology: Respiratory, Environmental and Exercise Physiology*, 53: 1175–1182.

CHAPTER 29

CRICKET

Richard G. Smith, Robert A. Harley and
Nigel P. Stockill

INTRODUCTION

Recent research has more clearly defined the physiological demands of cricket, and better informed those who design conditioning training programmes for cricketers to meet these demands (see Noakes and Durandt (2000) for overview). Modern, particularly one-day cricket, requires high over rates, athleticism in the field, high scoring rates and explosive running between wickets. The nature of the game and the environment in which it is often played, particularly at elite level, make demands on players. They have to have a sound athletic base, nutritional and hydration awareness and specific role conditioning to delay the onset of fatigue and accompanying deterioration in skill or concentration and prevent injury. Individually targeted and monitored, sport-specific conditioning contributes to the development of these qualities.

Regular fitness assessments at appropriate stages of training cycles provide valuable information both on players' strengths and weaknesses, conditioning status and the effectiveness of prescribed training and mainte-nance programmes. In addition, such assessments, combined with effective medical and musculo-skeletal screening, can identify potential functional imbalances or mechanical irregularities that could contribute to future acute or chronic injury and allow the introduction of preventative strategies.

The following protocols provide guidelines on the rationale, monitoring procedures and assessments used to profile men and women cricketers.

TEST ORDER, ENVIRONMENT AND PREPARATION OF PLAYERS

The assessment protocols are described in the order in which ideally they should be carried out to: (1) maximise reproducibility and reliability, (2) establish an effectively administered test battery and (3) minimise any influence of fatigue from preceding tests. The test battery should preferably be split over two days, commencing testing at the same time each day with only minimal controlled recovery activity in between. Athletes should be fully briefed on test procedures and how to prepare physically, nutritionally and mentally for testing. The anthropometry, jump and strength tests should be completed on day 1, with the sprint, agility and multi-stage fitness tests on day 2 (if assessment schedules are limited, testing may be completed on the same day, however interpretation of data should reference a potential increased fatigue profile).

As the test battery is predominately field-based, every attempt should be made to standardise all aspects of the assessment environment such as timings, test administration and player preparation. The coefficients of variation are reported for the sport-specific sprint assessments.

ANTHROPOMETRY

Standard anthropometric variables are assessed. Please refer to Chapter 9.

RANGE OF MOTION

Players National Cricket Centre undergo a rigorous muscular skeletal/range of motion screening. The description of which is outside the scope of this sport-specific testing chapter. Readers are referred to Chapter 10.

UPPER BODY STRENGTH – OLYMPIC BENCH PRESS

Rationale

Strength is an important component of fitness for cricketers as it underpins good technique and dynamic movement and it helps to prevent injury. The bench press is a common upper body strengthening exercise, which conditions most of the major muscle groups of the anterior upper torso. As well as a training exercise, bench press can be used as an exercise test to assess general upper body strength.

Test procedure

The test involves performing the Olympic bench press, lifting the maximum amount of weight, with correct technique (bar held with shoulder width grip, bar moving from arm extension down to lightly touching chest at bottom to full arm extension at top of movement and hips/lower back remaining flat on bench) for a set number of repetitions. The test should be supervised and spotted for by an appropriately qualified instructor.

Athletes should ensure that they complete a sub-maximal warm-up set of 10 repetitions at ~50% bodyweight. They may then progress into a further sub-maximal set with rest or attempt their maximal lift.

Athletes who are competent weight lifters should complete a 3-rep or 1-rep maximum lift with total weight lifted in kilograms and repetitions recorded. Weight lifted as a relative value to bodyweight, that is, percentage bodyweight for a given number of reps should also be reported.

Athletes who are less experienced weight lifters should complete a 10-rep maximum lift with total weight lifted in kilograms and repetitions recorded. As for competent lifters, weight lifted as a relative value to bodyweight, that is, percentage bodyweight for a given number of reps should also be reported.

LOWER BODY STRENGTH – OLYMPIC SQUAT (OR 45° LEG PRESS)

Rationale

Lower limb strength is an essential component of fitness for cricketers as it provides the athletic base for all skills involved in the game and is fundamental to injury prevention. The Olympic Squat and variations of it, is a principal conditioning exercise for leg strength, a foundation for other strength and power exercises and functional training modalities. As well as a conditioning exercise, the Olympic Squat can be used as an exercise test to assess general lower body strength and functional movement. (For younger players or players inexperienced in lifting techniques the Olympic Squat test can be replaced with an equivalent test performed on a 45° plate loaded leg press.)

Test procedure

The test involves performing the Olympic Squat, lifting the maximum amount of weight, with correct technique (bar held across shoulders and upper back), squatting to a knee angle of 90°, returning to full leg extension for a set number of repetitions. The test should be supervised and spotted by an appropriately qualified instructor.

Athletes should ensure that they complete a sub-maximal warm-up set of 10 repetitions at ~50% bodyweight. They may then progress into a further sub-maximal set with rest or attempt their maximal lift.

Recorded values for competent and less experienced lifters should follow the patterns for the bench press.

SPEED-SINGLE SPRINT TEST (17.7 m)

Rationale

The single sprint test is used to assess initial acceleration and straight line speed. In cricket, players must accelerate over short distances when running between wickets and in the field both to stop and retrieve the ball.

Preparation

Testing should be completed indoors on a non-slip surface, ideally in an indoor cricket hall with cricket-specific floor markings. If cricket floor markings are not available, line markings should be measured accurately with a tape measure and marked as illustrated in Figure 29.1. Each line should be marked 2 m across the run course, with cones and tape, with the timing gates set up directly above the cones at 1m height, creating gates for the athletes to run through. (To save time later, measure and mark out 505 agility test lines at the same time Figure 29.1.) For safety make sure that there is an adequate run off at the end of the course with appropriate crash pads in front of hard walls.

Test procedure

Athletes should complete a standardised dynamic warm-up and two sub-maximal runs over the course at 75% and 90% of maximal effort with a walk back recovery, before their first maximal attempt. Athletes should be reminded that this is a maximal effort test and that they should focus on their start, and accelerate as fast as they can through and beyond the end of the course.

Instruct the athlete to assume the start position, with their back foot on the wicket line ensuring front foot remains behind the crease line facing forward. After a ready signal is given (when timing gates are set), the athlete may start their sprint when ready, accelerating maximally through the second set of timing gates at 17.7 m, and de-accelerating only having passed through the second timing gate.

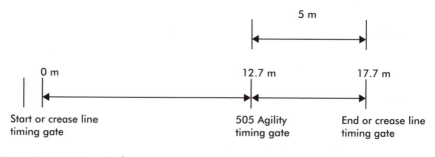

Figure 29.1 Sprint protocol

Each athlete should complete three attempts with full recovery between sprints, with the best sprint time being recorded to the nearest 100th of a second. Additional timing gates can be placed at 5 m and 10 m intervals to review first step speed and acceleration.

SPEED-AGILITY: RUN THREE 505 AGILITY TEST

Rationale

Speed-agility is an important part of cricket performance, and is a combination of neurological processing (responding to stimuli or anticipation), speed, power and co-ordination mechanics. The combined run three and 505 agility test provides a cricket-specific context and if testing is recorded on suitable media, provides both quantitative and technical information about a player's movement. Completing the test with and without protective batting equipment allows an assessment of the functionality of the athlete's equipment and the impact this has on his or her performance.

Preparation

As described for the single sprint test and illustrated in Figure 29.1. An additional set of timing gates are set at 12.7 m from the start line (5 m from the end line). The end timing gates are set at their lowest height to record the bat crossing the crease line as in a match-play (note this is different from the single sprint where the end gates are set at 1 m).

Test procedure

Instruct the athlete to assume the start position holding their bat, with their back foot on the wicket line and front foot behind the crease line facing forward. After a ready signal is given (when timing gates are set), the athlete may start their sprint run three when ready, accelerating maximally through to turn at the end crease line, return to the start line and then complete the third run back through the end crease line running the bat in along or just above the ground (this would only be implemented for safety reasons as sometimes the bat does not slide smoothly across the artificial surface).

Each athlete should complete two maximal run 3's: one attempt with a left-handed 505 turn and one with a right-handed 505 turn with their bat, but without protective equipment and with full recovery between sprints. Players then repeat both run 3/505 sprints wearing full protective clothing. Run 3 and 505 sprint times are recorded to the nearest 100th of a second.

REPRODUCIBILITY OF SPRINT TESTING

Coefficients of variation for 14 senior County Cricketers, calculated using the test–retest method (McDougal *et al.*, 1991), were

- Single sprint with no protective clothing = 1.2%
- Single sprint with full protective clothing (Pads, helmet etc.) = 1.5%
- Running a 3 with no protective clothing = 0.8%

VERTICAL JUMP TEST

Rationale

The ability to accelerate between the wicket and in the field depends on the impulse generating capability of the lower limb. To assess training-induced changes, a vertical jump test is used.

Test procedure

Standard countermovement jump and squat jump protocol are employed. Readers are referred to the appropriate chapter of the book.

Table 29.1 Mean data for senior men (England Test/ODI and National Academy Squads) and senior women (England Test/ODI Squad) 2005

	Senior men mean	Senior men range	Senior women mean	Senior women range
Body mass (kg)	85.3	79.1–96.5	65.9	52.4–82.8
Stature (cm)	182.4	176.5–189.8	167.9	152.3–181.2
Body fat sum of 7 (mm)	57.3	32.1–85.8	93.2	49.1–211.7
Bench press 10-rep max (kg)	77.6	65–95	—	—
45° leg press 10-rep max (kg)	297	210–360	—	—
Single sprint no pads (s)	2.75	2.55–2.88	2.99	2.86–3.31
Run three with pads and bat (s)	9.55	9.33–9.98	10.85	10.28–11.79
Run three with bat, without pads (s)	9.47	9.15–10.04	10.51	9.99–11.36
505 (s)	2.03	1.91–2.26	2.36	2.11–2.53
MST (level:shuttle)	12:4	11:1–14:8	10:6	8:2–12:8
Predicted $\dot{V}O_{2max}$ ($ml \cdot kg^{-1} \cdot min^{-1}$)	54.8	50.6–62.7	48.7	40.5–56.0

AEROBIC FITNESS

Rationale

The Multi-stage Fitness Test (MSFT) is used as a screening test to establish a marker of basic underpinning aerobic power. The test is inexpensive and easy to administer with squads of players. The test is performed over 20 m with turns (similar to distances run in cricket) and has been found to provide an adequate estimate of aerobic power (Ramsbottom *et al.*, 1988).

Test procedure

The full description of the 20 m multi-stage shuttle run test is described elsewhere in Chapter 33, page 261. Heart rate monitors are used to assess peak heart rates at the point of volitional exhaustion (Table 29.1).

REFERENCES

MacDougall, J.D., Wenger, H.A. and Green H.J. (1991). *Physiological Testing of the High-Performance Athlete*. Champaign, IL: Human Kinetics.

Noakes, T.D. and Durandt, J.J. (2000). Physiology demands of cricket. *Journal of Sports Science*, 18(12): 919–929.

Ramsbottom, R., Brewer, J. and Williams, S.C. (1988). A progressive shuttle run test to estimate maximal oxygen uptake. *British Journal of Sports Medicine*, 22(4): 141–144.

CHAPTER 30

BASKETBALL

Robert A. Harley, Jo Doust and Stuart H. Mills

This chapter outlines a battery of field-based fitness tests which reflect both the movement patterns of basketball and the main energy systems used during the sport. Each test has high ecological validity and data are reported in comparison to laboratory tests. The battery has been designed for ease of use with squads in the field and to keep assessment time to a minimum. The appropriateness of laboratory-based tests and tests to assess other important components of basketball fitness, for example, ranges of motion and strength should not be neglected but these areas are well described in Chapters 10 and 13 and should be used to complement this field-testing battery. The battery of tests will be sensitive to changes that take place with training since it has strong specificity (i.e. employing movement patterns and energy systems required during basketball performance), validity and reliability. Although basketball is essentially a repeated-sprint sport, no basketball-specific test of repeated-sprint ability has been included in the proposed field-testing battery. At present this topic requires further research into the underlying physiology to generate a valid field-based assessment protocol.

ANTHROPOMETRY

Standard anthropometric variables are assessed. Stature, body mass and prediction of body fat percentage using sum of four skinfolds (Durnin and Womersley, 1974).

FLEXIBILITY AND STRENGTH ASSESSMENT

It is recognised that flexibility and strength are important components of fitness for basketball performance. Assessment techniques of these variables are discussed in detail in Chapters 10 and 13.

SINGLE SPRINT TEST (ACCELERATION AND SPEED)

Rationale

Speed is a key asset of the basketball player. In fast breaks, the player who is quickest down the court will be in the most commanding tactical position, whether attacking or defending. Also acceleration is of vital importance to the basketball player as it is often the first few steps that determine whether a player can get past their defender. Times in the sprint test are recorded over the first 5 m and 20 m.

Test procedure

- Mark a starting line (0 m) and a finishing line (20 m) with tape.
- Photo electric cells (accurate to 0.01 s) are set at a height of 1 m and at 0, 5 and 20 m intervals.
- Starting position for the sprints is with the front foot placed behind a piece of tape set 0.5 m back from the start line (first timing gate).
- Players start when ready and the time commences when they break the start line beam. This removes reaction time from their sprint times.
- Players sprint as fast as possible through to the finish line making sure not to slow down before braking the finishing beam.
- The best 5 m and 20 m times should be recorded from at least two attempts at the test. Two trials were concluded to be sufficient to attain a reproducible score which reflects a player's sprint capability (Harley, 1997a).

Reproducibility results for 20 m sprint

Mean $\pm$ sd sprint times (s) for trial one (3.41 ± 0.21) and two (3.39 ± 0.23)

Test–retest correlation of $r = 0.96$, $p < 0.001$, $n = 20$

Test–retest method error assessment of CV = 1.6% (Harley, 1997a)

VERTICAL JUMP TEST

Rationale

Of paramount importance to the basketball player is the ability to jump high. During a game, jumping during rebounding or to block or take a jump shot will depend on both the skill of the player and their leg extensor impulse. It is recognised that there is a variety of protocols which could be employed but the one used in this battery of tests is the countermovement jump (CMJ) with arm swing using a switch mat and timer system.

Test procedure for a CMJ with arm swing using a switch mat and timer system

- Players step onto a switch mat connected to a timer. It is recommended that the particular switch mat system used is validated against a force platform as some commercially available systems have been found to give invalid results.
- Players start in an upright position.
- In one continuous movement players bend their knees to an angle of approximately 90°, swinging the arms back as they sink down (counter-movement) and with no delay, jump in the air as high as they can and land back down on the mat.
- Players should be given three attempts and the best score recorded.
- Height jumped can be calculated from time in the air using the formulae:

Height in cm=[(time in air)^2 × gravitational acceleration/8] × 100. For the test to be valid you have to ensure the subject centre of gravity is in the same position on landing as it is in take off.

Reproducibility results

Significant test–retest correlation of $r = 0.91$, $p < 0.001$, $n = 10$
Test–retest of CV = 2.4 ± 0.2%
The CV value of 2.4% is the mean CV calculated from 27 subjects performing 10 CMJ with arm swing (Harley, 1997a).

TEST OF LATERAL QUICKNESS

Rationale

An important element in basketball is the ability to move sideways at speed (defensive slide). This is important to ensure that the defender can position their body between the attacking player and the basket therefore not allowing the attacker an easy lay-up. This test requires a player to move across the base of the key, employing the typical movement patterns performed during defence, over a distance of 36 m with six reversals of direction. Data from Miller (1993) found that players were spending ~6 s each minute in a defensive slide move-ment. The total distance for this test was selected due to the results gained from a small pilot study which found that male players were taking between 10 and 12 s to complete the test. This ensured that it overloaded the requirements of the game situation and that the duration of ~11 s reduced the percentage error due to the hand held timing being employed compared to tests of shorter duration.

Test procedure

- On a full size court the distance of the base of the key (where it touches the base line) is exactly 6 m. Place a piece of tape on the base line mid-way between where the two key lines cross the base of the key (see diagram).
- To optimise player performance the area of the court being used for the test should be wiped with a damp cloth to remove dust from the surface of the court and therefore reduce risk of slipping.
- A player starts with their left foot in the middle of the key on the base line facing into the court.
- On 'go' they move sideways (without crossing their feet) to their left to touch the outside foot on the edge of the key, changing direction so they slide back across the base line again touching their outside foot on the far corner of the key, and so on until they have completed three full cycles across the base line (middle to left side of the key, changing direction across to the right side and then changing direction again back to middle is one complete cycle, see Figure 30.1). The cycles are performed continuously without stopping.
- Throughout the test players face into the court.
- Timing is stopped as the centre of the player's body crosses the middle of the key. Photo electric cells could be employed to enhance the accuracy of the timing procedures.
- Players should be given two trials and the better of the two is recorded. Two trials were concluded to be sufficient to attain a reproducible score which reflects a player's lateral quickness ability (Harley, 1997a).

Reproducibility results

Test–retest correlation $r = 0.68$, $p < 0.05$, $n = 11$
Test–retest method error assessment of CV = 2.8%

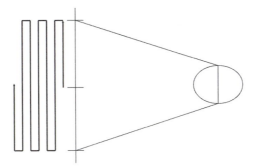

Figure 30.1 Lateral movement test protocol

THE LINE DRILL

Rationale

The ability to sprint repeatedly without fatigue is essential to the basketball player. The line drill proposed by Seminick (1990) requires players to sprint maximally up and down the court over a set distance using the same movement patterns they would perform during a game of basketball. The test lasts ~23 to 33 s depending on an individual's condition and therefore stresses the ATP-CP and anaerobic glycolytic energy systems. The protocol outlined here is a slight modification to that proposed by Seminick (1990).

Test procedure

- Players start with their front foot behind the base line and on the command 'go' run to the near free throw line and back to the base line (see Figure 30.2), then to the half way line and back, then the far free throw line and back, and to the far base line and back (touching a foot on each line).
- The stop watch is stopped as the player's chest crosses the *near free throw line*, not the base line, ensuring that there is room to decelerate.
- If all the test criteria are not fulfilled the test result is discounted.
- The total distance covered on the court (full length court being 26 m) during the sprint should be 124.2 m. The total distance should be measured and scores converted to speed, that is, $m \cdot s^{-1}$ to compare results of scores measured on different length courts.
- To increase motivation players can be paired up and race each other.
- Prior to testing the court should be wiped with a damp cloth to remove dust.
- Photo electric cells should be used if available to increase the accuracy of the timing of the test. The first set of timing lights should be positioned on the base line with the second set of lights positioned on the near free throw line. This test would involve players starting 0.5 m back from the base line so as to remove the effect of reaction time as in the 20 m sprint.
- Two players can perform the test down the centre of the court (one on the left-hand side and one on the right) within the width of the circles and move up and back on the same plane to reduce turning distance.

Reproducibility results

Test–retest correlation $r = 0.93$, $p < 0.01$, $n = 10$
Test–retest method error assessment of CV = 1.3% (Harley, 1997a)

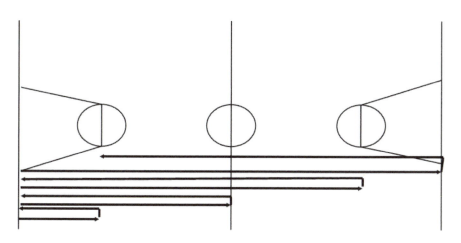

Figure 30.2 Modified line drill protocol

Validity results

Correlation $r = -0.79$, $p < 0.01$, $n = 10$, was found between scores obtained on a 30-s Wingate test (calculated as mean power expressed as $W \cdot kg^{-1}$) and the modified line drill (Harley 1997a).

AEROBIC POWER TESTS

Rationale

Energy production via aerobic pathways is of critical importance to the basketball player with games and practices lasting up to 2 h. During a game, players are exercising at a high proportion of maximal oxygen uptake ($\dot{V}O_{2max}$) as indicated by the high mean hearts rates recorded during game and match play (Ramsay *et al.*, 1970). The 20 m multistage shuttle run test (MST) (Ramsbottom *et al.*, 1988) is the most widely used test for predicting $\dot{V}O_{2max}$ in the field.

MST reproducibility results in basketball players

Test–retest correlation $r = 0.92$, $p < 0.05$, $n = 11$
Test–retest method error assessment of CV = 2.8% (Harley, 1997a)

Validity of the MST in predicting aerobic power of basketball players

There was no significant difference between the two sets of peak oxygen uptake values ($t = 1.53$, $p < 0.05$) determined under laboratory conditions (increasing speed protocol on a treadmill using offline gas analysis) and predicted in the field using the MST for 11 male University basketball players. A significant correlation ($r = 0.70$, $p < 0.05$) was also found to exist between the two sets of peak oxygen uptake values (Harley, 1997a).

SPORTS-SPECIFIC AEROBIC POWER TEST

MST performed dribbling a basketball

Protocol

- The procedure is exactly the same as that outlined for the MST with the exception that subjects are required to dribble a basketball throughout the duration of the test.
- If the player looses control of the ball at any time they need to be fed a new ball immediately.

Reproducibility of the MST dribbling test

The test–retest correlation is $r = 0.95$ with a coefficient of variation of 1.1%.

Validity of the MST dribbling test

A difference ($p < 0.001$) of 3.8 ml·kg^{-1}·min^{-1} equating to ~14 shuttles existed between maximal oxygen uptake predicted from the MST performed with and without the ball, as might be anticipated, with the value being greater without the ball. A high correlation ($r = 0.99$) was found between the two tests.

Maximal oxygen uptake may be able to be predicted from the original results of Ramsbottom *et al.* (1988) if results were corrected using the equation $y = 3.57 + 0.992x$ (x = predicted max with ball, y = predicted max without ball). However, these data are based on a sample of 11 players so a much larger sample size and further analysis of the energy cost of dribbling a basketball with players of different skill would be required before such a procedure could be recommended. It is likely that lesser skilled dribblers would have a greater difference between the scores predicted from the MST performed with and without the ball.

Suggested testing order/warm-up issues

Table 30.1 outlines the suggested order and time intervals to optimise player performance if the full field testing battery is employed. Table 30.2 provides mean and range data for English national league players.

Table 30.1 Suggested order of testing

Testing sequence	Number of attempts	Time between attempts (min)	Time between tests (min)
Body mass			
Height			
Skinfolds			
Warm-up			
CMJ with arm swing	3	1	3
20 m sprint	2/3	3	3
Lateral movement	2	6	6
Line drill	1	—	15
MST	1	—	—

Table 30.2 Mean data for 50 male and 30 female national league players (Harley and Doust 1997b)

	Senior men mean	Senior men range	Senior women mean	Senior women range
Height (cm)	189	171.6–2250	176	169–190
Body mass (kg)	83.3	73.4–1160	74.6	61–84
Body fat (%)	13.8	8–25		
Lateral movement (s)	10.80	9.91–12.31	12.08	11.20–13.60
CMJ with arm swing (using the switch mat and timer method) (cm)	47	24–59	33	21–41
5-m sprint (s)	1.04	0.92–1.18	1.15	1.05–1.28
20-m sprint (s)	3.12	3.86–2.88	3.45	3.80–3.19
Line drill (s)	26.3	34.6–23.8	30.6	33.4–28.6
MST (level:shuttle)	11:5	6:9–15:2	9:6	7:1–12:0
Predicted $\dot{V}O_{2max}$ (ml·kg^{-1}·min^{-1})	51.7	36.0–64.6	45.2	36.8–54.0

REFERENCES

Durnin, J.V.G.A. and Wormesley, J. (1974). Body fat assessed from total body density and estimation from skinfold thickness. *British Journal of Nutrition*, 32: 77–92.

Harley, R.A. (1997a). The development, evaluation and implementation of sports specific fitness monitoring in basketball. Unpublished Master of Philosophy Thesis. University of Brighton.

Harley, R.A. and Doust, J.H. (1997b). *Strength and Fitness Training for Basketball: A Sports Science Manual*. National Coaching Foundation, Leeds.

Miller, S. (1993). English Basketball Association: Match Analysis Report. *Sports Science Support Programme*.

Ramsay, J.D., Ayoub, M.M., Dudek, R.A. and Edgar, H.S. (1970). Heart rate recovery during a college basketball match. *Research Quarterly for Exercise and Sport*, 41: 528–535.

Ramsbottom, R., Brewer, J. and Williams, S.C. (1988). A progressive shuttle run test to estimate maximal oxygen uptake. *British Journal of Sports Medicine*, 22(4): 141–144.

Seminick, D. (1990). The line drill test. *National Strength and Conditioning Association Journal*, 12(2): 47–49.

CHAPTER 31

SOCCER

Chris Barnes

INTRODUCTION

There is a general agreement that certain physiological attributes are a pre-requisite to success in the sport of soccer, and should thus form the basis of any battery of tests performed. Contributory attributes include; cardiorespiratory endurance, speed (linear), agility (multi-directional speed), speed endurance, strength and muscular power (Reilly, 1993: 371–372). The complex nature of the sport means that no single test exists which is capable of assessing all factors concurrently, and that several tests must usually be employed. Despite the large number of studies, which investigate the physical fitness of soccer players of all levels, there exist no agreed national or international battery of tests or associated performance standards.

The fact that soccer is a team sport often results in testing sessions being conducted on a large number of players (sometimes up to 25) simultaneously. With this in mind, this section will focus on field tests of performance, as they are the most viable means of collecting information on large numbers in a relatively short time frame. Due to the fact that field tests of performance are generally more sport-specific than laboratory measures, they are also thought to be a more valid tool (MacDougall and Wenger, 1991).

Issues and protocols for the laboratory-based assessment of aerobic and anaerobic capacity are dealt with elsewhere in this document, as is the assessment of general strength and power.

General issues to be considered when testing soccer players include:

Organisation

The testing session will usually involve subjects performing a number of protocols, and test order and timings need to be planned ahead, and standardised.

Because time is often at a premium for routine testing sessions, sessions must be structured to obtain the maximum amount of information in the minimum amount of time. To minimise the effect of circadian variations, testing sessions should be scheduled at the same time of day (Reilly, 1986: 346–366).

Communication

Test instructions need to be communicated effectively to all parties. For most of the tests there will be one opportunity to collect data, and this should not be compromised by ineffective communication. Similarly, results from tests should be communicated back to coaches and players at the earliest possible convenience post-test.

Environment

To maximise test repeatability, attempts should be made to standardise the test environment. Tests should preferably be performed indoors, on a surface which displays performance characteristics similar to that of turf (e.g. Fieldturf®).

Habituation

In our experience, at least two familiarisation sessions are required on most field tests of performance before acceptable test repeatability is achieved.

The tests described here have been shown previously to be valid and reliable (Hood et al., 2002). Data, where presented, is taken from English Premier League standard players, and youth players associated with a professional club. Data is for outfield players and is not sub-divided according to playing position. Also presented are indicators of the level of reliability achieved, including a statistic for Coefficient of Variation (CV%) and Typical (or Standard) Error of Measurement (Hopkins, 2000) which can be used on an ongoing basis to determine whether changes in performance are meaningful or merely an artefact of the test itself.

SOCCER-SPECIFIC ENDURANCE

Two tests of soccer-specific endurance which have been widely reported in the literature are those developed by Bangsbo (1994: 88–89) and Ekblom (1989). Both incorporate soccer-specific movements and have been demonstrated to produce similar heart rate and lactate responses to soccer match play.

A widely used test of intermittent endurance capacity is the Intermittent Yo-Yo Endurance Test (IYET, Bangsbo, 1994: 89–98). The main difference between the IYET and the more widely known Multistage Fitness Test (MFT) is the inclusion of a 5 s period of active recovery after each pair of 20 m shuttles.

The test can be performed at two levels, the difference between levels being the entry speed. It is appropriate to use level 2 for soccer players of a reasonable standard. The test is performed according to the same principles as the MFT with time to perform shuttles decreasing as directed by an audible bleep. Failure to maintain the pace results in exclusion from the test and maximum distance covered being recorded.

To ensure standardisation of test administration, the following points should be considered:

Subjects should be fully briefed prior to the test.

Pre-test warm-up should be standardized.

Subjects should be required to run round cones at each end of the shuttle to ensure distances covered are standardised.

Cone height should be standardised between tests.

Test administrators should be consistent in determining end point of the test for subjects. We find that a system where two warnings are given prior to termination gives good repeatability.

Data for Youth and Senior professional soccer players are presented in Table 31.1. Copies of the IYET are available from Forsport Ltd (www. forsport.co.uk).

Other tests widely used for the assessment of soccer-specific endurance include the Yo-Yo Intermittent Recovery Test (IYRT, Bangsbo, 1994: 97), which has been shown to be valid and reliable (test–retest CV 4.9%, Krustrup *et al.*, 2003), and the Loughborough Intermittent Shuttle Test (LIST), which has also demonstrated good reliability (Nicholas *et al.*, 2000).

LINEAR SPEED

Ideally, tests of linear speed should be performed indoors, preferably on a simulated turf surface (e.g. Fieldturf®). Results for distances ranging from 5 to 50 m have been reported in the literature (e.g. Kollath and Quade, 1993; Strudwick *et al.*, 2002). More important than the actual distances to be measured is the fact that protocols are standardised. It is suggested that times for one shorter sprint (5 or 10 m) and one longer sprint (30 m or above) are recorded.

To record data for 10 and 30 m sprints, timing gates should be positioned at 0, 10 and 30 m. A mark should be made 1 m behind the 0 m gates. Subjects start the test with preferred foot on this mark. Subject runs at maximal effort through all three sets of gates. Minimum 2 min recovery is allowed before the test is repeated. Three runs are undertaken with data from the fastest 30 m run being recorded (along with the corresponding 10 m time).

Key administrative points include:

Subjects need to be fully warmed up prior to commencing maximal activities.

Feedback should only be given after all tests have been completed.

Data for Youth and Senior English professional soccer players are presented in Table 31.1. Similar normative data on elite young players attending Football Academies attached to English professional clubs was presented by Balmer and Franks (2000).

MULTI-DIRECTIONAL SPEED (AGILITY)

It is without question that the capacity to change quickly the direction of movement is a key performance determinant in the sport of soccer. Several tests of agility have been presented in the literature (e.g. Cable, 1998), although little normative data has been provided. The two tests described later have been evaluated in terms of reliability, and are sensitive to changes in performance over time.

The modified 'Balsom Run'

The test is based on a protocol described by Balsom (1994, 112), and is illustrated in Figure 31.1(a). Following one run through the course at 50% and one run at 75% of maximal effort, the subject starts with their preferred foot on a mark at the start point, 4 m behind the first set of timing gates. They run to and round cone B and back round cone A, round the outside of cone C, round cone D and back to cone C, round the outside of cone B and through the second set of timing gates. Three runs should be completed with a minimum of 2 min separating runs. The fastest time should be noted.

The 'M' run test

The test involves subjects navigating a course of cones laid out as in Figure 31.1(b). Following one run through the course at 50% and one run at 75% effort, the subject starts with their preferred foot on a mark at the start point, 1 m behind the first set of timing gates. They then run forwards to cone A, backwards from cone A to cone B and then forwards to cone C and through the second set of timing gates. As with the modified Balsom run, three runs are completed with a minimum of 2 min between runs. The fastest time is recorded.

Key administrative points for multi-directional tests include:

Warm-up prior to tests should be standardised.
Time between runs should be standardised.
Cone height should be standardised between tests.
The test should ideally be performed on artificial turf in the subjects training boot of choice (usually moulded sole boots).

Data for Youth and Senior professional soccer players are presented in Table 31.1.

Table 31.1 Selected performance data on battery of soccer-specific field tests for English Premier League Senior and Youth players

Parameter	Test	Senior (n = 28)			Under 19 (n = 14)			Under 17 (n = 12)			Repeatability	
		Mean	SD	Range	Mean	SD	Range	Mean	SD	Range	CV(%)	TEM
Soccer-specific endurance	YIFT (m)	2020	366	1340–2600	1960	188	1620–2380	1840	290	1480–2420	3.90	260 m
Linear speed	10 m (s)	1.71	0.06	1.63–1.86	1.70	0.10	1.65–1.89	1.74	0.08	1.62–1.79	4.85	0.09 s
	30 m (s)	4.14	0.12	3.94–4.34	4.13	0.17	3.94–4.39	4.19	0.19	3.92–4.33	2.12	0.10 s
Multi-directional	Modified Balsom run (s)	11.70	0.36	11.20–12.66	11.75	0.23	11.49–12.15	11.91	0.32	11.59–12.41	0.81	0.09 s
Speed (agility)	M run (s)	10.52	0.29	10.03–11.08	10.61	0.26	10.14–11.20	No data available			0.77	0.08 s
Repeated sprint ability	Repeated Balsom run (s)	60.03	0.94	57.10–61.75	60.38	1.17	59.35–62.38	61.13	1.47	59.16–63.2	0.88	0.25 s
	Repeated M run (s)	53.65	1.40	51.65–56.70	53.92	1.28	51.74–56.66	No data available			0.91	0.22 s

REPEATED SPRINT ABILITY

The sport of soccer at any level is characterised by repeated bouts of high-intensity actions interspersed with more prolonged periods of low-intensity activities. The capacity to perform these high-intensity actions effectively impacts significantly on individual and team performance. As such, a number of field tests of repeated sprint ability have been developed (e.g. Fitzsimmons *et al.*, 1993; Aziz *et al.*, 2000). The following tests evaluate individual capacity in this respect. Both are derived from protocols presented earlier to evaluate multi-directional speed.

Repeated Balsom run

Three subjects can be tested simultaneously. Five bouts of the Modified Balsom Run (Figure 31.1(a)) are performed by each subject.

When Subject A commences his first run a stopwatch is started. Time for the run is taken from the timing gates and recorded. Subject B commences his run at $t = 20$ s on the stopwatch, and time for the run is recorded. Subject C commences at $t = 40$ s. Run 2 for Subject A commences at $t = 60$ s, Subject B at $t = 80$ s and Subject C at $t = 100$ s. Runs are performed at 20 s intervals until all three subjects have performed five runs ($\sim.6$ min). Fastest and total times are recorded (Figure 31.1).

Repeated M run

Timings for runs are the same as for the repeated Balsom run, three subjects can be tested simultaneously, with subjects performing five runs each at 20 s intervals. Fastest and total times are recorded for each subject. Key administrative points include:

Subjects should be clearly informed that the total time for five runs will be noted.
This instruction helps to maximise compliance for each run.
No feedback on run times should be given until all runs have been completed.
If a subject trips or fails to navigate a run correctly, the times for the other four runs should be averaged, and this time added to the total for the four remaining runs.

TEST ORDER

It is often the case when testing professional footballers that limited time is allocated to profile a whole squad. This means that full recovery between tests may not be possible and the issue of test order becomes important. When running the above battery of tests we always apply the following sequence.

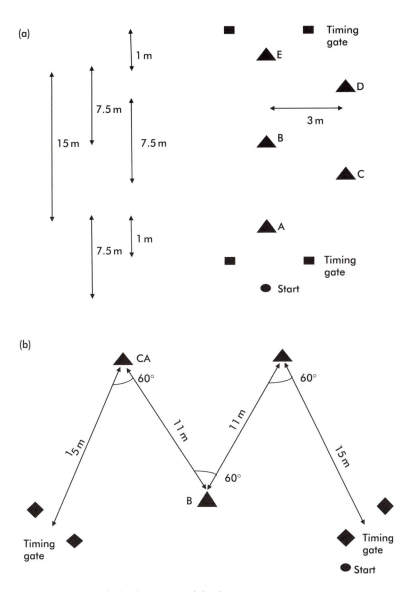

Figure 31.1 (a) The Modified Balsom Run and (b) The 'M' Run

1 Yo–Yo intermittent test, followed by 60 min recovery.
2 Linear speed tests followed by 15 min recovery.
3 Multi-directional speed test coupled with Speed Endurance test (i.e. all five runs undertaken – fastest and total times recorded).

In all instances appropriate warm-up should be undertaken prior to each test protocol.

REFERENCES

Aziz, A.R., Chia, M. and Teh, K.C. (2000). The relationship between maximal oxygen uptake and repeated sprint performance indices in field hockey and soccer players. *Journal of Sports Medicine and Physical Fitness*, 40: 195–200.

Balmer, N. and Franks, A. (2000). Normative values for 5 and 15 metre sprint times in young elite footballers. *Insight*, 1(4): 24–25.

Balsom, P. (1994). Evaluation of physical performance. In B. Ekblom (ed.), *Football (Soccer)*, p. 112. Oxford: Blackwell Scientific Publications.

Bangsbo, J. (1994). *Fitness Training in Football – a Scientific Approach*, pp. 88–99. Bagsvaerd: HO+Storm.

Cable, T. (1998). Agility in football. *Insight*, 1(2): 42.

Ekblom, B. (1989). A field test for soccer players. *Science and Football*, 1: 13–15.

Fitzsimmons, M., Dawson, B., Ward. D. and Wilkinson. A. (1993). Cycling and running tests of repeated sprint ability. *Australian Journal of Science and Medicine in Sport*, 25: 82–87.

Hood, P.E., Barnes, C.A. and Gregson, W. (2002). Reliability of a battery of soccer-specific field tests. *Journal of Sport Sciences*, 20: 20–21.

Hopkins, W.G. (2000). Measures of reliability in sports medicine and science: correspondence. *Sports Medicine*, 30: 375–381.

Kollath, F. and Quade K. (1993). Measurement of sprinting speed in professional and amateur soccer players. In T. Reilly, J. Clarys and A. Stibbe (eds), *Science and Soccer II*, pp. 31–36. London: E. & F.N. Spon.

Krustrup, P., Mohr, M., Amstrup, T., Rysgaard, T., Johansen, J., Steensberg, A., Pedersen, P.K. and Bangsbo, J. (2003). The Yo-Yo Intermittent Recovery Test: physiological response, reliability, and validity. *Medicine and Science in Sports and Exercise*, 34(4): 687–705.

MacDougall, J.D. and Wenger, H.A. (1991). The purpose of physiological testing. In J.D. MacDougall and H.A. Wenger (eds), *Physiological Testing of the High-Performance Athlete*, pp. 1–5. Champaign, IL: Human Kinetics.

Nicholas, C.W., Nuttall, F.E. and Williams, C. (2000). The Loughborough Intermittent Shuttle Test: a field test that simulates the activity pattern of soccer. *Journal of Sports Sciences*, 18: 97–104.

Reilly, T. (1986). Circadian rhythms and exercise. In D. Macleod, R. Maughan, M. Nimmo, T. Reilly and C. Williams (eds), *Exercise: Benefits, Limits and Adaptations*, pp. 346–366. London: E. and F.N. Spon.

Reilly, T. (1993). Football. In T. Reilly, N. Secher, P. Snell and C. Williams (eds), *Physiology of Sports*, pp. 371–372. London: E. and F.N. Spon.

Strudwick, A., Reilly, T. and Doran, D. (2002). Anthropometric and fitness characteristics of elite players in two football codes. *Journal of Sports Medicine and Physical Fitness*, 42: 239–242.

CHAPTER 32

NETBALL

Nick Grantham

BACKGROUND

Netball is an intense and physically demanding sport. Elite players competing today can no longer rely on skill alone and successful international netball players must develop essential physical qualities if they are to perform at the highest level. Individualised physical preparation combined with regular fitness monitoring provides the cornerstones of any long-term development programme (Ellis and Smith, 2000; Harman and Pandorf, 2000).

The tests in this chapter have been used by English National squads (All England Netball Association, 2002) and comprise field tests that can be performed on court. The on-court field tests allow for the effective monitoring of the major physical attributes that contribute to successful performance in netball: speed, acceleration and agility, lower limb strength and power, and short-and long-term endurance (Steele and Chad, 1992).

EQUIPMENT CHECKLIST

This section lists the appropriate equipment required for each of the tests.

Vertical jump

Vertec jump device

10-m sprint test

Timing gates, dual beam, accurate to 0.01 s
Measuring Tape

777 agility test

Timing gates, dual beam, accurate to 0.01 s
Measuring tape

Multistage Fitness Test (MSFT)

Measuring tape
MSFT CD
CD player
Cones
Heart rate monitors

Upper body performance test

Medicine ball (4 kg)
Chair
Pressure pad
Measuring tape

Speed-endurance test

Timing gates, dual beam, accurate to 0.01 s
Measuring tape

ORDER OF TESTS

To ensure reliability and validity of the test protocols each test should be completed in the same order. This will allow valid test–retest comparisons. The recommended order is as follows:

1   Vertical jump
2   10-m sprint test
3   777 agility test
4   Multistage Fitness Test
5   Speed-endurance test

PROTOCOLS

Impulse and jump ability

The aim of this test is to assess the player's ability to perform vertical jumps and examine differences when reaching with dominant and non-dominant hands. The test evaluates both lower limb impulse generating and jumping capabilities.

Vertical jump test procedure

The test should be preceded by an appropriate warm-up followed by at least two sub-maximal practice jumps. The players' standing height should be measured while they stand flat-footed with their back against a wall and reach vertically with both hands. The tester should adjust the height of the vertical column so that it is at a distance that accommodates the jumping ability of the player. Ask the athlete to stand side on to the device so that, when the dominant hand reaches straight upward, it is directly below the centre of the vanes. Instruct the player to jump as high as possible using arm swing and counter-movement. At the peak of the jump, the athlete should attempt to tap the highest possible vane with their hand. The athlete should complete at least three jumps on each side. Vertical jump height can be calculated by subtracting the players standing reach from the highest absolute height jumped (vane tapped at the highest point of the jump).

An electronic jump mat could also be used with caveats about body posture at take off and landing as indicated elsewhere in this text.

Speed and acceleration

Netball players must be able to accelerate over short distances to make position, evade opponents and intercept passes. The aim of this test is to assess the player's ability to produce such acceleration.

10-m sprint test procedure

The 10-m sprint test should be completed on a sprung wooden floor. The test should be preceded by an appropriate warm-up followed by at least two sub-maximal practice runs before the completion of several maximal sprint and start efforts.

Players start from a static position with one foot on the start line (1 m away from the first timing gate). The player can start when ready (so eliminating reaction time). The player sprints through the first timing gate, through to the final timing gate, making sure not to slow down before the finish gate. Each player must attempt at least four sprints with a minimum of 1–2 min recovery between each one. The fastest time is recorded.

Agility

Successful Netball players must be able to accelerate, decelerate and change direction effectively (Steele and Chad, 1992). The aim of this test is to assess a player's ability to sprint, change direction and sprint again. The test involves three 7-m sprints with two 90° turns.

777 agility test procedure

The 777 agility test sprint test should be completed on a sprung wooden floor. The test should be preceded by an appropriate warm-up and each player must complete 2–3 practice runs at sub-maximal intensity to habituate to the test (Figure 32.1).

Players start from a static position with one foot on the line (2 m from the first timing gate). The player can start when ready (eliminating reaction time). The player sprints through the first timing gate towards the first target area (40 cm^2), they then turn and sprint to the second target area (40 cm^2), turn and sprint through the final timing gate. Each player must complete a minimum of four sprints. During all the trials players must touch the targets with one foot (any part of the foot must touch or cross into the target area). If a player fails to touch the target area the time is not counted. The fastest time is used as the final result.

Acceleration

Endurance

The contribution of aerobic pathways to energy production during a netball match is marked. The aim of this test is to assess the aerobic capacity of the player while performing similar movement patterns to those in netball (repeated acceleration, deceleration and changes of direction).

Multistage fitness test procedure

Prior to starting the test each player is fitted with a heart rate monitor. Ensure that players listen carefully to the test instructions on the CD. Begin the test on level 1. The player runs in time with a series of bleeps until they are no longer able to sustain the pace. The player must attempt to be at the opposite end of the 20 m course by the time the next bleep sounds. If the player fails to reach the line at the sound of the bleep they will receive a warning. Once a player has received three successive warnings they are withdrawn from the test. The score is the level and shuttle at which the player was stopped. Maximum heart rate should be recorded to confirm that a maximal effort had been achieved.

Speed endurance

The contribution of anaerobic mechanisms during a netball match is also marked. The aim if this test is to assess the player's ability to maintain performance during repeated bouts of high-intensity exercise over an extended period of time while performing similar movement patterns to those in netball (repeated acceleration, deceleration and changes of direction).

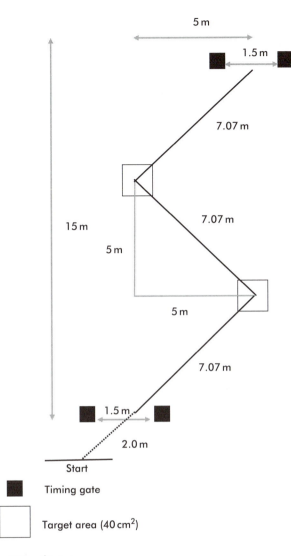

Figure 32.1   The 777 agility test

The speed endurance test should be completed on a sprung wooden floor. The test should be preceded by an appropriate warm-up and each player must complete 2–3 practice runs at sub-maximal intensity to habituate to the test (Figure 32.2).

Players start from a static position with one foot on the line (1 m from the first timing gate). The first repetition is started at the right-hand side timing gate, the second start at the left hand timing gate. Players continue to alternate the direction of the run until all 10 runs have been completed. Prior to the start of each repetition, players are given a 5 s countdown. For each repetition, players sprint to the first marker (directly opposite the inner timing gate), turn and sprint to the second marker (between the two timing gates), turn and sprint to the third marker (directly opposite the inner timing gate), turn and sprint through the second timing gate. The player has 30 s from the start of one repetition to the start of the next repetition to complete the course and recover, before the next repetition begins. A total of 10 repetitions are completed. Players must touch the line at the marker with one foot. If the player fails to touch the line, the time is taken, but not used as their best time. Fastest and slowest times are recorded as well as the average and total time taken to complete all 10 runs.

Examples of results from these tests are illustrated in Table 32.1.

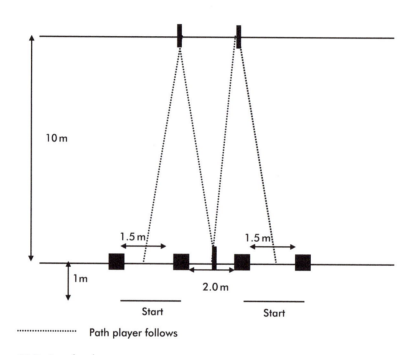

Figure 32.2 Speed endurance test

Table 32.1 Fitness tests and National Squad Performance Standards

Fitness test	National Squad Performance Standard
Vertical jump	58 cm
10 m sprint test	<1.77 s
777 agility test	<4.3 s
Multistage Fitness Test	12–5 (55.1 ml·kg^{-1}·min^{-1})
Speed-endurance test (mean)	<10 s

ACKNOWLEDGEMENTS

Nuala Byrne, Claire Palmer, Andy Borrie.

REFERENCES

All England Netball Association. (2002). *World Class Performance Plan Physiological Fitness Monitoring Protocols*. Hitchin, Hertfordshire: All England Netball Association.

Ellis, L. and Smith, P. (2000). Protocols for the physiological assessment of netball players. In C.J. Gorre (ed.), *Physiological Tests for Elite Athletes*, Chapter 20. Australian Sports Commission. Leeds: Human Kinetics.

Harman, E. and Pandorf, C. (2000). Principles of test selection and administration. In T.R. Baeche and R.W. Earle (eds), *Essentials of Strength Training and Conditioning (2nd edn)*, Chapter 14. National Strength and Conditioning Association. Champaign, IL: Human Kinetics.

Steele, J. and Chad, K. (1992). An analysis of the movement of netball players during matchplay: implications for designing training programmes. *Sports Coach*, 15(1): 21–28.

CHAPTER 33

RUGBY LEAGUE

Simon L. Breivik

INTRODUCTION

Rugby League is an intermittent 'collision' sport that comprises teams of 13 players: six 'Forwards' (those involved in the scrum) and seven 'Backs' (those not involved in the scrum). Performance relies on several factors, including: tactical decision-making, player skill levels, interplay between players, mental toughness and physical fitness. Given the 'multiple-sprint' nature of the game, the physiological demands placed on its players are complex and comprise several fitness components. Hence it is advantageous for the physiologist to possess a thorough understanding of the sport prior to working with a squad of Rugby League players.

During a match, players are frequently engaged in brief bouts of high intensity, anaerobic activity (sprinting and tackling), interspersed with low-intensity periods of aerobic 'recovery' (standing, walking and jogging) (Gabbett, 2005). Recent alterations, such as the 10-m rule and unlimited player interchange have affected the dynamics of the game and made it increasingly physical (Meir *et al.*, 2001). A recent study has determined mean match intensities of 81% maximum oxygen uptake (VO_{2max}) with mean blood lactate concentrations of 7.2 mmol·l^{-1} (Coutts *et al.*, 2003).

Physiological demands in Rugby League are specific to a player's role on the pitch. Forwards are generally involved in more physical collisions and tackles than backs and are subjected to a higher ratio of high- to low-intensity activity. Backs are required to run at speed, using their agility to catch and tackle opposing players and to avoid being tackled themselves (Gabbett, 2005). The physiologist should take player position into account when analysing fitness test results and prescribing subsequent training programmes.

Although Rugby League game shares many similarities with its cousin, Rugby Union, time-motion analyses have revealed notable differences between the two codes in physiological demand placed on its participants. Because of the absence of rucks, mauls and lineouts in Rugby League, there are fewer stoppages and the ball is typically in play for 50 out of the 80 min of match duration (Brewer and Davis, 1995), compared with less than the 40 min or so seen in Rugby Union. As well as League players being actively engaged in play for longer, there are fewer players available to cover the pitch (13 players compared to 15 for a Union) and this places players under further physical strain. The combination of these factors implies that Rugby League is a more physically demanding sport than Rugby Union.

ASSESSMENTS

Players should be subjected to a battery of tests designed to reflect the overall demands of the game. Physiological testing should include anthropometry (stature, body mass and skinfold thickness), assessments of maximal intensity exercise (vertical jump/Wingate), muscular strength (1-RM), speed (10, 20, 40-m sprint), speed agility ('L' test) and aerobic capacity (MultiStage Fitness Test).

BODY COMPOSITION

Excess body fat is detrimental to athletic performance because it acts as 'dead weight' that has to be transported and slows a performer. In Rugby League, backs are required to be quick and agile (Gabbett, 2005) and should avoid carrying excess fat and also excess lean body mass. Forwards, who are involved with most of the physical contact, should be strong and powerful and thus carry more lean body mass than backs. Forwards typically possess greater amounts of body fat than backs, which has been suggested to provide them with a level of protection from injury (Brewer and Davis, 1995). However, carrying excess body fat will adversely affect player mobility and speed so a suitable balance should be met.

Skinfold thickness

Measurements of skinfold thickness should be performed using Harpenden Skinfold callipers at either four (biceps, triceps, subscapular, suprailiac) or seven (chest, abdomen, triceps, subscapular, suprailiac, thigh and midaxilla) sites. The results can be summed and displayed in millimetres or with due care, converted to a percentage body fat.

MAXIMAL INTENSITY EXERCISE

A player's ability to change speed quickly and perform pushing, tackling and lifting tasks during a match relies on their capacity to generate high levels of lower-body muscular impulse and power. Muscular power is of particular importance to forward players, who generally display higher test scores than backs, especially when results are reported relative to body mass (Brewer and Davis, 1995).

The Sargent Test or Vertical Jump Test, are effective methods of evaluating the maximal intensity exercise capability of the legs (Gabbett, 2005).

The Sargent test

The Sargent jump score reflects the vertical difference in centimetres between a person's maximum standing reach height and their maximum jump-and-reach height. Participants should stand side-on to a wall with their feet flat on the ground, reach as high up as they can and place a mark on the wall with chalk. The participant should then place the chalk between the tips of their index and middle finger, assume a crouch position and spring maximally upward, marking the wall with the chalk at maximum jump height. The vertical difference between the two marks is measured to the nearest 0.1 cm and recorded with the highest value from three trials used as their vertical jump score.

Modified vertical jump test

Recently, 'timing mats', that calculate vertical jump height by measuring the time participants' feet are off the mat, have become widely available. Participants should be prohibited from falsely increasing jump time by flexing at the knees mid-flight – legs should remain extended throughout the jump. Jump height is displayed to the nearest 0.1 cm. The highest score achieved from three attempts should be recorded.

MUSCULAR STRENGTH

Rugby League players require strength to perform optimally in scrums, tackles and to penetrate the opposition's defence. Forward players generally score more highly than backs in measurements of muscular strength due to the demands of their playing position.

One repetition maximum

To maximise test validity, the exercises selected should adequately replicate sport-specific movement patterns. The 'Bench Press' and 'Back Squat' activate the upper and lower body 'pushing muscles', thus adequately replicating the demands placed upon the players during a match (Meir, 1993).

Tests should be carried out in a technically sound manner with the assistance of an appropriately qualified instructor who should: (1) ensure correct technique, (2) spot and (3) offer encouragement. An appropriate warm-up, consisting of between two and three sets of the test exercise using a relatively light weight, should be performed prior to the test.

The first attempt should be at ~50% of the player's estimated maximum, and subsequent lifts should gradually increase in difficulty with a maximal lift attempted within approximately five attempts. The participant should receive sufficient recovery between lifts (2–5 min) (Harman *et al.*, 2000) and one repetition maximum should be measured to the nearest kilogram.

SPEED

Movement with the ball in Rugby League is executed at speed. Players, particularly backs, need to be quick to outrun their opposition, track back to defend during counter-attacks and support team members in set plays.

Time-motion analysis of Rugby League matches has shown that players rarely sprint maximally for more than 40 m in a single bout (Gabbett, 2005), while mean sprints range from 10 to 20 m. This highlights the importance of assessing a player's acceleration as well as their maximum running speed.

40-m sprint test

Electronic timing gates should be placed at 0, 10, 20 and 40 m (Brewer and Davis, 1995) in a straight line and on a well-grassed, flat surface preferably in the absence of a head or tail wind. Following the completion of a suitable warm-up, participants should set off from a stationary position 1 m before the 0 m timing gates and complete the 40-m course as quickly as possible.

Timing will start once the first beam is broken and stop when the participant passes through the 40-m beam. Speed should be measured to the nearest 0.01 s with the quicker result from two trials recorded as the score. Split times, recorded at 10 and 20 m can be used to determine a player's ability to accelerate.

AGILITY

The ability to accelerate, decelerate and change direction at speed is an important quality in Rugby League, particularly for backs, who use forward acceleration and rapid lateral changes in direction to escape being tackled.

The 'L' run

The 'L' run reflects the movement pattern of Rugby League (Webb and Lander, 1983). For safety reasons, tests of agility should be performed on a non-slip surface with players dressed in appropriate footwear.

Three cones are placed 5 m apart, in the shape of an 'L' (see Figure 33.1). Following the completion of an appropriate warm-up, players are requested to

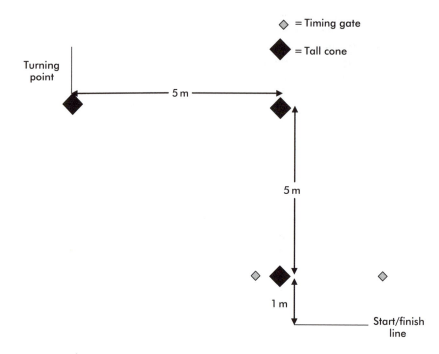

Figure 33.1 The 'L' run

run (from a standing position) forward 5 m, change direction 90° to their left, run forward a further 5 m, turn 180° and return as quickly as possible following the same route to the start/finish line. Cones should be tall enough to discourage short cuts being taken. Time taken to complete the course is recorded to the nearest 0.01 s and the quicker attempt from two is recorded.

AEROBIC POWER

Players should possess aerobic fitness sufficient to sustain activity throughout an 80-min match. During a match, players will cover between 5,000 and 10,000 m, with most energy being supplied aerobically. High aerobic fitness is also required to aid recovery from intermittent bouts of anaerobic activity.

Although laboratory testing provides a direct and accurate measure of aerobic fitness (VO_{2max}), treadmill protocols do not adequately replicate the movement patterns associated with Rugby League.

Multistage Fitness Test

The Multistage Fitness Test, commonly referred to as the 'bleep test', is a valid means to estimate VO_{2max} (Ramsbottom *et al.*, 1988) as it partly replicates movement patterns performed in multiple sprint sports. Additionally, it allows a test administrator to assess several players simultaneously.

Table 33.1   Mean Fitness Test results for Elite Rugby League players

Test	Forward	Back
Body mass (kg)	92	82
Predicted body fat (%)	15	12.5
Vertical jump (cm)	49	51
Bench press (kg)	119	113
Speed to 10 m (s)	2.0	1.9
Speed to 20 m (s)	3.1	2.9
Speed to 40 m (s)	5.6	5.3
Agility (s)	6	5.8
Predicted VO_{2max} (ml·kg^{-1}·min^{-1})	60	62

Source: Brewer and Davis, 1995; Gabbett, 2002

Participants are requested to run back and forth between cones (placed 20 m apart), whilst keeping in time with the pace, which is dictated by a series of 'bleeps' on a CD or cassette tape. The frequency of the bleeps increases progressively as the levels elapse with participants running to volitional exhaustion. The test is ended when the participant can no longer maintain the required pace. The level and shuttle should be recorded and converted to predicted VO_{2max} (Table 33.1).

REFERENCES

Brewer, J. and Davis, J. (1995). Applied physiology of Rugby League. *Sports Medicine*, 20(3): 129–135.

Coutts, A., Reaburn, P. and Abt, G. (2003). Heart rate, blood lactate concentration and estimated energy expenditure in a semi-professional rugby league team during a match: a case study. *Journal of Sports Sciences*, 21: 97–103.

Gabbett, T.J. (2002). Physiological characteristics of junior and senior rugby league players. *British Journal of Sports Medicine*, 36: 334–339.

Gabbett, T.J. (2005). A comparison of physiological and anthropometric characteristics among playing positions in junior rugby league players. *British Journal of Sports Medicine*, 39: 675–680.

Harman, E., Garhammer, J. and Pendorf, C. (2000). Administration, scoring, and interpretation of selected tests. In T.R. Baechle, and R.W. Earle (eds), *Essentials of Strength and Conditioning*, pp. 287–318. Champaign, IL: Human Kinetics.

Meir, R. (1993). Evaluating player's fitness in Rugby League: reducing subjectivity. *Strength and Conditioning Coach*, 1(4): 11–17.

Meir, R., Colla, P. and Milligan, C. (2001). Impact of the 10-m rule change on professional rugby league: implications for training. *Strength and Conditioning Journal*, 23: 42–46.

Ramsbottom, R., and Brewer, J. and Williams, C. (1988). A progressive shuttle run test to estimate maximal oxygen uptake. *British Journal of Sports Medicine*, 22(4): 141–144.

Webb, P. and Lander, J. (1983). An economical fitness testing battery for high school and college rugby teams. *Sports Coach*, 7: 44–46.

CHAPTER 34

RUGBY UNION

Richard J. Tong and Huw D. Wiltshire

INTRODUCTION

The physical demands of international Rugby Union have increased in recent years due to a combination of factors including; the move to professionalism, changes in laws/game plans, and the increase in the number of high-profile competitions and international matches. This increased demand is supported by the increase in ball-in-play time in the Six Nations Championship from 42% in 2003 to 44% in 2004 (Thomas, 2004). To meet these increased demands sports scientists need to review the fitness requirements of the sport so that players can achieve optimal performance for the duration of matches. It is now universally accepted that success in rugby union requires high levels of skill combined with equally high levels of aerobic and anaerobic fitness (Duthie *et al.*, 2003). Although all players need to possess a good level of all-round fitness positional differences exist within rugby union (Quarrie *et al.*, 1996; Nicholas, 1997) which have implications for training prescription and testing procedures. Duthie *et al.* (2003) summarise these differences stating that:

- Front row positions demand strength and power.
- Locks should be tall with a large body mass.
- Loose forwards require strength and power with excellent speed, acceleration and endurance.
- Inside backs require strength, speed and power.
- Outside backs require considerable speed to out-manoeuvre their opponents.

Most rugby playing nations have a team of sports scientists which analyse every aspect of the game and the players. The fitness advisors now provide individualised training schedules and closely monitor the fitness levels of their players.

However, although there is general agreement about which fitness components are required for success in rugby, there is considerable variation in the testing protocols adopted by different countries. Therefore, the purpose of this chapter is to explain which components of fitness should be measured and to provide a protocol for each component of fitness with desirable performance standards for international players. Appropriate references will be provided where alternative tests are described in more detail. Test protocols presented in this chapter will be field-based, as field-testing in rugby union is normally sensitive enough to monitor training adaptations and identify individual differences. However, if specific weaknesses exist in a player's fitness profile laboratory procedures may be necessary. These generic laboratory protocols are provided elsewhere in this book.

TEST PREPARATION

In an international season there might be as many as five testing sessions with some measures being taken at all sessions whilst others may only occur once or twice a season. Table 34.1 provides a guide for the scheduling of the fitness tests. The general preparation/conditions for all testing sessions should be similar, including:

- Players being rested, injury free and accustomed to the test procedures.
- Tests occurring at the same time of day under similar environmental conditions using comparable equipment and facilities.
- Appropriate and consistent pre-test preparation, order of tests, warm-up and recovery between tests.

Table 34.1 Order of testing protocols

Test	Timing	Order
Stature	Day 1 a.m.	1
Body mass	Day 1 a.m.	2
Body composition	Day 1 a.m.	3
Leg power	Day 1 a.m	4
Speed	Day 2 a.m.	5
Multiple sprint test	Day 2 a.m.	6
3RM bench press	Day 3 a.m.	7
3RM half squat	Day 3 a.m.	8
3RM bench pull	Day 3 a.m.	9
Chin ups	Day 4 a.m.	10
3 km run	Day 4 a.m.	11

AEROBIC FITNESS

Introduction

A range of tests and distances have been used to assess the aerobic fitness levels of international rugby players during the last decade. Most major rugby playing countries now adopt a timed distance run of between 1 and 5 km.

Test procedure: 3 km timed run

- Complete 7.5 laps (3 km) of a 400 m synthetic running track.
- Ensure the weather conditions are favourable.
- Where appropriate, test players in small groups according to previous test result or playing position and be consistent in this group selection as this can affect the motivation (and hence test results) of some players.
- Provide a clear indication of the lap the players are completing and the split lap times.

Results

There can be a range of results for endurance performance within playing categories in rugby union. Some of the variation may be due to genetic endowment (including body size, shape and composition) but other variation can be due to training status. Hence when looking at the results of aerobic tests it is important to consider any changes in the player's body composition. Table 34.2 provides a series of desired performance standards for the 3 km timed run.

Alternative tests

A 20-m multistage shuttle run test, ASC Fitness Test for Rugby (Nance, 1998), 1–6 km timed runs, directly determined VO_{2max} test.

Table 34.2 Desired performance standards for the 3 km run (target lap times)

	Time min:s	Lap number						
		7	6	5	4	3	2	1
Front row	<12:15	11:26	9:48	7:70	6:32	4:54	3:16	1:38
Locks	<12:00	11:12	9:36	7:60	6:24	4:48	3:12	1:36
Back row	<11:30	10:44	9:12	7:40	6:08	4:36	3:04	1:32
Inside backs	<11:15	10:30	9:00	7:30	6:00	4:30	3:00	1:30
Outside backs	<11:45	10:58	9:24	7:50	6:16	4:42	3:08	1:34

SPEED

Introduction

Acceleration tends to outweigh maximal speed in terms of a pre-requisite for success in rugby union (Jenkins and Raeburn, 2000). The distances selected for testing speed and acceleration vary according to country and in our experience are not always based on sound scientific principles, but rather the size of the indoor venue (although there is close agreement between speeds achieved over the distances used). In general photoelectric timing of distances of between 5 and 60 m are utilised with 10 and 40 m being the most common.

Test procedure: 10 and 40 m timed sprints

- Use an indoor synthetic track with an adequate run down area.
- Set up the photoelectric cells at 10 and 40 m at a height of 1.2 m.
- Players should run in normal running shoes (and not spikes) and start from a crouched/standing position. Blocks should not be used.
- Players should start from a line 30 cm behind the first cell.
- Allow players to perform three timed sprints with the fastest split times being recorded.

Results

These results can vary depending on running surface, footwear and position of the start line. Players should be adequately warmed-up and complete the 10 and 40 m distances in the same sprint. Table 34.3 provides some desired performance times by playing position.

Alternative tests

These include sprints over 5–60 m.

Table 34.3 Desired performance standards for 10, 30 and 40 m sprints

	Standing 10 m (s)	Standing 30 m (s)	Standing 40 m (s)
Front row	<1.82	<4.40	<5.60
Locks	<1.75	<4.35	<5.50
Back row	<1.70	<4.25	<5.20
Inside backs	<1.65	<4.10	<4.90
Outside backs	<1.65	<4.00	<4.85

ANAEROBIC ASSESSMENT

Introduction

To assess the anaerobic fitness of a player a test should be able to provide a maximum score, a mean value for all sprints and a fatigue index. The maximum score provides a useful indication of pure speed, the mean value will provide an estimate of the anaerobic capacity of the individual and the fatigue index will indicate the player's ability to resist fatigue. Different rugby playing countries have selected a range of tests to assess this aspect of fitness but they are all based on the principles. Many sports scientists consider this to be one of the most relevant tests for rugby players and therefore if there is only time for one test then this test is often the test they select.

Test procedure: 40 m multiple sprint test

- Set the photoelectric cells up at 0 and 40 m with a starting line marked 30 cm behind these two cells.
- Players start from 30 cm behind the light beam at 0 m.
- The player completes 10 sprints every 30 s from the start of the first sprint using a running stop-watch (i.e. players get ~23–25 s recovery between sprints).
- Players alternate the ends from which they start (i.e. sprint 1 from 0 to 40 m, sprint 2 from 40 to 0 m, sprint 3 from 0 to 40 m).
- Players are provided with a 5 s warning (so they can prepare themselves for the next sprint) and then a count down of 2, 1, GO.

Results

The 10 sprint times are recorded to the nearest 0.01 s. The maximum score is the fastest time, the mean is the sum of all the 10 sprints divided by 10 and the fatigue index is the slowest speed divided by the fastest speed multiplied by 100% minus 100.

For example,

Fastest speed = 5.60 s; slowest speed = 6.00 s;
Fatigue index = [(6.00 − 5.60) / 6.00] × 100%
(%)   = 6.7%

These three results can be considered individually or together. It is also important to compare the criterion result with the player's 40 m speed time to ensure that the player is performing maximally in the multiple sprint test. Table 34.4 provides some desirable performance scores.

Table 34.4   Desired performance standards for the 10 m × 40 m multiple sprint test

	Fastest sprint (s)	Mean value (s)	Fatigue index (%)
Front row	<1.82	<5.90	<8
Locks	<1.75	<5.80	<8
Back row	<1.70	<5.65	<8
Inside backs	<1.65	<5.55	<10
Outside backs	<1.65	<5.50	<10

Alternative tests

These include a range of multiple sprint tests over between 35 and 150 m with recovery periods of between 24 and 40 s. Examples include the Australian phosphate recovery anaerobic capacity test (8 × 35 m with 24 s recovery), South African 150 m Shuttle (6 × 150 m with 30 s recovery), Phosphate Decrement Test (Dawson et al., 1991).

STRENGTH ASSESSMENT

Introduction

Different countries have adopted a range of strength-based measures but these are all based on a variation of 1–5 repetition maximum (RM) testing using large muscle groups. As rugby involves the upper body in both pushing and pulling activities it is important that this is replicated when testing rugby players. The bench press, bench pull and chin ups provide a good measure of upper body strength, both in terms of absolute terms and relative to a player's body mass. It is universally accepted that lower body strength can be assessed using some variation of a free weight squat.

Test procedure

Bench press

- Players should warm-up and be tested on a flat bench.
- The bar should be gripped shoulder width apart with their feet on the floor.
- Players should lower the bar to touch the chest and then fully extend the bar.
- The player continues to increase the load until their 3RM is reached. There should be a 5-min rest between attempts.

Bench pull

- The player lies in a prone position on a raised bench that allows the player to fully extend the elbows without the weights touching the floor.
- Knees are bent at 90% with the feet kept in the air.
- Players must lift the bar to touch the underside of the bench.
- The player continues to increase the load until their 3RM is reached. There should be a 5-min rest between attempts.

Half squat

- The bar should be gripped shoulder width apart with the player's toes pointing forwards.
- Players are required to complete a repetition with just the bar to determine the height that they need to squat to ensure their thighs are parallel with the floor.
- An adjustable bench is situated below the player to act as a target and ensure the required range of movement (standing upright to touching the bench).
- The player continues to increase the load until their 3RM is reached. There should be a 5-min rest between attempts.

Chin ups

- The players adopt an underhand grip shoulder width apart.
- Players start with the arms fully extended and knees bent to 90°.
- The player completes as many chin ups (chin above the bar and the arms fully extended) without stopping.

Results

Desirable performance scores by playing position are given in Table 34.5. These scores can also be presented relative to body mass.

Table 34.5 Desired performance standards for the strength tests

	Bench (kg)	Prone row (kg)	Squat (kg)	Chins
Front row	>155	>125	>240	>10
Locks	>145	>120	>210	>10
Back row	>150	>115	>220	>12
Inside backs	>140	>115	>200	>15
Outside backs	>135	>110	>200	>15

Alternative tests

Different countries use alternative numbers of repetitions between 1RM–5RM and further test protocols are sometimes adopted for particular positions. These include power clean, press-pull and dead lift. Some test batteries include a test to record the number of repetitions against a set load, normally relative to body mass.

BODY COMPOSITION

Introduction

Within the last decade the use of predictive equations to estimate percentage body fat in rugby has largely been replaced by reporting the sum of skinfold thicknesses. Body composition measurement is now more readily acceptable by rugby players as previously the percentage body fat of a player would change significantly when they reached 30 years of age as a different regression equation was used to calculate body density and to predict percentage body fat. All international rugby players are now routinely assessed for skinfold thickness but there is no universally accepted view on the number and location of sites. This ranges from 4 to 11 skinfold sites. However, what is more important than the number and anatomical site is the accuracy of the measurements.

Test procedure

Seven sites – triceps, biceps, subscapular, suprailiac, forearm, mid-thigh, mid-calf. These measurement techniques will be covered in Chapter 9.

Results

Due to the variation in site selection and technique and the range of values recorded per player normative data is of limited use. It is more important to monitor the absolute change of each individual relative to their baseline measurement.

Alternative tests

Body Stat, Hydrostatic weighing, Body Scanning.

LEG POWER

Introduction

A variety of jump tests are used in rugby fitness testing to assess explosive leg power. These protocols include two-leg and one-leg jumps, countermovement jumps and depth jumps. Countermovement and vertical jumps are regularly used for assessing leg power in rugby players although power tests have been used with forwards pushing in a horizontal position against a scrummaging machine or force platform.

Test procedure

The two jumps that are described are a countermovement jump and a vertical jump. These protocols were selected as they do not require a force platform or photoelectric jump mat.

Countermovement jump

- A jump meter is attached to the player's waist and all measurements taken are recorded to the nearest centimetre.
- Players are instructed to jump for maximum height following a dip or prior squatting movement keeping their hands on their hips, which minimises the skill element of the measure and concentrates on the functional capacity of the leg extensor muscles.
- Players are allowed three attempts and the best score is recorded.

Vertical jump

- The vertical jump protocol is identical to the countermovement jump but the subjects are allowed to use their arms to assist in jumping.

Results

The equipment available will determine the presentation of the results which can be expressed in height jumped, time in the air or as power output (Table 34.6).

Table 34.6 Desired performance standards for countermovement jump and vertical jump

	Countermovement jump (cm)	Vertical jump (cm)
Front row	40	48
Locks	45	53
Back row	47	55
Inside backs	52	60
Outside backs	52	60

Alternative tests

Modified Sargeant jumps, horizontal push tests against scrummaging machines, vertical jump test using force platforms or photoelectric jump mats.

REFERENCES

Dawson, B., Ackland, T., Roberts, C. and Lawrence, S. (1991). The phosphate recovery test revisited. *Sports Coach*, 14(3): 41.

Duthie, G., Pyne, D. and Hooper, S. (2003). Applied physiology and game analysis of Rugby Union. *Sports Medicine*, 33(13): 973–991.

Jenkins, D. and Raeburn, G. (2000). Protocols for the physiological assessment of Rugby Union Players. In C. Gore (ed.), *Physiological Tests for Elite Athletes*, Champaign, IL: Human Kinetics.

Nance, S. (1998). *Australian Rugby Union Rugby Specific Fitness Test*. Australian Rugby Union Ltd. Sydney, Australia: ASC.

Nicholas, C.W. (1997). Anthropometric and physiological characteristics of Rugby Union football Players. *Sports Medicine*, 23(6): 375–396.

Quarrie, K.L., Handcock, P. and Toomey, M.J. (1996). The New Zealand Rugby injury and performance project: IV. Anthropometric and physical performance comparisons between positional categories of Senior A rugby players. *British Journal of Sports Medicine*, 30: 53–56.

Thomas, C. (2004). http://www.irb.com/Playing/Game+analysis/ accessed 280405.

CHAPTER 35

JUDO

Andy Harrison, Jeremy A. Moody and
Kevin Thompson

Judo is characterised by short-duration, high-intensity, intermittent exercise (NCCP, 1990; Callister *et al.*, 1991; Ebine *et al.*, 1991; Takahashi, 1992; Nunes, 1997) followed by periods of constant pulling, pushing, lifting, grappling and gripping movements in preparation for the next explosive effort. The movement patterns primarily consist of grappling the costume via lapel, collar and/or sleeve in order to gain dominance so that subsequent short bursts of effort can exploit this advantageous position. Such sparring can occur either in the standing or ground positions depending on the situation, tactics or individual strengths and weaknesses of the judo player. Sikorski *et al.* (1987) categorised judo work into four separate phases 0–10 s, 11–20 s, 21–30 s and greater than 30 s. The highest frequency of activity (39%) was in the 11–20 s range with 80% of rest/breaks in the 0–10 s range. NCCP (1990) reported periods of 10–30 s work with a 10–15 s periods of active recovery. In another study, Sikorski *et al.* (1987) reported that the mean period of work time did not exceed 25 s and rest periods were no more than 10 s in duration. Their research illustrated that attacks occurred every 10–15 s.

Both the aerobic and anaerobic energy systems are taxed during judo competition. The anaerobic system provides the short and explosive all-out bursts of maximal power and strength that characterise this sport, whereas the aerobic system contributes more to the sustained effort necessary for a potentially 5 min match. Therefore in order to offset fatigue and maintain technique high levels of physical fitness are necessary. Research has highlighted the following parameters as influencing judo performance: body mass and composition, muscular strength, muscular endurance, muscular power, flexibility, anaerobic power and cardiovascular fitness. However, it must be remembered that judo is a technical sport and that these characteristics form only the platform upon which the athlete must base their technical skill and strategy.

Competitive judo is organised by weight division (women <48 to 78+ kg, men <56 to 100+ kg) yet due to the heterogeneity of body types a wide range in body mass, height and body fat is often evident both within and between divisions (Marchocka et al., 1984). Generally, low adiposity is a desirable characteristic irrespective of weight category. A possible exception is the open class heavy weight division where additional non-force producing body mass may provide an advantage. The measurement of body composition thus provides a valuable monitoring tool not only to indicate when a reduction in adiposity is warranted and possible, but also to indicate when a shift to a higher weight division may be necessary owing to the development of a higher lean body mass.

Muscular strength, endurance and power are all essential performance capabilities owing to the nature of attacking and defensive manoeuvres as well as the duration of a match. Such actions require judoka to have all round physical development of these capabilities, although some coaches place particular emphasis upon grip strength. Due to the sports weight division structure a strong relationship exists between strength and total body mass (Thomas et al., 1989; Fagerlund and Hakkinen, 1991; Borowski et al., 2001). Studies have typically measured strength using isokinetic dynamometers (Thomas et al., 1989; Fagerlund and Hakkinen, 1991; Kraemer et al., 2001; Harrison et al., 2003), although recent research has rightly begun to consider isometric contractions (Kraemer et al., 2001; Utter et al., 2002) due to their particular importance within judo groundwork technique (Reilly and Secher, 1990).

The technical requirement of judo involves both offensive and defensive movements through a wide range of joint motion. Consequently, coaches consider flexibility important for certain sport-specific movements and the possible reduction of injury (Horswill, 1992). Despite this judoka are typically reported to possess only 'normal' flexibility.

Judoka exhibit a high degree of aerobic and anaerobic development in the upper and lower body (Thomas et al., 1989). Although, research has also shown a considerable variation in maximum oxygen uptake and anaerobic capacity across judo athletes, which may be partly attributable to body size (Thomas et al., 1989; Callister et al., 1991). Subjective observation suggests that a judoka's level of anaerobic vs. aerobic fitness may influence their competitive strategy, such that those with a high anaerobic capacity fight offensively, while those with a high aerobic capacity adopt a more defensive fighting style (Fagerlund et al., 1991; Horswill, 1992; Callan et al., 2000).

When considering the physiological aspects of judo it must be remembered that in the period preceding competition it is common practice for the majority of its athletes to attempt to rapidly lose body mass in order to 'make weight'. Unfortunately, such strategies are often instigated with little consideration or knowledge of the potential impact upon performance and/or health. Many judoka believe that they possess an advantage over an opponent competing at their natural weight if they reduce their body mass to a minimum to qualify for a lower weight division. The premise is that the period after the weigh-in provides an opportunity to consume sufficient fluid and food to return to the 'training weight' prior to the beginning of the competition. It is not uncommon for athletes to be 2–5 kg heavier during competition than at their weigh-in. The timing of the weigh-in, which varies from the day before to a few hours before competition, is consequently a crucial factor. Actual weight loss

Table 35.1 Physical and physiological characteristics of judo players

			Range of mean values or mean ±SD	
Anthropometry				
Body composition	Skinfold (%)	Male	3–16.0	Thomas et al. (1989), Fagerlund et al. (1991), Horswill (1992), Callan et al. (2000), Degoutte et al. (2002), Harrison et al. (2003)
	Skinfold (%)	Female	16.8 ± 0.8	Harrison et al. (2003)
	Hydrostatic weighing (%)	Male	5.8 ± 0.3	Utter et al. (2002)
Isokinetic strength				
Grip	Hand Grip Dynamometer (kgf)	Male	53–61	Thomas et al. (1989), Harrison et al (2003)
	Hand Grip Dynamometer (kgf)	Female	44–45	Harrison et al. (2003)
Upper body	1RM Bench Press (kg)	Male	97–100	Thomas et al. (1989), Harrison et al. (2003)
	1RM Bench Press (kg)	Female	59 ± 9	Harrison et al. (2003)
	Bench Press Dynamometer (kgf)	Male	96 ± 20	Fagerlund et al. (1991)
	Back Pull Dynamometer (kgf)	Male	145 ± 23	Harrison et al. (2003)
	Back Pull Dynamometer (kgf)	Female	106 ± 15	Harrison et al. (2003)
Lower body	Hip and lower back Dynamometer (kgf)	Male	160.3–185	Fagerlund et al. (1991), Kraemer et al. (2001)
Isometric strength				
Lower body	Lower body pull (N·kg^{-1})	Male	52.7 ± 1.1	Utter et al. (2002)
Upper body	Upper body pull (kgf)	Male	98.2 ± 4.1	Kraemer et al. (2001)
Srength endurance				
Upper body	Bench Press at 70% 1RM (max. reps.)	Male	16 ± 3	Thomas et al. (1989)
Power				
Lower body	Vertical countermovement jump (cm)	Male	43.2–60	Claessens et al. (1986), Tumilty (1986), Thomas et al. (1989), Callan et al. (2000), Utter et al. (2002), Harrison et al. (2003),

Category	References	Sex	Test	Value
	Thomas et al. (1989), Kraemer et al. (2001), Utter et al. (2002),		Vertical countermovement jump (W)	4,318.5–5,304
	Harrison et al. (2003)	Female	Vertical countermovement jump (cm)	34 ± 5
Flexibility *Lower body*	Claessens et al. (1986), Thomas et al. (1989), Little (1991), Callan et al. (2000), Harrison et al. (2003)	Male	Sit and reach (cm)	3.8–43.2
	Harrison et al. (2003)	Female	Sit and reach (cm)	38 ± 4
Aerobic capacity – oxygen uptake measurements	Tumilty et al. (1986), Thomas et al. (1989), Horswill et al. (1992), Callan et al. (2000), Franchini et al. (2000), Utter et al. (2002), Degoutte et al. (2003)	Male	Continuous, graded treadmill Protocol (ml·kg·min)	50.4–69.9
	Horswill et al. (1992), Starczewska et al. (1999), Callan et al. (2000), Borkowski et al. (2001)	Male	Continuous, incremental cycle Protocol (ml·kg·min)	41.2–64.0
	Borkowski et al. (2001)	Female		49.8 ± 4.8
Aerobic capacity – heart rate measurements	Thomas et al. (1989), Callan et al. (2000), Utter et al. (2002), Degoutte et al. (2003)	Male	Treadmill VO_{2max} test (b·min^{-1})	174–191
	Callan et al. (2000), Degoutte et al. (2003)		Cycle VO_{2max} test (b·min^{-1})	176–198
			Mean during 5 min judo match (198.2 ± 0.7 max b·min^{-1})	182.4 ± 0.4
Anaerobic capacity *Upper body peak power*	Little (1991)	Male	Modified standing arm crank Wingate (W·kg^{-1}·BM^{-1})	8.5 ± 0.7
	Thomas et al. (1989), Horswill et al. 1990), Starczewska et al. (1999), Callan et al. (2000)		Modified seated arm crank Wingate (W·kg^{-1}·min^{-1})	3.5–11.3

(continued)

Table 35.1 Continued

			Range of mean values or mean ±SD
Lower body peak power	Tumilty et al. (1986), Thomas et al. (1989), Horswill et al. (1992), Starczewska et al. (1999), Borkowski et al. (2001)	Male — Cycle Wingate (W·kg⁻¹·BM⁻¹)	11–19.9
	Borkowski et al. (2001)	Female	10.6 ± 0.9
Peak lactate measurements	Callan et al. (2000)	Male — Treadmill VO_{2max} test (mmol·l)⁻¹	$15.1 \pm >$
	Callan et al. (2000), Borkowski et al. (2001)	Cycle VO_{2max} test (mmol·l)⁻¹	$10.6–13.1$
	Thomas et al. (1989)	Modified arm crank Wingate (mmol·l)⁻¹	14.5 ± 1.7
	Thomas et al. (1989)	Cycle Wingate (mmol·l)⁻¹	15.2 ± 1.8
	Degoutte et al. (2003)	3 min post a 5 min judo match (mmol·l)⁻¹	12.3 ± 0.8

Table 35.2 Suggested testing protocols

Tests	Equipment	Protocol
Strength endurance, strength and power		
Handgrip strength	Handgrip Dynamometer	Complete a 1RM followed by one maximal repetition every 20 s × 10. (Total 10 repetitions)
		Complete for left and right hands
Bench press	Bench press station, olympic bar and plates	3RM and no of reps at 70% of 1RM
Bench press (1:1 ratio with body mass)	Bench press station, olympic bar and plates	Isometric hold 5 cm above chest (bar to be retained within 0–10 cm range of chest). Measure time held in position
Back pull	Back pull station, olympic bar and plates	1RM and no of reps at % of BW (50% males; 40% females)
Back pull	KMS/BMS Strength Diagnostic System	Isometric hold at % of BM (50% males; 40% females). Bar to be retained within 0–10 cm range of under surface of bench. Measure time held in position
Pull up and hold (pronated and supinated grip)	Wall mounted pull up station; Stopwatch	Raise body into position where chin is above bar, execute hold. Finish timing when the top of the head is below the bar
Rope climb	Secured gymnastics rope facility	Standardise rope length, count absolute total repetitions or number over a set time period
Pull ups with towel for grip	Wall mounted pull up station; towel	Total number of pull ups with chin obtaining height equal to horizontal line of hands
Countermovement vertical jump	Jump belt	Hands to be positioned on the hips throughout the jump
Countermovement vertical jump	KMS/BMS Strength Diagnostic System	Hands to be positioned on the hips throughout the jump
Drop jump (0.3 m)	0.3-m box; KMS/BMS Strength Diagnostic System	Measurements of ground contact time and vertical jump height
Power Cleans (1RM)	Olympic bar and plates, KMS/BMS Strength Diagnostic System	Following warm-up and technical repetitions at 60–80% attempt to achieve the 1RM within three attempts
Power Cleans @ body mass	Olympic bar and plates, KMS/BMS Strength Diagnostic System	Total number of repetitions. 15 s between lifts (tester to count 3,2,1, lift to maintain timing)
Speed and anaerobic performance		
30 m sprint (hand timed)	Stopwatch; cones	Starting: Set…Go. Finish: When torso crosses the 30 m line

(continued)

Table 35.2 Continued

Tests	Equipment	Protocol
30 m sprint (electronic timing)	Brower electronic timing system	Start: use pressure release mechanism from '3 point' start position. Split times at 5, 10, 20 and 30 m
Anaerobic power and capacity Wingate Anaerobic Test (Bar-Or, 1996) (lower body and upper body)	Monark Cycle Ergometer and associated software Monark Arm Ergometer and associated software	
Repeated sprint test	Stopwatch and calculator	E.g. Complete 8 × 40 m shuttle (10:20:10) with 30 s recovery between repetitions
Aerobic capacity Coopers Test (Cooper, 1968)	400 m (or 200 m) running track	Run the greatest distance possible in 12 min.
Maximal oxygen uptake	Multistage fitness test (Brewer *et al.*, 1988) Laboratory protocol (see protocols elsewhere in guidelines)	Complete test to volitional exhaustion

methods are often sourced from coaches or peers and include: dehydration, fasting and, for a few, even vomiting, laxatives and diuretics (Steen and Brownell, 1990). Such strategies may undermine performance if they lead to dehydration, depleted glycogen stores and a reduction in lean muscle mass. Both research and anecdotal evidence suggests that dehydration is the most common method of 'making weight' with urine osmolality measurements as high as 1,000 mOsmols·kg^{-1} not uncommon (Harrison *et al.*, 2003). Studies have shown that rapid weight loss is associated with concurrent decrements in performance however research regarding the specific impact of rapid weight loss on combative sports performance is equivocal (Horswill, 1992; Filaire *et al.*, 2001; Kraemer *et al.*, 2001). Although, it has been suggested that coupling stressors (i.e. the degree of absolute weight loss combined with the tournament schedule) may adversely affect performance as a tournament progresses (Kraemer *et al.*, 2001). Judoka who participate in weight loss techniques might be doing so at the expense of peak physical performance, particularly in the later rounds of a tournament.

In summary, the physiological and physical profiles of judo athletes differ markedly and the factors responsible for success may be specific to each weight division. Ultimately, successful performance may be dependent upon the correct compromise between 'making weight' and maximising physiological capacities (Callister *et al.*, 1991).

The data in Table 35.1 provides some physical characteristics of judoka and amateur wrestlers. The application of the data with regard to sports science and strength and conditioning support programmes remains highly contentious and further research into the performance requirements and physical capabilities of judoka is clearly warranted. Table 35.2 provides a number of possibilities for the physiological and physical assessment of judoka. These tests have either been reported in the literature or have been used within sports science and strength and conditioning support programmes with elite judoka.

REFERENCES

Borowski, L., Faff, J. and Starczewska-Czapowska, J. (2001). Evaluation of the aerobic and anaerobic fitness in judoists from the Polish National Team. *Biology of Sport*, 18(2): 107–117.

Brewer, J., Ramsbottom, R. and Williams, C. (1995). Multistage Fitness Test. In Crisfield (ed.), *Measuring Performance: A Guide to Field Based Fitness Testing*, pp. 112–113. Leeds: National Coaching Foundation.

Callan, S., Brunner, D., Devolve, K., Mulligan, S., Hesson, J., Wilber, R. and Kearney, J. (2000). Physiological profiles of elite freestyle wrestlers. *Journal of Strength and Conditioning Research*, 14(2): 162–169.

Callister, R., Callister, R.J., Fleck, S.J. and Dudley, G.A. (1990). Physiological and performance responses to overtraining in elite judo athletes. *Medicine and Science in Sports and Exercise*, 22: 816–824.

Callister, R., Callister, R.J., Staron, R.S., Fleck, S.J., Tesch, P. and Dudley, G.A. (1991). Physiological characteristics of elite judo athletes. *International Journal of Sports Medicine*, 12(2): 196–203.

Castarlenas, J.L. and Planas, A. (1997). A study of temporal structure of judo combat. *Physical Education and Sports*, 47: 32–39.

Claessens, A., Beunen, G., Lefevre, J., Mertens, G. and Wellens, R. (1986). A physiological profile of well trained male judo players. In J. Watkins, T. Reilly and L. Burwitz (eds), *Proceedings of the VII Commonwealth and International Conference on Sport, Physical Education, Dance, Recreation and Health*. London: E & FN Spon.

Cooper, K.H. (1968). A means of assessing maximal oxygen uptake. *Journal of American Medical Association*, 203: 135–138.

Degoutte, F., Jouanel, P. and Filaire, E. (2003). Energy demands during a judo match and recovery. *British Journal of Sports Medicine*, 37: 245–249.

Ebine, K., Yoneda, I. and Hase, H. (1991). Physiological characteristics of exercise and findings of laboratory tests in Japanese elite judo athletes. *Medicine in Sport*, 65(2): 183–192.

Fagurlund, R. and Hakkinen, K. (1991). Strength profile of Finnish Judo players: measurement and evaluation. *Biology of Sport*, 8(3): 143–149.

Filaire, E., Maso, F., Degoutte, F., Jouanel, P. and Lac, G. (2001). Food restriction, performance, psychological state and lipid values in judo athletes. *International Journal of Sports Medicine*, 22(6): 454–459.

Franchini, E., Takito, M., Matsushigue, K. and Kiss, M. (2000). Influences of aerobic fitness on blood lactate concentration and on intermittent anaerobic exercise after a judo fight (abstract). In *Sports Medicine Australia 2000 Pre Olympic Congress: International Congress on Sport Science, Sport Medicine and Physical Education*, p. 363. Brisbane, Australia.

Harrison, A., Thompson, K., Cosgrove, M., Hardman, S. and Dietzig, B. (2003). Physical characteristics and body weight management of international judo players (abstract). *Journal of Sports Science*, 21(4): 275.

Horswill, C. (1992). Applied physiology of amateur wrestling. *Sports Medicine*, 14: 114–143.

Horswill, C., Hickner, R., Scott, J., Costill, D. and Gould, D. (1990). Weight loss, dietary carbohydrate modifications, and high intensity, physical performance. *Medicine and Science in Sports and Exercise*, 22: 470–476.

Kraemer, W., Fry, A., Rubin, M., Triplett-McBride, T., Gordon, S., Koziris, L., Lynch, J., Volek, J., Meuffels, D., Newton, R. and Fleck, S. (2001). Physiological and performance responses to tournament wrestling. *Medicine and Science in Sports and Exercise*, 33(8): 1367–1378.

Little, N. (1991). Physical performance attributes of junior, and senior women, juvenile, junior and senior men judokas. *Journal of Sports Medicine and Physical Fitness*, 31: 510–520.

Marchocka, M., Nowacha, E. and Sikorski, W. (1984). Specific body build of judo athletes depending on the fighting technique used. *Biology of Sport*, 1: 261–264.

National Coaching Certification Programme (NCCP). (1990). Level 3: Judo Technical Manual. Glouster, Ontario: Judo Canada.

Nunes, A.V. (1997). The difficult of evaluation in high level judo athletes. In *Proceeding of International Symposium of Science and Technology in Sport*, pp. 39–40. Porto Alegre: November 19–22.

Pulkkinen, W.J. (2001). *The Sport Science of Elite Judo Athletes*. Ontario. Canada: Pulkinetics.

Reilly, T. and Secher, N. (1990). *Physiology of Sports: An Overview*. In T. Reilly, N. Secher, P. Snell and C. Williams. (eds), *Physiology of Sports*, p. 467. E & FN Spon.

Sharp, N.C. and Koutedakis, Y. (1987). Anaerobic power and capacity measurements of the upper body in elite judo players, gymnasts and rowers. *The Australian Journal of Science and Medicine in Sport*, 19(3): 9–13.

Sikorski, W.G., Mickiewicz, G., Maole, B. and Laska, C. (1987). *Structure of the Contest and Work Capacity of the Judoist*. Polish Judo Association. Warsaw, Poland: Institute of Sport.

Starczewaska-Czapowska, J., Faff, J. and Borkowski, L. (1999). Comparison of the physical fitness of the successful and less successful elite wrestlers. *Biology of Sport*, 16(4): 225–232.

Steen, S.N. and Brownell, K.D. (1990). Patterns of weight loss and regain in wrestlers: has the tradition changed. *Medicine and Science in Sports and Exercise*, 22(6): 762–768.

Takahashi, R. (1992). *Power Training for Judo: Plyometric Training with Medicine Balls. National Strength and Conditioning Association Journal*, 14(2): 66–71.

Thomas, S., Cox, M.H., Legal, Y.M., Smith, H.K. and Verde, T.J. (1989). Physiological profiles of the Canadian judo team. *Canadian Journal of Sports Science*, 14(3): 142–147.

Tumilty, D., Hahn, A. and Telford, R. (1986). A physiological profile of well trained male judo players. In J. Watkins, T. Reilly and L. Burwitz (eds), *Proceedings of the VII Commonwealth and International conference on Sport, Physical Education, Dance, Recreation and Health*. London: E & FN Spon.

Utter, A., O'Bryant, H., Hafe, G. and Trone, G. (2002). Physiological profile of an elite freestyle wrestler preparing for competition; a case study. *Journal of Strength and Conditioning Research*, 16(2): 308–315.

CHAPTER 36

WINTER SPORTS

Richard J. Godfrey, David Macutkiewicz,
Kate Owen and Gregory P. Whyte

Collectively the Winter Sports are characterised as those sports that traditionally appear as part of the Winter Olympic Games. There are a number of Winter Olympic Sports in which Great Britain has had no representation in recent Olympic Games including: long track speed skating, cross-country skiing and ice hockey. Further, several Winter Olympic sports have GB representation, however these sports have a limited physiological support programme, for example, curling.

Accordingly, the following chapter focuses on the physiological support for GB Winter Olympic sports that represent Team GB and have a significant physiological support structure, including Nordic skiing, bobsleigh and bob skeleton.

NORDIC SKIING

Nordic skiing comprises three disciplines; biathlon and cross-country skiing and 'Nordic combined' (ski jumping and cross-country skiing). In this chapter only biathlon and cross-country skiing will be addressed.

In modern competition cross-country skiing encompasses the two major technique-specific disciplines of 'traditional' and 'freestyle' technique. Traditional is also referred to as 'classic' and is the diagonal technique where opposite arm and leg work together in providing locomotion. This technique is also often referred to as 'Langlauf'. Freestyle most commonly refers to skating technique. For competition these two techniques are separated and racing occurs over distances of 5–30 km for women and 10–50 km for men. In general, the longer distances are often those using traditional technique (classic) and the shorter distances often use freestyle. Classic technique is generally far more aerobic, whilst freestyle is more dynamic and has the athlete working at a relatively higher physiological intensity.

Biathlon competition involves cross-country skiing (exclusively freestyle technique) and shooting. There is a greater variety of physiological demand in using freestyle technique as the individual is required to work at a fairly high intensity for protracted periods with heart rates commonly approaching 180 bpm, and then shoot at five targets. This requires the athlete to be able to reduce heart rate in approaching the shooting range (generally by slowing down) and to fire between beats (through psychophysiological cues which are not yet fully understood).

In training for cross-country skiing large training volumes have traditionally been the norm with injury rates being lower than for running as foot-ground impact forces are very much lower. In winter most training occurs on snow but in summer, athletes commonly train using roller skis (skis which are c. 45 cm long with wheels at each end). Training involves a substantial amount of work that combines technique improvement with maintenance of a large aerobic power. For biathletes, training often utilises classic technique to establish and maintain a good aerobic endurance base and freestyle as it is specific to competition. Freestyle technique requires superior balance as the athlete must be able to drive with one leg and then glide on the other. Increasingly, it is recognised that to be good at freestyle technique requires, in addition to endurance training, some high-intensity and explosive training which should, initially, be developed in the weights room during the non-competitive phase. This recognises the biomechanics of the muscle–tendon complex where, when the ski is first in contact with the snow there is a slow build up of force as the active muscle contracts relatively slowly to transfer energy to the tendon. This being in preparation for the later, 'kick away', explosive drive prior to the driving foot leaving the snow. In both freestyle and classic technique the arms are used, not just to provide balance but to also contribute substantial force to propulsion.

Physiology testing

Generally, Nordic skiers require a large aerobic power and to be competitive at elite level it is estimated that, for men, an excess of 85 ml·kg^{-1}·min^{-1} is a prerequisite. In racing an exercise intensity of 80–90% $\dot{V}O_{2max}$ is the norm and hence elite skiers not only require a large aerobic power they also need a high fractional utilisation and lactate threshold should occur at a high percentage of $\dot{V}O_{2max}$.

In Scandinavian countries $\dot{V}O_{2max}$ and lactate profile tests are carried out, on both classic and freestyle techniques, on a motorised treadmill specifically designed for the task (2 m wide and 4 m long). In the United Kingdom, $\dot{V}O_{2max}$ and lactate profiles of cross-country skiers and biathletes have generally been assessed on modes of activity which relate more specifically to the numerous modes that are used in training and hence motorised treadmill running and Kingcycle testing has been the norm. In addition, it is important to test upper body $\dot{V}O_{2max}$, strength and power. Over many years (since 1988 in the United Kingdom) lab-based testing has been supplemented with field-based data collection with athletes rollerskiing on dry roads in Summer or on snow, skiing in Winter.

In the lab and field one generalised test protocol has been implemented to assess the current level of endurance conditioning; a discontinuous incremental protocol (Figure 36.1). In the lab this involves 4-min work stages alternating with 30-s rest intervals during which capillary (earlobe) blood samples are

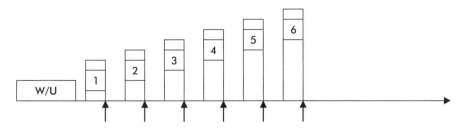

Figure 36.1 Generalised test protocol used in the lab and field. W/U is warm-up. Blocks 1–6 are 4-min efforts, with each successive block being at higher exercise intensity. There is a rest interval of 30 s between each block for the collection of a capillary blood sample for lactate (BLa) analysis and here represented by vertical arrows. Heart rate (HR) data is also collected and the relationship between HR and BLa allows the training intensity associated with 'threshold'/'tempo' to be identified

taken for subsequent analysis of lactate concentration. On the treadmill this generally involves calculating the pace for a recent 10 km run (preferably a race), setting this as the pace for the third or fourth stage. Higher stages are calculated by successively adding 0.6 kph to that 10 km pace, lower stages by successive subtraction of 0.6 kph. In the field the same generalised protocol is used but stages are generally longer (6–10 min). Athletes wear heart rate monitors and are instructed to start the first stage at a really easy intensity (c. 40–50% of HRM). Each successive stage is then conducted at, on average, 5–10 bpm higher than the one before. Again, 30-s rest is given between each stage and a lactate profile constructed. From the lactate profile, heart rate ranges can be derived for 'threshold' training and for base endurance.

Additional tests for biathletes, and for cross-country skiers specialising in freestyle races, should include measures of leg power such as standing vertical jump (SVJ), Bosco test or similar.

In summary the following tests are recommended for biathlon and cross-country skiers:

- A measure of body composition (namely, body mass and skinfold data).
- A measure of lower body $\dot{V}O_{2max}$ and lactate profile (where possible on roller skis on a treadmill or, as a poorer option, using cycle ergometery or running on a motorised treadmill).
- A measure of upper body $\dot{V}O_{2max}$ and lactate profile (this is best done using a double poling action and a pulley system which can be calibrated and the resistance varied in increments of power output).
- SVJ or similar.

REFERENCES

Martin, D.T., Skog, C., Gillam, I., Ellis, L. and Cameron, K. (2000). Protocols for assessment of cross-country skiers and biathletes. In C.J. Gore (ed.), *Physiological Tests for Elite Athletes*. Australian Sports Commission. Champaign, IL: Human Kinetics.

Rusko, H. (ed.) (2003). *Cross Country Skiing*. IOC Medical Commission. Oxford: Blackwell Publishing.

BOBSLEIGH

Bobsleigh developed as a modification of tobogganing and was introduced at the 1924 Olympic Winter Games in Chamonix. The aerodynamic sleighs race down the course (at least 1,500 m long) with at least 15 banked curves with speeds up to 143 km·h^{-1}.

Teams consist of either two or four men. A two-man team consists of a driver and a brakeman. Before mounting the bob, momentum needs to be initiated. Athletes begin with a standing start and push the bob over a distance of between 30 m and 40 m to achieve maximal acceleration within around 6 s. Athletes must then jump into the bob at speed. Once in the bob, they work together transferring their bodyweight around corners for maximum speed. Athletes stay as low as possible to achieve maximum aerodynamic performance. The driver must take the fastest line possible down the track from top to bottom by steering using ropes that are attached to the runner-bearing front axle. The brakeman slows the bob by a harrow-type brake that digs into the ice. The maximum weight for a two-man team and bob is 390 kg. The maximum weight for a four-man team and its bob is 630 kg. The additional two members in a four-man team work as the side pushers.

To achieve optimal performance, the athlete needs speed, strength, power, hand-eye coordination and spatial awareness. The test battery should be designed to test those factors most important for performance.

PHYSIOLOGICAL TESTING PROCEDURES

Lower body power

To assess this parameter the British Bobsleigh Association uses results from a vertical jump test and a 1 repetition maximum power clean with squad members. A modified Wingate test could also be used where athletes sprint maximally from a standing start for 6 s against a load of 8% or 6% body mass for males and females, respectively. This test is used to assess the ability of the athlete to push against and accelerate a load from a standing start. Results could be used in conjunction with those recorded from the 30 s sprint test which measures maximal sprint speed.

VERTICAL JUMP TEST

Proposed protocol

A 5-min warm-up should precede at least three practise jumps. Prior to these practises the tester must make it clear which 'Vertical Jump' technique is required, the countermovement or static, the athlete should be fully habituated to the technique. At least 4 min rest should then be allowed prior to the test

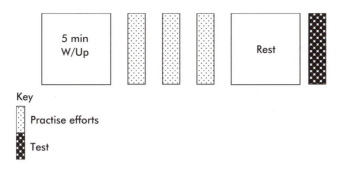

Key

Practise efforts

Test

Figure 36.2 Vertical jump test

effort. Flight time and contact time is measured by a microcomputer digital timer connected to a jumping platform. Height attained during the jump is the value recorded (Figure 36.2).

Qualifying standards

The British National and International team requires athletes to achieve the following standards for vertical; men = 0.70 m, women = 0.50 m (National) and men = 0.80 m and women = 0.60 m (International) (www.british-bobsleigh.com).

POWERCLEAN 1RM

Proposed protocol

The athlete warms-up by performing 5–10 reps using a light weight. After a rest of 1 min ~10–20% of the warm-up weight is added and another 5–10 reps are performed (Figure 36.3). Following another 2 min rest, 2–3 reps of a heavy weight are performed. After this, the athlete should rest for 2–4 min and then attempt a 1 rep maximum (1RM) lift. If this is successful the athlete should have another 4 min rest before attempting a heavier weight for a 1RM. This process should continue until the athlete fails to lift the weight successfully. The athlete should attempt no more than five 1RM lifts for assessment due to avoid the effects of fatigue influencing the result. The last successful weight lifted should be counted as the 1RM (adapted from Earle, 1999).

Qualifying standards

The British National and International team requires athletes to achieve the following standards for power clean; men = 110 kg, women = 70 kg (National) and men = 120 kg and women = 80 kg (International) (www.british-bobsleigh.com).

MODIFIED WINGATE TEST: 6 s SPRINT

Proposed protocol

The test can be performed on a Monark cycle ergometer. A 5-min warm-up should be completed after which the athlete should perform a practise start. This allows the athlete to experience the inertia of the fly wheel in preparation for the test. A load equivalent to 8% and 6% bodyweight should be added to the cycle ergometer for males and females, respectively. For the test, the athlete should be instructed to cycle as hard and as fast as possible for the full 6 s (Figure 36.4).

Measurements

Peak power (W), peak power relative to bodyweight (W·kg^{-1}) and time taken to reach peak power (s) achieved during the test should be recorded. Ideally, elite male athletes should reach a peak power of 1,400–1,500 W (approx. equivalent to 15–16 W·kg^{-1}) and reach this within 2.5 s. Elite females should reach a peak power of around 800–900 W (approx. equivalent to 12 W·kg^{-1}) within 3.0–3.5 s.

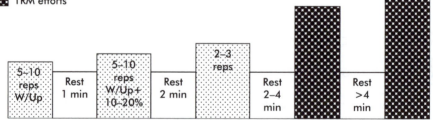

Figure 36.3   Powerclean 1RM

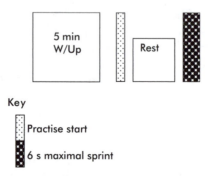

Figure 36.4   Modified Wingate test: 6 s sprint

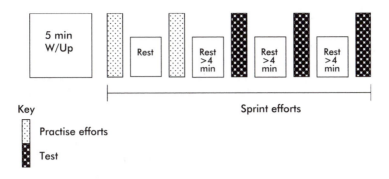

Figure 36.5 A 30 m maximal sprint test

SPEED

Excellent speed is needed for a bobsleigh athlete. The ability to accelerate maximally from a standing start and achieve an aerodynamic position in the bob quickly is essential.

Proposed protocol

Sprint tests can be carried out using timing gates positioned on an athletics track 30 m apart. Athletes are then assessed on the time it takes to break the beams between the timing gates. At least 5 min warm-up should be completed followed by two practice sprints with adequate rest in between. At least 4 min rest should be given following these practice efforts before athletes perform a maximal 30 m sprint. Athletes could be given opportunity to repeat this test more than once with the best effort being counted. In this scenario, at least 4 min rest time should be given between each sprint (Figure 36.5).

Qualifying standards

The British National and International team requires athletes to achieve the following standard for 30 m sprint; Men = 3.90 s, women = 4.25 s (National) and men = 3.80 s and women = 4.15 s (International) (www.british-bobsleigh.com).

REFERENCES

British Bobsleigh Association website (www.british-bobsleigh.com).

Earle, R.W. (1999). Weight training exercise prescription. In Roger W. Earle and Thomas R. Baechle, *Essentials of Personal Training Workbook*; Lincoln, NE: NSCA Certification Commission.

BOB SKELETON

Skeleton is the fastest growing of all the bobsleigh sports. The sport was re-introduced into the Olympics at the Winter Games in Salt Lake City in 2002 after a 54 year absence. The athlete adopts a face down, head first, minimal drag riding position on board a sled. The ice tracks range in length between 1,000 m and 1,500 m and speeds of up to 135 km·hr^{-1} have been recorded. The athlete has to sprint 20–30 m, accelerating the sled before diving on board. The athlete continues to accelerate under gravity, aiming to adopt the most energy efficient line possible while negotiating up to 15 curves. The aim is to achieve the fastest possible descent of the track. The sled has no brakes or mechanical steering and has minimal protection. Steering is induced by shifting the rider's body weight. As the athlete descends the track, 'G forces' of plus 5 G's are experienced.

The physical demands of bob skeleton require athletes to have speed, strength, power, hand-eye coordination and spatial awareness. The British Bob Skeleton Association also highlights the importance of mental toughness, 'nerves of steel' and hard work.

Minimum sprint standards for selection are set for men and women of different age groups. Generally, skeleton athletes are smaller in height and weight than bobsleigh athletes and this is said to be due to the advantage of maintaining a lower centre of gravity when pushing the bob and a faster cadence in the bent over position whist in the acceleration zone. It is also easier to maintain an efficient aerodynamic posture on the sled. The max weight of a bob-skeleton athlete and sled is 110 kg.

As with 'Bobsleigh' the acceleration of the sled is important to achieve the fastest start as possible. Athletes need to be fast and powerful and so the same tests may be implemented.

PHYSIOLOGICAL TESTING PROCEDURES

Lower body power

There are a number of tests that could be used to assess this parameter; vertical jump e.g., 1RM power clean or a modified Wingate test. In this last test, athletes sprint maximally from a standing start for 6 s against a load of 8% or 6% body mass for males and females, respectively. This assesses the ability of the athlete to push against and accelerate a load from a standing start. Results could be used in conjunction with those recorded from the 30 s sprint test which measures maximal sprint speed.

VERTICAL JUMP TEST

Proposed protocol

A 5 min warm-up should precede at least three practise jumps. Prior to these practises the tester must make it clear which 'Vertical Jump' technique is required, the countermovement or static, and coach the athlete accordingly until the tester is satisfied. At least 4 min rest should then be allowed prior to the test effort. Flight time and contact time is measured by a microcomputer digital timer connected to a jumping platform. Height attained during the jump serves as the end result (Figure 36.6).

POWERCLEAN 1RM

Proposed protocol

The athlete warms up by performing 5–10 reps using a light weight. After a rest of 1 min ~10–20% of the warm up weight is added and another 5–10 reps are performed (Figure 36.7). Following another 2 min rest, 2–3 reps of a heavy weight are performed. After this, the athlete should rest for 2–4 min and then attempt a 1 RM lift. If this is successful the athlete should have another 4 min rest before attempting a heavier weight for a 1RM. This process should continue until the athlete fails to lift the weight successfully. The athlete should attempt a maximum of five 1RM lifts for assessment due to inevitable effects of fatigue. The last successful weight lifted should be counted as the 1RM (adapted from Earle, 1999).

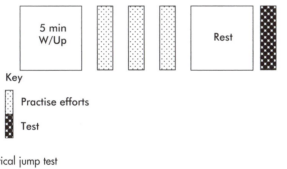

Key

Practise efforts

Test

Figure 36.6   Vertical jump test

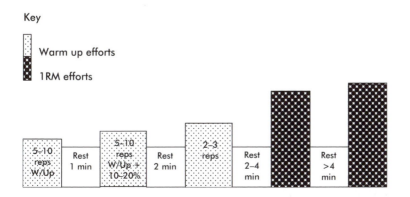

Key

Warm up efforts

1RM efforts

Figure 36.7   Powerclean 1RM

MODIFIED WINGATE TEST: 6 s SPRINT

Proposed protocol

The test can be performed on a Monark cycle ergometer. A 5-min warm-up should be completed after which the athlete should perform a practise start. This allows the athlete to experience the inertia of the fly wheel in preparation for the test. A load equivalent to 8% and 6% bodyweight should be added to the cycle ergometer for males and females, respectively. For the test, the athlete should be instructed to cycle maximally for the full 6 s (Figure 36.8).

Measurements

Peak power (W), peak power relative to bodyweight (W·kg^{-1}) and time taken to reach peak power (s) achieved during the test should be recorded.

SPEED

As for bobsleigh, excellent speed is needed for a skeleton athlete. The ability to accelerate maximally from a standing start and achieve an aerodynamic position quickly is essential.

Proposed protocol

Sprint tests can be carried out using timing gates positioned on an athletics track 30 m apart. Athletes are then assessed on the time it takes to break the beams between the timing gates. At least 5-min warm-up should be completed followed by two practice sprints with adequate rest in between. At least 4 min rest should be given following these practice efforts before athletes perform a maximal 30 m sprint. Athletes could be given opportunity to repeat this test more than once with the best effort being counted. In this scenario, at least 4 min rest time should be given between each sprint (Figure 36.9).

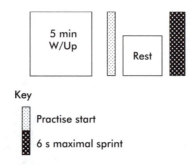

Figure 36.8 Modified Wingate test: 6 s sprint

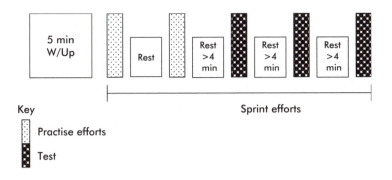

Key

Practise efforts

Test

Figure 36.9 A 30 m maximal sprint test

REFERENCES

Earle, R.W. (1999). Weight training exercise prescription. In Roger W. Earle and Thomas R. Baechle, *Essentials of Personal Training Workbook*. Lincoln, NE: NSCA Certification Commission.

ALPINE SKIING

Alpine skiing is made up of four separate disciplines: downhill, super giant slalom, Giant slalom and slalom. The slalom features tight technical turns, quick foot-to-foot movements and high level of agility. Giant slalom involves a series of medium radius turns with the emphasis upon higher velocities than the slalom and smooth transitions between foot movements (Atkins and Hagerman, 1984). The slalom and giant slalom typically lasts between 60 s and 80 s. The downhill and super giant slalom emphasises very high speeds (in excess of 100 MPH), large radius turns, various terrain changes and typically lasts between 90 s and 120 s. Whilst the downhill comprises a single run of the course, the slalom and giant slalom comprise of two timed runs.

Snow conditions, weather conditions, terrain, visibility and course setting, and starting position all affect the competitors ability to perform a fast and successful run. A competitors starting position is determined by their Federation International de Ski (FIS) ranking.

Oxygen uptake has been measured during skiing, showing that during the last part of the giant slalom oxygen uptake typically exceeds 75–80% of maximal oxygen uptake ($\dot{V}O_2$ max) and may also reach near maximal level (Tesch, 1995). The oxygen consumption may be proportionally substantial, the energy requirements have been shown to exceed 120% of the expected intensity of $\dot{V}O_{2max}$ (Saibene *et al.*, 1985). Veicsteinas *et al.* (1984) suggested that the aerobic demand of alpine skiing is only about 40%. The anaerobic component is also illustrated by the pronounced lactate accumulation of 7–15 mM demonstrated after SL and GS races (Veicsteinas *et al.*, 1984; Astrand and Rodahl, 1986, pp. 646–682). A greater reliance upon anaerobic energy production has also been suggested to differences in joint angle/posture and associated per cent maximal voluntary contraction (MVC) patterns of the quadriceps during GS and SL reducing blood flow and oxygen delivery to the working muscles (Lewis et al., 1985; Berg et al., 1995; Hintermeister et al., 1995; Szmedra et al., 2001).

Muscular strength of the lower limb musculature has been suggested to be an important factor for alpine skiing. A combination of isometric, eccentric and concentric muscular actions is involved during alpine skiing events. Holding a crouched or tuck position requires force development through isometric muscle action of the quadriceps femoris muscle group. Concentric muscle action is required for the elevation of the centre of gravity during the initiation of a turn and the tightness of the turn dictates the degree of torsion and the forces required by eccentric muscle action to oppose both gravity and centrifugal forces (Leach et al., 1994). Muscle activation can reach near maximal levels during the course of every turn (Berg and Eiken, 1999) however, Berg et al. (1994) estimated maximal mean electromyographic (EMG) activity of the vastus group during GS at 74% of MVC in the late eccentric phase and 67% MVC during concentric activity.

PHYSIOLOGICAL TESTING PROCEDURES

Proposed protocol

Athletes should begin with a warm-up after which a lactate sample should be collected. This quantifies the starting lactate and should be under 2 mMol. The athlete is provided with a heart rate monitor and the connection to the online gas analysis set up using a face mask or mouthpiece and nose clip with head gear for stability. Gas analysis continues throughout the test. Capillary blood samples for blood lactate (BLa) analysis are taken at the end of each 4 increment until a clear break point can be observed (BLa >4 mM). Following a 15-min rest period a ramp test of increments of 15 W every 15 s is performed until volitional exhaustion. At the end of the test another lactate sample is taken and maximal heart rate recorded. $\dot{V}O_{2max}$ is calculated as the $\dot{V}O_2$ during the last 60 s of the test according to criteria set out in the previous section 'The assessment of maximal oxygen uptake'. The maximum minute power is calculated as the average power maintained over the final 60 s of the test (Figure 36.10).

The importance of collecting these lactate values is that the effectiveness of training can be assessed by comparing responses at standardised work loads (Wilmore and Costill, 1988) and blood lactate and heart rate profiling also allows the establishment of training zones (Martin and Coe, 1997).

ANAEROBIC POWER

The contribution of the anaerobic energy system in alpine skiing is considerable and has been estimated to contribute to 60% of total energy expenditure

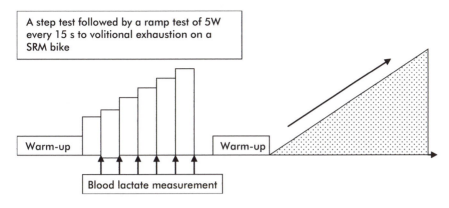

Figure 36.10   Aerobic power

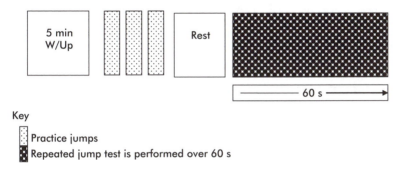

Key

Practice jumps

Repeated jump test is performed over 60 s

Figure 36.11   Vertical jump test: repeated efforts over 60 s

(Veicsteinas *et al.*, 1984). The associated % MVC patterns of the quadriceps, in order to oppose both gravity and centrifugal forces and joint angle/posture, during a turn reduces blood flow and oxygen delivery to the muscles, as a consequence, anaerobic metabolism is increased (Lewis *et al.*, 1985; Anderson and Montgomery, 1988; Berg *et al.*, 1995; Hintermeister *et al.*, 1995; Szmedra *et al.*, 2001).

Proposed protocol

Repeated jump test: Athletes should still perform a warm-up, practise jumps and rest as above. Following this, the athlete must jump maximally and repeatedly for 60 s with good technique. After this test a fatigue index may be calculated from the results by calculating the mean difference of the first three jumps and last three jumps, or, alternatively, the per cent decrease in jump height between the best and worst jump. Flight time and contact time can be measured by a microcomputer digital timer connected to a jumping platform and used to calculate average power (Figure 36.11).

Laboratory tests for maximal anaerobic power and capacity are most relevant to the athlete when they simulate the actual mode of exercise and involve the specific muscle groups used in the sport (MacDougall *et al.*, 1991). The specific nature of repeated jumping test utilising elastic energy and the stretch-shortening cycle, a characteristic of alpine skiing, gives a justification for its inclusion in the testing battery.

MODIFIED WINGATE TEST: 60 S SPRINT

Proposed protocol

The test can be performed on a Monark cycle ergometer. A 5-min warm-up should be completed after which the athlete should perform a practise start. This allows the athlete to experience the inertia of the fly wheel in preparation for the test. A load equivalent to 8% and 6% bodyweight should be added to the cycle ergometer for males and females, respectively. For the test, the athlete should be instructed to cycle maximally for the full 60 s (Figure 36.12).

Measurements

Peak power (W), peak power relative to bodymass (W·kg^{-1}) and time taken to reach peak power (s) and fatigue index (FI) achieved during the test should be recorded.

Although the mode of exercise during the wingate is not specific to skiing, the musculature involved is specific to skiing and has been used widely for the assessment of alpine skiers.

Lower body strength/power

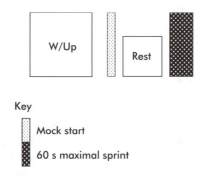

Figure 36.12 Modified Wingate test: 60 s sprint

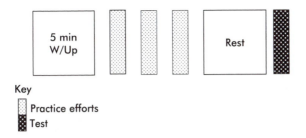

Figure 36.13 Vertical jump test: single effort

ISOKINETIC STRENGTH/POWER

Isokinetic strength of the knee extensor and flexor muscles could also be assessed using an isokinetic dynamometer such as the Biodex or Cybex system. Concentric muscular strength is measured at three different angular velocities (60°s, 180°s and 300°s) over four repetitions with a rest interval of 60 s between speeds.

VERTICAL JUMP TEST: SINGLE EFFORT

Proposed protocol

Single vertical jump test: A 5-min warm-up would precede three practise jumps. Prior to these practises the tester must make it clear which 'Vertical Jump' technique is required (the static, countermovement or countermovement with arms) and coach the athlete accordingly until the tester is satisfied. At least 4 min rest should then be allowed prior to the test effort. Flight time and contact time is measured by a microcomputer digital timer connected to a jumping platform and can be used to calculate power (W). Height attained during the jump and power (W·kg^{-1}) serves as the end result (Figure 36.13).

SQUAT 1RM

Proposed protocol

The athlete warms-up by performing 5–10 reps using a light weight. After a rest of 1 min ~10–20% of the warm-up weight is added and another 5–10 reps are performed (Figure 36.14). Following another 2-min rest, 2–3 reps of a heavy weight are performed. After this, the athlete should rest for 2–4 min and then attempt a repetition maximum (1RM) lift. If this is successful the athlete should have another 4 minutes rest before attempting a heavier weight for a 1RM.

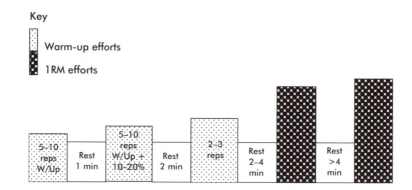

Key

Warm-up efforts

1RM efforts

Figure 36.14 Squat 1RM

This process should continue until the athlete fails to lift the weight successfully. The athlete should attempt a maximum of five 1RM lifts for assessment due to inevitable effects of fatigue. The last successful weight lifted should be counted as the 1RM (adapted from Earle, 1999).

LUGE

Luge is a sport in which its competitors lie supine on a four-foot by two-foot luge and slide, feet first, on an ice-coated track negotiating 13–15 high banking curves at speeds in excess of 90 mph. The individual with the lowest cumulative time (over 2 or 4 runs) wins the competition.

As with most sports involved with high speeds, technology and aerodynamics play a large part in an athletes pursuit of excellence. In order for the 'sliders' to complete the track in the fastest possible time good 'starting', 'riding' and 'driving' skills are vital.

The 'Start' is an explosive effort that is designed to propel the sled down the start ramp and into the first curve with the highest velocity possible. Force during the 'start' is produced in two phases, the first phase is the Pull phase where the slider holds two handles and with an explosive effort pulls the body and the sled past the handles and down the start ramp. This movement requires the sequential use of shoulder flexion and hyperextension, hip flexion followed by hip extension. The power-orientated movement is controlled as the thighs isometrically contract resulting in foot pressure on the front of the sled, all from a seated position.

The second or 'paddle' phase is initiated once the sled is in motion and past the starting handles. The slider then use spiked gloves to paw or paddle the ice in order to increase sled velocity. The supine riding position that is adopted reduces drag and maximises aerodynamics. The slider controls the body in order to negotiate the high banking curves and allow the sled to run freely on the straights.

Good driving and riding involves muscles of both the upper and lower extremities. Short or more sustained contraction of the hamstrings, latisimus dorsi and deltoids are performed for driving the sled and action of the abdominal and neck flexors give control the supine position on the sled. The movements required to ride and drive successfully can be forceful depending on the amount of steering required, however the movements are subtle and give the appearance of an effortless run.

Aerobic power

A sliders aerobic capacity can be assessed using a standard ramp test on a cycle or treadmill ergometer. Although not an essential parameter for performance, poor aerobic capacity will have implications on general help and the athlete's ability to perform the volume of training required and endurance the rigors or a competitive season.

Proposed protocol

Athletes should begin with a warm-up after which a lactate sample should be collected. This quantifies the starting lactate and should be under 2 mMol. The athlete is provided with a heart rate monitor and the connection to the online gas analysis set up using a face mask or mouthpiece and nose clip with head gear for stability. Gas analysis continues throughout the test until volitional exhaustion. At the end of the test another lactate sample is taken and maximal heart rate recorded. VO_{2max} is calculated as the VO_2 during the last 60 s of the test according to criteria set out in the previous section 'The assessment of maximal oxygen uptake'. The maximum speed and incline reached should also be noted and used to compare in future tests (Figure 36.15).

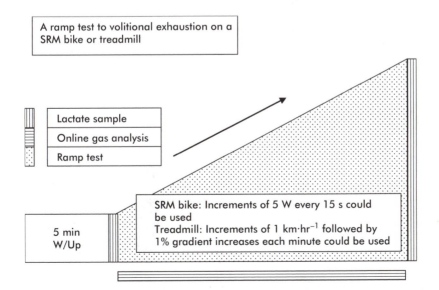

A ramp test to volitional exhaustion on a SRM bike or treadmill

Lactate sample

Online gas analysis

Ramp test

5 min
W/Up

SRM bike: Increments of 5 W every 15 s could be used
Treadmill: Increments of 1 km·hr^{-1} followed by 1% gradient increases each minute could be used

Figure 36.15 Aerobic power

UPPER/LOWER-BODY STRENGTH

The muscular strength and power of the upper and lower body are fundamental to the successful execution of the luge start and in controlling the movement of the sled during a run.

SQUAT/POWER CLEAN/BENCH PULL/ BENCH PRESS 1RM

Proposed protocol

The athlete warms-up by performing 5–10 reps using a light weight. After a rest of 1 min ~10–20% of the warm-up weight is added and another 5–10 reps are performed (Figure 36.16). Following another 2-min rest, 2–3 reps of a heavy weight are performed. After this, the athlete should rest for 2–4 min and then attempt a 1 RM lift. If this is successful the athlete should have another 4 min rest before attempting a heavier weight for a 1RM. This process should continue until the athlete fails to lift the weight successfully. The athlete should attempt a maximum of five 1RM lifts for assessment due to inevitable effects of fatigue. The last successful weight lifted should be counted as the 1RM (adapted from Earle, 1999).

10 m MAXIMAL PADDLE TEST

Proposed protocol

In order to assess the paddle phase of the start a 10 m maximal paddle tests using a wheeled luge sled can be carried out using timing gates positioned on an athletics track 10 m apart. Athletes are then assessed on the time it takes to break the beams between the timing gates. At least 5-min warm-up should be

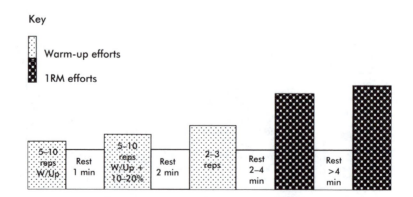

Figure 36.16 Squat/power clean/bench pull/bench press 1RM

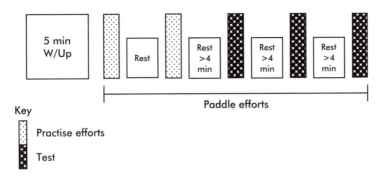

Figure 36.17 A 10 m maximal paddle test

completed followed by two practice paddle with adequate rest in between. At least 4 min rest should be given following these practice efforts before athletes perform a maximal 10 m paddle. Athletes could be given opportunity to repeat this test more than once with the best effort being counted. In this scenario, at least 4 min rest time should be given between each sprint (Figure 36.17).

FLEXIBILITY

Good flexibility within the lower back, hips and hamstrings are essential to allow the slider to achieve the correct posture during the compression, pull and paddle phases of the start.

SPINAL FORWARD FLEXION (SIT-AND-REACH)-BACK AND HAMSTRING TEST

Proposed protocol

The athlete begins by sitting with their feet on the sit and reach box with their feet together touching the scale on the front of the box. Hands should be together, with the index fingers parallel and fully extended and thumbs locked. The athlete reaches forward and touches the scale on the front of the box as far down as possible in a smooth motion. The knees must remain fully extended throughout the test. Athletes should be given opportunity to repeat this test three times with the best effort being counted (Figure 36.18).

SNOWBOARDING

Snowboarding made its World Cup debut in 1994 and then into the Olympic programme for the Nagano Games in 1998. The events are the Half Pipe and the Parallel Giant Slalom, with the Snowboard Cross being introduced into the up and coming Torino 2006 Olympic Games. The Half Pipe measures

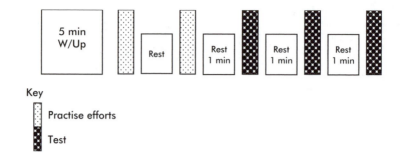

Key

Practise efforts

Test

Figure 36.18   Spinal forward flexion (sit-and-reach)-back and hamstring test

the athletes' ability to jump over the lip of a half-cylindrical field, and to cope with the side walls and then land back whilst performing manoeuvres and tricks. The entire performance is accompanied by the athletes' choice of music track. The Parallel Giant Slalom is a test of the athletes' alpine skills. The competition requires two rides to race down two identical parallel slopes. The vertical drop between the start and the finish is 120–200 m. There are two qualifying runs that lead to a 16-person tournament. The new event, Snowboard Cross, requires the athletes to be skilled in both alpine techniques and acrobatic techniques. The course is made up of a variety of terrains and obstacles with blue and red flags indicating the entrance to the obstacles. There are two qualifying runs in which 32 athlete's progress to the finals.

To date, there has been no research into the physiology and testing of snowboarders, however it can be suggested from factors involved in performance that the physiological demands of the Half Pipe and Snowboard Cross event are; agility, spatial awareness, leg strength, aerobic power, anaerobic power. The physiological demands of the Parallel Giant Slalom are agility, aerobic power, anaerobic power and leg strength. The test battery for Snowboarders is based on the most important parameters needed for performance as mentioned earlier.

Agility

Core stability

Poor core stability affects take off, landing, rotation, air positions and efficiency of movement. A test to monitor the development of boarder's abdominal and lower back muscles is to assume the 'plank' position which takes on the basic press up position with elbows on the ground. To analyse progress, duration held can be compared and a video analysis system utilised to assess position.

Aerobic power

Although limited research has been conducted looking at the cardiovascular demands of snowboard cross and the halfpipe, there are similarities in the demands of snowboard cross with that of alpine skiing's super G and giant slalom which have been demonstrated to require a good aerobic power.

Proposed protocol

Athletes should begin with a warm-up after which a lactate sample should be collected. This quantifies the starting lactate and should be under 2 mMol. The athlete is provided with a heart rate monitor and the connection to the online gas analysis set up using a face mask or mouthpiece and nose clip with head gear for stability. Gas analysis continues throughout the test until volitional exhaustion. At the end of the test another lactate sample is taken and maximal heart rate recorded. VO_{2max} is calculated as the VO_2 during the last 60 s of the test according to criteria set out in the previous section 'The assessment of maximal oxygen uptake'. The maximum speed and incline reached should also be noted and used to compare in future tests (Figure 36.19).

ANAEROBIC POWER

In events such as the halfpipe and snowboard cross, where the ability to repeatedly produce high levels of force is a fundamental component of performance, measures of anaerobic power provide valuable insight and information into an athletes physiological profile.

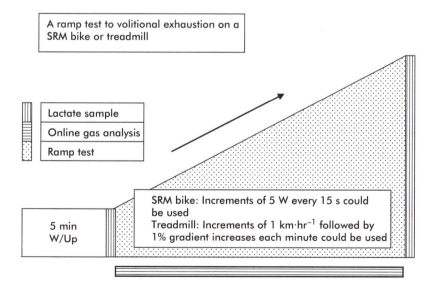

Figure 36.19   Aerobic power

VERTICAL JUMP TEST: REPEATED EFFORTS OVER 30 s

Proposed protocol

Repeated jump test: Athletes should perform a warm-up, practise jumps and rest as discussed earlier. Following this, the athlete must jump maximally and repeatedly for 60 s with good technique. After this test a fatigue index may be calculated from the results by calculating the mean difference of the first three jumps and last three jumps, or, alternatively, the per cent decrease in jump height between the best and worst jump. Flight time and contact time can be measured by a microcomputer digital timer connected to a jumping platform and used to calculate average power (Figure 36.20).

Laboratory tests for maximal anaerobic power and capacity are most relevant to the athlete when they simulate the actual mode of exercise and involve the specific muscle groups used in the sport (MacDougall *et al.*, 1991). The specific nature of repeated jumping test utilising elastic energy and the stretch-shortening cycle, a characteristic of board slalom and the halfpipe.

WINGATE TEST: 30 s SPRINT

Proposed protocol

The test can be performed on a Monark cycle ergometer. A 5-min warm-up should be completed after which the athlete should perform a practise start. This allows the athlete to experience the inertia of the fly wheel in preparation for the test. A load equivalent to 8% and 6% bodyweight should be added to the cycle ergometer for males and females, respectively. For the test, the athlete should be instructed to cycle maximally for the full 30 s (Figure 36.21).

Measurements

Peak power (W), peak power relative to bodyweight (W·kg^{-1}) and time taken to reach peak power (s) and fatigue index (FI) achieved during the test should be recorded.

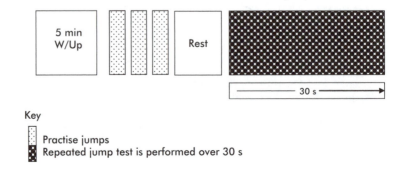

Key

Practise jumps
Repeated jump test is performed over 30 s

Figure 36.20   Vertical jump test: repeated efforts over 30 s

Although the mode of exercise during the Wingate is not specific to skiing, the musculature involved is specific to skiing and has been used widely for the assessment of alpine skiers.

LOWER BODY STRENGTH/POWER

In order to perform the radial turns during snowboard cross and achieve the vertical heights necessary to perform the aerial manoeuvres in the halfpipe event, high levels of muscular force must be generated by the lower extremities.

ISOKINETIC STRENGTH

Isokinetic strength of the knee extensor and flexor muscles could also be assessed using an isokinetic dynamometer such as the Biodex or Cybex system. Concentric muscular strength is measured at three different angular velocities ($60°$s, $180°$s and $300°$s) over four repetitions with a rest interval of 60 s between speeds.

VERTICAL JUMP TEST: SINGLE EFFORT

Proposed protocol

Single vertical jump test: A 5-min warm-up would precede three practise jumps. Prior to these practises the tester must make it clear which 'Vertical Jump' technique is required (the static, countermovement or countermovement with arms) and coach the athlete accordingly until the tester is satisfied. At least 4 min rest should then be allowed prior to the test effort. Flight time and contact time is measured by a microcomputer digital timer connected to a jumping platform

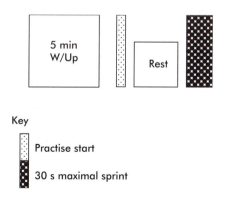

Key

Practise start

30 s maximal sprint

Figure 36.21 Wingate test: 30 s sprint

and can be used to calculate power (W). Height attained during the jump and power (W·kg^{-1}) serves as the end result (Figure 36.22).

SQUAT 1 RM

Proposed protocol

The athlete warms-up by performing 5–10 reps using a light weight. After a rest of 1 min ~10–20% of the warm-up weight is added and another 5–10 reps are performed (Figure 36.23). Following another 2-min rest, 2–3 reps of a heavy weight are performed. After this, the athlete should rest for 2–4 min and then attempt a 1RM lift. If this is successful the athlete should have another 4 min rest before attempting a heavier weight for a 1RM. This process should continue until the athlete fails to lift the weight successfully. The athlete should attempt a maximum of five 1RM lifts for assessment due to inevitable effects of fatigue. The last successful weight lifted should be counted as the 1RM (adapted from Earle, 1999).

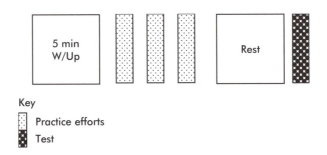

Figure 36.22 Vertical jump test: single effort

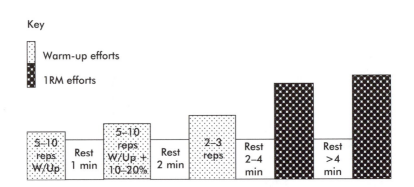

Figure 36.23 Squat 1RM

SPEED SKATING

Short-track speed skating was first introduced at the 1992 Alberville Winter Olympic Games as an exhibition event. Competition event distance in short track consist of 500 m, 1,000 m, 1,500 m and 3,000 m, athlete's generally compete in all of the distances, although they may specialise in a particular distance.

Short-track speed skating is a very technical sport, thus skill and technique play a vital role in performance, along with a high level of aerobic endurance, strength, power muscular endurance and anaerobic capacity. It is suggested that the high lactate values observed are in part due to the restricted muscle blood flow when skating due to the high intramuscular forces intrinsic to the unique posture assumed by speed skaters and/or to the prolonged duty cycle of the skating stroke (Foster *et al.*, 1999).

Laboratory-based assessments for aerobic capacity are generally performed on cycle ergometers, as off-ice conditioning is performed on road and mountain bicycles and researchers have suggested that the correlation between cycling and skating performance is closer than other forms of laboratory assessments, for example treadmill running.

AEROBIC POWER

Proposed protocol

Athletes should begin with a warm-up after which a lactate sample should be collected. This quantifies the starting lactate and should be under 2 mMol. The athlete is provided with a heart rate monitor and the connection to the online gas analysis set up using a face mask or mouthpiece and nose clip with head gear for stability. Gas analysis continues throughout the test. Capillary blood samples for blood lactate (BLa) analysis are taken at the end of each 4 increment until a clear break point can be observed (BLa >4 mM). Following a 15-min rest period a ramp test of increments of 15 W every 15 s is performed until volitional exhaustion. At the end of the test another lactate sample is taken and maximal heart rate recorded. VO_{2max} is calculated as the VO_2 during the last 60 s of the test according to criteria set out in the previous section 'The assessment of maximal oxygen uptake'. The maximum minute power is calculated as the average power maintained over the final 60 s of the test (Figure 36.24).

The importance of collecting these lactate values is that the effectiveness of training can be assessed by comparing responses at standardised work loads (Wilmore and Costill, 1988) and blood lactate and heart rate profiling also allows the establishment of training zones (Martin and Coe, 1997).

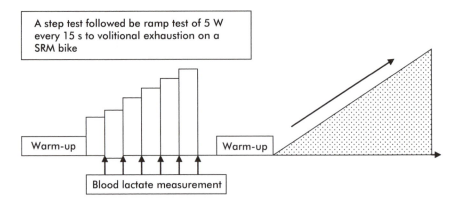

A step test followed be ramp test of 5 W every 15 s to volitional exhaustion on a SRM bike

Warm-up

Warm-up

Blood lactate measurement

Figure 36.24 Aerobic power

ANAEROBIC CAPACITY

The high level of muscular forces required during the prolonged skating stroke and the reduced blood flow from the skating posture, maintaining a high force output utilising anaerobic energy systems is fundamental.

VERTICAL JUMP TEST: REPEATED EFFORTS OVER 60 s

Proposed protocol

Repeated jump test: Athletes should perform a warm-up, practise jumps and rest as discussed earlier. Following this, the athlete must jump maximally and repeatedly for 60 s with good technique. After this test a fatigue index may be calculated from the results by calculating the mean difference of the first three jumps and last three jumps, or, alternatively, the per cent decrease in jump height between the best and worst jump. Flight time and contact time can be measured by a microcomputer digital timer connected to a jumping platform and used to calculate average power (Figure 36.25).

Laboratory tests for maximal anaerobic power and capacity are most relevant to the athlete when they simulate the actual mode of exercise and involve the specific muscle groups used in the sport (MacDougall *et al.*, 1991).

WINGATE TEST: 60 s SPRINT

Proposed protocol

The test can be performed on a Monark cycle ergometer. A 5-min warm-up should be completed after which the athlete should perform a practise start. This allows the athlete to experience the inertia of the fly wheel in preparation

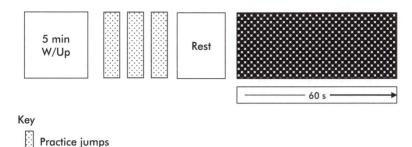

Key

Practice jumps
Repeated jump test is performed over 60 s

Figure 36.25   Vertical jump test: repeated efforts over 60 s

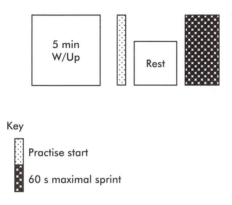

Key

Practise start

60 s maximal sprint

Figure 36.26   Wingate test: 60 s sprint

for the test. A load equivalent to 8% and 6% bodyweight should be added to the cycle ergometer for males and females, respectively. For the test, the athlete should be instructed to cycle maximally for the full 60 s (Figure 36.26).

Measurements

Peak power (W), peak power relative to bodyweight ($W \cdot kg^{-1}$) and time taken to reach peak power (s) and fatigue index (FI) achieved during the test should be recorded.

Although the mode of exercise during the Wingate is not specific to skating, the musculature involved is specific to skating and has been used previously in the assessment of short-track speed skaters.

LOWER BODY STRENGTH/POWER

In order to produce the speed necessary during races and oppose the large centrifugal forces imposed, the musculature of the body must be able to produce a high level of force and produce that power explosively.

ISOKINETIC STRENGTH

Isokinetic strength of the knee extensor and flexor muscles could also be assessed using an isokinetic dynamometer such as the Biodex or Cybex system. Concentric muscular strength is measured at three different angular velocities (60°s, 180°s and 300°s) over four repetitions with a rest interval of 60 s between speeds.

VERTICAL JUMP TEST: SINGLE EFFORT

Proposed protocol

Single vertical jump test: A 5-min warm-up would precede three practice jumps. Prior to these practices the tester must make it clear which 'Vertical Jump' technique is required (the static, countermovement or countermovement with arms) and coach the athlete accordingly until the tester is satisfied. At least 4 min rest should then be allowed prior to the test effort. Flight time and contact time is measured by a microcomputer digital timer connected to a jumping platform and can be used to calculate power (W). Height attained during the jump and power (W·kg^{-1}) serves as the end result (Figure 36.27).

SQUAT 1RM

Proposed protocol

The athlete warms up by performing 5–10 reps using a light weight. After a rest of 1 min ~10–20% of the warm-up weight is added and another 5–10 reps are performed (Figure 36.28). Following another 2 min rest, 2–3 reps of a heavy weight are performed. After this, the athlete should rest for 2–4 min and then attempt a 1RM lift. If this is successful the athlete should have another 4 min rest before attempting a heavier weight for a 1RM. This process should continue until the athlete fails to lift the weight successfully. The athlete should attempt a maximum of five 1RM lifts for assessment due to inevitable effects of fatigue. The last successful weight lifted should be counted as the 1RM (adapted from Earle, 1999).

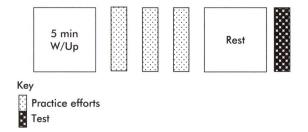

Figure 36.27 Vertical jump test: single effort

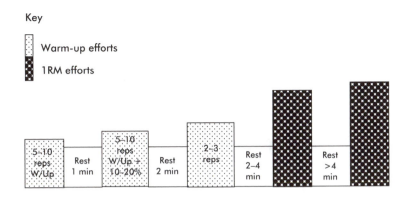

Figure 36.28 Squat 1RM

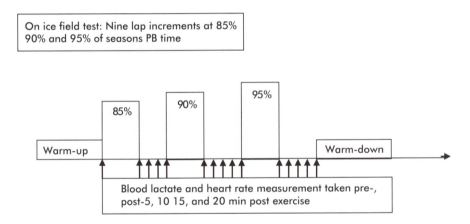

Figure 36.29 On-ice assessment

ON-ICE ASSESSMENT

On-ice assessments have been developed in order to assess the skater's sports-specific fitness and the training progress of the individual athletes throughout the season as well as from year to year.

Proposed protocol

Developed in association with the British Short track speed skating coaching staff, heart rate and lactate are measured in response to performing nine laps of the track at 80%, 90% and 95% of individuals seasons personal best time. Recovery profile is measured at 5, 10, 15 and 20 min post-exercise (Figure 36.29).

CHAPTER 37

FIGURE SKATING

Chris Byrne and Roger Eston

BACKGROUND

Figure skating disciplines

The International Skating Union (ISU, 2004a,b) recognises the following branches of international figure skating: (1) single skating; (2) pair skating; (3) ice dancing; and (4) synchronised skating. The former three branches are separate disciplines in all ISU Senior and Junior Championships and the Olympic Winter Games, and will therefore be the focus of these testing guidelines.

Duration and technical requirements of competitive figure skating

The duration and technical composition of competitive figure skating is dependent on the skating discipline and the programme being performed. In single and pair skating, competitors are required to undertake a short programme with eight required elements (e.g. single jumps, combination jumps, spins and connecting steps) in a maximum duration of 2-min 50-s and, on a separate occasion, a free skating programme (i.e. skating to unspecified movements to music chosen by the skater) within ±10-s of 4-min 30-s for senior men and pairs, and 4-min for senior women (ISU, 2004a). Both males and females are required to incorporate triple rotation jumps, with males having the option of incorporating quadruple rotation jumps. In addition, pair skating requires throw-jumps and overhead lifts and holds. Ice dancing requires competitors to complete three skating programmes (i.e. compulsory dance, original dance and free dance) on separate occasions, whose duration and technical composition will be dependent on the prescribed or chosen music. The duration of the free dance programme is fixed at 4-min for seniors and 3-min for juniors, whereas the compulsory and original dance duration are designated annually by the ISU (ISU, 2004b).

Physiological demands of competitive figure skating

Very little data exists quantifying the physiological demands of competitive figure skating and the physiological characteristics of elite skaters (Kjaer and Larsson, 1992). Our own measurements of heart rate during simulated on-ice competition on members of the UK National Ice Skating Association (NISA) senior squad in 1999 revealed that skaters achieved and maintained exercise intensities between 85% and 100% of maximal heart rate (%HR$_{max}$) in all skating disciplines. Figure 37.1 illustrates the mean response of six single skaters performing their free skating programme (4-min to 4-min 30-s duration). Their average exercise intensity was (mean ± SEM) 94 ± 1%HR$_{max}$. Equally high intensities were observed in single skaters performing their short programme (92 ± 2%, $n = 4$) and dancers performing their free dance programme (92 ± 2%, $n = 2$). These observations are in agreement with the results of Kjaer and Larsson (1992) who reported intensities of 92 ± 3% and 93 ± 3%HR$_{max}$ in eight elite Danish single skaters for short and free skating programmes, respectively. In addition, our measurements of blood lactate concentration taken 3 min after simulated competition revealed concentrations of 7.7 ± 0.4 ($n = 9$) and 8.6 ± 0.4 ($n = 7$) mmol·l^{-1} for single skaters short and free programmes, respectively. These observations are in close agreement with the results of Kjaer and Larsson (1992) who reported blood lactate

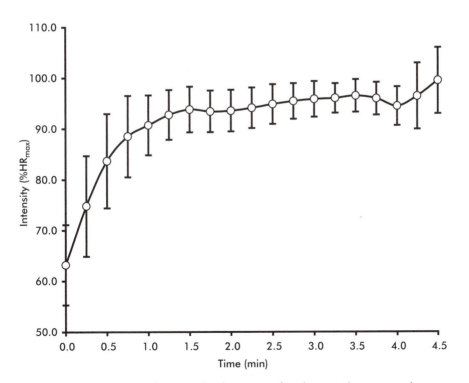

Figure 37.1 Exercise intensity during simulated competitive free skating. Values represent heart rate (mean ± SEM) of six (three males, three females) NISA senior single skaters expressed as a percentage of maximal heart rate achieved during cycle ergometer exercise testing in the laboratory

concentrations of 9.0 ± 1.3 and 7.4 ± 0.6 mmol·l^{-1}, 1-min after simulated competition free skating in males and females, respectively. For ice dancing we recorded blood lactate concentrations of 7.0 ± 0.8 ($n = 4$) and 8.1 ± 0.5 mmol·l^{-1} ($n = 4$) following original dance and free dance programmes, respectively.

Implications for performance, training and physiological testing

Although data quantifying the physiological demands of competitive figure skating are limited, the implications are clear: figure skaters are required to attain and maintain an intensity of exercise in the domain *severe sub-maximal* ($\leq$VO$_{2max}$) for up to 4–4.5-min, with intermittent periods of *supramaximal* ($>$VO$_{2max}$) intensity exercise (Mannix *et al.*, 1996; Jones and Doust, 2001). Therefore, training and physiological testing should focus on developing the power and fatigue resistance of the glycolytic and aerobic energy systems. Furthermore, the attainment of skating speed and jump height, the ability to lift and hold a skating partner in an artistic position or produce throw jumps requires significant muscle strength and power for successful execution. Therefore, training and physiological testing should also focus on developing muscle strength and power through the ATP-PC energy system. Finally, body composition will play a mediatory role in affecting the development of the above physiological qualities and, along with flexibility, will make an aesthetic contribution to performance in this subjectively judged sport.

PHYSIOLOGICAL TESTS FOR FIGURE SKATING

The physiological tests described in the following section are standard laboratory tests of physiological function, exercise performance and anthropometry relating to maximal oxygen uptake, maximal intensity exercise performance, vertical jump performance, body composition and flexibility. Table 37.1 contains normative data obtained from NISA senior squad members in 1999.

Maximal oxygen uptake, peak exercise intensity and lactate transition thresholds

Rationale

A high-level of aerobic power is an essential requirement for competitive figure skaters (Kjaer and Larsson, 1992): allowing the maintenance of sub-maximal exercise intensities with less fatigue throughout a skating programme and the attainment of a high power output at VO$_{2max}$ (Mannix *et al.*, 1996). An incremental test to exhaustion combining measurements of VO$_2$, blood lactate and heart rate at each power output (exercise intensity) will serve as an indicator of training adaptation and will also indicate optimal training intensities.

Table 37.1   Normative data obtained from laboratory testing of NISA senior squad members in 1999. Values are means ± SEM

	Males (n = 8)	Females (n = 8)
Anthropometry		
Age range (yr)	17–22	14–22
Height (m)	1.75 ± 0.02	1.60 ± 0.02
Mass (kg)	66.4 ± 2.4	51.5 ± 2.0
Sum of skinfolds (mm)	32 ± 4	47 ± 7
Fat (%)	8.5 ± 1.3	18.7 ± 2.2
Fat mass (kg)	5.8 ± 1.0	10.1 ± 1.5
Lean body mass (kg)	60.6 ± 1.8	42.8 ± 1.3
Aerobic power		
VO_{2max} (ml·kg·min^{-1})	52.6 ± 2.7	50.4 ± 3.5
Peak work rate (W)	288 ± 8	226 ± 16
Wingate 30 s cycle test		
Peak power (W)	814 ± 31	511 ± 31
Relative peak power (W·kg^{-1})	12.3 ± 0.4	10.0 ± 0.5
Mean power (W)	697 ± 28	419 ± 18
Relative mean power (W·kg^{-1})	10.5 ± 0.3	8.2 ± 0.3
Fatigue (%)	30 ± 3	31 ± 2
Vertical jump height (cm)		
Without use of arms		
Two-legged	39.4 ± 1.9	31.9 ± 2.6
Left-leg	21.8 ± 1.1	20.0 ± 2.1
Right-leg	23.1 ± 1.1	20.5 ± 2.8
With use of arms		
Two-legged	47.3 ± 1.6	37.0 ± 2.5
Left-leg	26.7 ± 1.6	24.5 ± 2.2
Right-leg	27.0 ± 0.9	23.2 ± 2.1
Flexibility (degrees)		
Hip flexion		
Left	118 ± 6	140 ± 8
Right	114 ± 5	140 ± 8
Hip extension		
Left	57 ± 2	75 ± 4
Right	55 ± 2	71 ± 6

Notes
Sum of skinfold sites: males = pectoral, abdomen, anterior mid-thigh (Jackson and Pollock, 1978); females = suprailium, triceps, anterior mid-thigh (Jackson et al., 1980)

Test procedure

Following familiarisation and a 5-min warm-up, the athlete completes a continuous incremental test of 3-min stages to exhaustion on a cycle ergometer. Oxygen uptake is measured continuously or over the final minute of each stage with heart rate and blood lactate measured in the last 30-s of each stage. The starting power output is 25 W with 25–50 W increments and a fixed pedal cadence of 60 revolutions per minute.

Data analysis

Figure 37.2 illustrates the power output–blood lactate response of an individual female skater to the incremental cycle test before and after the prescription of off-ice moderate intensity continuous and high-intensity interval training. A shift down and/or to the right in the power output-blood lactate plot is representative of an increase in sub-maximal exercise economy and/or aerobic power. After training, the test provides objective evidence of endurance training adaptations: the skater produces less blood lactate for a given power output

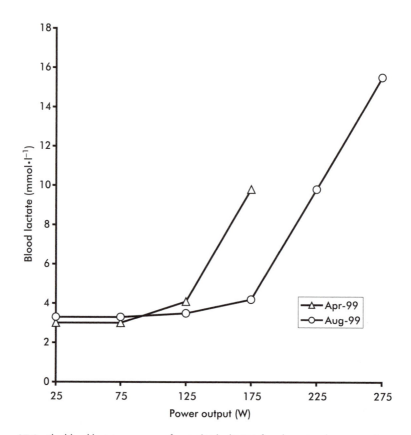

Figure 37.2 The blood lactate response of an individual NISA female senior skater to an incremental cycle ergometer test to exhaustion before and after a three-month off-ice moderate-intensity continuous and high-intensity interval training programme

(exercise intensity) and is able to attain greater power outputs (exercise intensities) during the latter stages of the test, resulting in a peak power output at VO_{2max} of 275 W after training Vs. 175 W before training.

Normative data

Table 37.1 provides normative data for VO_{2max} and peak power output in NISA senior squad members gained from the continuous incremental cycle ergometer test described previously.

Maximal intensity exercise: Wingate 30 s cycle test

Rationale

The Wingate test is a maximal-intensity cycle test of 30-s duration, which assesses power output of the legs. Explosive leg power is essential to provide elements of skating performance such as skating speed and jump height. The all-out 30-s test lacks specificity to skating performance and a more sport-specific test reflecting the intermittent nature of maximal intensity exercise in figure skating may be the 5×6-s cycle ergometer test involving five repetitions of 6-s sprinting with 24-s recovery separating each sprint. The standard Wingate test may also fail to satisfy muscle force–velocity relationships resulting in sub-optimal load, pedalling rate and peak power output. Alternative methods of assessing maximal intensity exercise are described by Winter and Maclaren (2001), see Chapter 14.

Test procedure

A basket-loading cycle ergometer linked to a computer for flywheel data logging is recommended (e.g. Monark 814E). Athletes undertake a 5-min cycling warm-up at 100 W, with a flat-out sprint at 3 min for 5-s, followed by 5 min of rest. An external load corresponding to 7.5% of body mass is applied to the basket. With the load supported, athletes begin pedalling at 50–60 revolutions per minute: upon the command '3, 2, 1, Go', the load is applied abruptly and the athlete pedals flat-out for 30-s. The test finishes with a warm-down of 2-min cycling at 100 W.

Data analysis

Three main variables are extracted from the test (see Table 37.1). Peak power represents the highest power output attained during the test; typically in the first 5 s. Mean power is the average power over the 30 s and fatigue represents the percentage difference between peak power and the lowest power output recorded during the test. Training-induced increases in the ability to attain and maintain high power outputs are likely to be reflected in the peak and mean power variables. However, fatigue may increase due to an increase in peak power without a corresponding increase in the minimum power recorded during the test.

Normative data

Table 37.1 illustrates Wingate test data from NISA senior squad figure skaters expressed in absolute values and relative to body mass.

Vertical jump performance

Rationale

Jump height represents a cornerstone of figure skating performance for single and pair skaters with potential to affect technical merit and artistic impression (Podolsky *et al.*, 1990). The test also represents a simple measure of explosive leg power in dancers.

Test procedure

The following instructions refer to the measurement of countermovement jump height. Commercial electronic timing mat systems (e.g. Ergo Tester, Globus, Italy) conveniently assess jump height by measuring flight time, which is triggered by the feet of the athlete leaving the mat and stopped at the instant of contact upon landing. Timing mat computations assume the jumper's position on the mat is the same at take-off and landing, therefore athletes are instructed to avoid deviating from the vertical plane (i.e. no horizontal movement), to keep their body erect throughout the jump, and to make initial contact with the toes upon landing. Upon the verbal command 'Go' athletes move from an erect standing position and make a downward countermovement to a knee angle of 90°, followed in one continuous movement by a vertical jump for maximum height. The countermovement jump may be performed with or without arm involvement and with single or double leg take-off. Arm involvement can simply be removed by asking athletes to place their hands on their hips throughout the jump. Single-legged jumps with arm involvement offer greater specificity to on-ice jumping.

Normative data

Table 37.1 provides countermovement jump height with and without arm involvement in a mixed sample of single skaters and dancers.

Body composition

Rationale

Body composition is of importance from both a functional and aesthetic standpoint. A high percentage of body fat will reduce relative aerobic power, relative maximal power output and jump height.

Test procedure

Skinfolds provide a valid indication of the levels of fatness located over the body in the subcutaneous storage areas. Methods and standards of the International Society for the Advancement of Kinathropometry (ISAK) should be adhered to (see Chapter 14).

Data analysis

Sum of skinfolds or calculation of body density and proportions of fat and lean mass can be used to track changes in body composition over time.

Normative data

Table 37.1 provides anthropometric data for NISA senior squad members. The results are similar to those reported by Kjaer and Larsson (1992) for elite Danish single skaters. Their percent body fat values were $7 \pm 1\%$ and $17 \pm 1\%$ for 3 males and 5 females, respectively.

Flexibility

Rationale

Flexibility is defined as the range of motion at a single joint or a series of joints (Borms and van Roy, 2001). Good flexibility is considered essential for sports such as figure skating, which require maximum amplitude of movement for technique to be executed optimally (Borms and van Roy, 2001).

Test procedure

Detailed methods for flexibility assessment are provided by Borms and van Roy (2001), see Chapter 14. A Leighton Flexometer was used to gain the active hip flexion and extension data presented in Table 37.1. For hip flexion, the subject lies supine on a firm bed or table and performs a complete active flexion in a pure saggital plane. The active leg may be flexed at the knee to avoid passive insufficiency of the hamstrings (Borms and van Roy, 2001); although a straight leg was used in our data collection to increase the sport-specificity of the test. Restraint is required for the opposing leg (e.g. Velcro strap or manual restraint) to ensure that the movement is not continued in the spine (Borms and van Roy, 2001). For hip extension, the subject lies prone at the end of a firm bed or table with their feet on the floor. A maximal active extension of the hip is performed with one leg whilst the opposite leg remains in position.

Data analysis

Readings are taken at the maximum movement amplitude.

Normative data

Table 37.1 provides hip range of motion data measured from NISA senior squad members in 1999.

REFERENCES

Borms, J. and van Roy, P. (2001). Flexibility. In R.G. Eston and T. Reilly (eds), *Kinanthropometry and Exercise Physiology Laboratory Manual: Tests, Procedures and Data*, 2nd edn, *Volume 1: Anthropometry*, pp. 117–147. Routledge, London.

International Skating Union (ISU) (2004a). Special Regulations. Single and Pair Skating. International Skating Union Official Website. Accessed 29/09/2005. http://www.isu.org/vsite/vfile/page/fileurl/0,11040,4844-160824-178039-80283-0-file,00.pdf.

International Skating Union (ISU) (2004b). Special Regulations. Ice Dancing. Accessed 29/09/2005. http://www.isu.org/vsite/vfile/page/fileurl/0,11040,4844-161655-178870-81791-0-file,00.pdf.

Jackson, A.S. and Pollock, M.L. (1978). Generalized equations for predicting body density of men. *British Journal of Nutrition*, 40: 497–504.

Jackson, A.S., Pollock, M.L. and Ward, A. (1980). Generalized equations for predicting body density of women. *Medicine and Science in Sports and Exercise*, 12: 175–182.

Jones, A.M. and Doust, J.H. (2001). Limitations to submaximal exercise performance. In R.G. Eston and T. Reilly. (eds), *Kinanthropometry and Exercise Physiology Laboratory Manual: Tests, Procedures and Data*, 2nd edn, *Volume 2: Exercise Physiology*, pp. 235–262. Routledge, London.

Kjaer, M. and Larsson, B. (1992). Physiological profile and incidence of injuries among elite figure skaters. *Journal of Sports Sciences*, 10: 29–36.

Mannix, E.T., Healy, A. and Farber, M.O. (1996). Aerobic power and supramaximal endurance of competitive figure skaters. *Journal of Sports Medicine and Physical Fitness*, 36: 161–168.

Podolsky, A., Kaufman, K.R., Cahalan, T.D., Aleshinsky, S.Y. and Chao, E.Y.S. (1990). The relationship of strength and jump height in figure skaters. *American Journal of Sports Medicine*, 18: 400–405.

Winter E.M. and Maclaren, D. (2001). Maximal intensity exercise. In R.G. Eston and T. Reilly (eds), *Kinanthropometry and Exercise Physiology Laboratory Manual: Tests, Procedures and Data*, 2nd edn, *Volume 2: Exercise Physiology*, pp. 263–288. Routledge, London.

PART 5

SPECIAL POPULATIONS

CHAPTER 38

CHILDREN AND FITNESS TESTING

Gareth Stratton and Craig A. Williams

RATIONALE

There are a number of reasons why guidelines specific for children should be created instead of adopting adult-based ones. These reasons include differences between children and adults in:

- ethics
- informed consent
- the physiological differences due to body size
- the impact of growth and maturation
- the need for a different laboratory environment.

In the last 20 years there has been a proliferation of testing protocols which has resulted in an increasing amount of data related to children's physiology. For the purposes of these guidelines we are delimiting the definition of a child as below 18 years. These guidelines are designed to recommend accurate techniques in measuring physical and physiological parameters in children and can be adopted for sporting or research purposes.

TESTING MODALITIES

Field and laboratory tests of fitness and performance represent the two main modalities available to the paediatric exercise scientist. Both are widely used, although field tests are commonly used as part of fitness education in schools. The choice of test depends on a number of factors such as cost, expedience, accuracy and tester experience. Field tests are limited because they provide no direct physiological data but more accurately assess 'motor performance'.

The advantage of field tests are that they require relatively inexpensive equipment, personnel involved in testing need less training and they can be performed with large sample sizes in readily available facilities, for example, sports halls. Hence field tests are convenient for large epidemiological studies of the population. Laboratory tests however are more expensive, require specialist facilities and staff, but produce more sensitive and precise 'physiological' data and a greater insight into biological mechanisms. Field tests are a useful tool available to coaches, teachers and allied health professionals, and although the tests have been criticised for their use in schools they are useful in tracking changes in whole population studies.

PARTICIPATION OF THE CHILD IN A PROJECT

The Medical Research Council (MRC) have stated that children may participate in research projects which have a therapeutic or a non-therapeutic benefit which does not necessarily benefit the child involved. However, children's involvement in testing must involve 'negligible risk' defined as no greater than risks of harm ordinarily encountered in daily life. The following test procedures are considered negligible risk:

- Observation of behaviour
- Non-invasive physiological monitoring
- Developmental assessments and physical examinations
- Changes in diet
- Obtaining blood and urine samples (Medical Research Council, 1991).

For exercise testing in England and Wales, children under the age of 18 cannot legally consent to participate in exercise tests and therefore parental or guardian's consent is essential. Recently, obtaining children's assent has become accepted practice alongside, but not in place of parental consent, as a safeguard to ensure the child is not being coerced into a project by their parents (Jago and Bailey, 1998; Williams, 2003). Explanation of the details of procedures should allow for the wide range of intellectual capabilities of the participating children. Common to all ethical procedures for research should be the emphasis that the child is free to withdraw at any time.

The Protection of Children Act 1999 (DoH, 2000) requires all activities involving children to have a responsibility to protect their welfare. Where children are involved in testing without the presence of a parent or guardian those responsible for testing are acting 'in loco parentis' (*in the place of the parent*). It is now common procedure that anyone without a Criminal Records Bureau Enhanced Disclosure should not be left in sole charge of children (DfES). It is also recommended that testers are not left in a one to one situation with children including during transport arrangements.

At the location of testing (field or laboratory) it is important to ensure such details as the following have been arranged:

- Participation health questionnaire including, if necessary, detailed information on medical, special educational, cultural and nutritional needs.
- Clear establishment of who is in charge.
- Check your insurance covers working with children.
- Establish acceptable conduct for children, that is, children cannot go off on their own anywhere, what are you going to do if a child becomes uncooperative.
- Details in the event of an emergency, for example, contact details of school, head teacher or classroom teacher, parent/guardian.

ASSESSING MATURATION

The assessment of physiological and physical changes during growth is essential for the valid interpretation of human performance.

Maturation can be assessed in a variety of ways and until recently these have been mainly invasive and ethically questionable. A number of scientists (Greulich and Pyle, 1959; Roche, 1988) developed similar methods for assessing, 'skeletal maturation'. These involved an X-ray of the left wrist and hand where constituent bones were assessed for their stage of ossification against a developmental atlas. Whilst this approach is the gold standard for skeletal maturation it is limited for two important reasons: (1) Measurement requires advanced technical expertise and (2) youngsters receive a dose of radiation during each assessment.

One non-invasive method to assess maturation is 'morphological age' which calculates the percentage of predicted adult stature. This is a simple technique to use but is limited by the need for the exact stature of both biological parents and it is not valid for children under 10 years of age. Furthermore, skeletal age is the only measure of maturation that can be applied from infant to adulthood whereas morphological age is only valid between 10 and 18 years of age. Unfortunately neither skeletal nor morphological age is able to adequately predict pubertal stage.

The most practical approach to assess maturation was developed by Tanner (1962). Tanner developed a 5-point scale to assess 'biological maturity' through observation of secondary sexual characteristics. The scales depicted five or more stages of breast and pubic hair (girls) pubic hair and genitalia development (boys). A limitation is that trained health professionals such as paediatricians and school nurses are typically employed to assess the scales. Subsequently Morris and Udry (1980) developed a self-assessment scale based on Tanner stages and found that children were able to accurately assess their own stage of maturation with correlation coefficients in the range of 60–70% (Matsudo and Matsudo, 1993).

Mirwald and colleagues (2001) developed a technique that uses anthropometrical data to calculate maturity offset and thus avoids ethical and technical complexities found in other techniques. This approach has gained widespread approval, as it only requires decimal age and simple measures of body mass, stature and sitting height. Leg length is also required but is calculated by subtracting sitting height from stature. These data are then substituted into a regression equation and distance in time before or after peak height velocity is calculated. This method has an accuracy of ±0.4 years.

Males

Maturity offset $= -9.236 + (0.0002708 \times$ (leg length $\times$ sitting height)) $+ (-0.001663 \times$ (age $\times$ leg length)) $+ (0.007216 \times$ (age $\times$ sitting height)) $+ (0.02292 \times$ (mass by stature ratio)).

Females

Maturity offset $= -9.376 + (0.0001882 \times$ (leg length $\times$ sitting height)) $+ (-0.0022 \times$ (age $\times$ leg length)) $+ (0.005841 \times$ (age $\times$ sitting height)) $+ (0.02658 \times$ (age $\times$ mass)) $+ (0.07693 \times$ (mass by stature ratio)).

In these equations age is measured in decimal years, lengths in centimetres and mass in kilograms.

The technique proposed by Mirwald and Bailey is to be recommended, although morphological age and self-assessment of maturity are also acceptable methods.

ANTHROPOMETRY AND BODY COMPOSITION

Anthropometrical measures are important during growth. Typical growth curves of stature and mass are widely used by health professionals to assess a child's growth status against normative values (Child Growth Foundation). The basic principles of anthropometrics are the same for children and adults. The key

Sex	Maturation level	Estimated per cent body fat
Male	Prepubertal	25.56 (log S_4) − 22.23
	Pubertal	18.70 (log S_4) − 11.91
	Post-pubertal	18.88 (log S_4) − 15.58
Female	Prepubertal	29.85 (log S_4) − 25.87
	Pubertal	23.94 (log S_4) − 18.89
	Post-pubertal	39.02 (log S_4) − 43.49

Note
Where S_4 is the log (base 10) of the sum of four skinfolds in mm

differences are related to the approaches of the measurement, analysis and interpretation of data. For example, special regression equations have been used for the conversion of the sum of four skinfolds (biceps, triceps, subscapular and iliac crest) to percentage body fat for circumpubertal children (Deurenburg et al., 1990).

These equations are different to those reported for use with adults (Durnin and Womersley, 1974) as children's body density changes with age from about 1.08 g·ml^{-1} at age 7 to 1.10 g·ml^{-1} at age 18 (Westrate and Deurenberg, 1989).

Clearly changes in body density will affect calculation of per cent body fat when using hydrostatic weighing. Lohman (1984) has reported an alternative to the Siri's equation for 8–12-year-old girls and boys that accounts for this change. Subsequently per cent body fat can be calculated using the following equations:

% fat $= (530/D) - 489$ (Lohman, 1984)
% fat $= ((5.62 - 4.2 \ (age - 2))/D) - (525 - 4.7 \ (age - 2))$ (Westrate and Deurenburg, 1989).

Skin folds are still a valid method for assessing adiposity in children. However, changes in the density of subcutaneous fat and other body tissues mitigate converting skinfold measures to percentage body fat. Therefore using the 'sum of skinfolds' technique is the most appropriate way of reporting adiposity data.

BODY MASS INDEX

The Body Mass Index (BMI) has taken on greater emphasis for reporting changes in whole population adiposity. Whereas this measure is not appropriate for use with individuals, its use in tracking population trends in adiposity has been controversial. There are a number of different interpretations of BMI cut points for UK children (Chinn and Rhona, 2002). These are calculated from adult cut points of 25 kg·m^{-2} (overweight) and 30 kg·m^{-2} (obese). BMI cut points for overweight and obese children are included in Table 38.1.

Body Mass Index is not without its critics, but the measure is still the most widely used to report changes in adiposity at a population level. Waist circumference is also being used in field studies where an estimate of visceral adiposity is required. Other more sophisticated measures such as bio-impedance, Dual Electron X-Ray Absorptiometry (DEXA), magnetic resonance imaging (MRI) and air displacement plethysmography (BodPod) are also available but their use is outside the scope of this chapter.

LABORATORY TESTS

Ergometry

Depending on the purpose of the testing, ergometers (treadmill or cycle) should be as child-friendly as possible. This should include either child-sized

Table 38.1 Cut off points for girls and boys between age 5 and 18 years defined to pass
through UK BMI 25 and 30 at age 19.5 years

Age (years)	Overweight		Obese	
	Boys	Girls	Boys	Girls
5	16.9	17.3	18.9	19.6
6	17	17.5	19.3	20.1
7	17.3	17.9	20	21
8	17.7	18.4	20.8	22
9	18.2	19.1	21.8	23
10	18.8	19.8	22.8	24
11	19.4	20.5	23.7	25
12	20.1	21.3	24.6	26
13	20.8	22.1	25.5	26.9
14	21.5	22.8	26.4	27.7
15	22.3	23.4	27.2	28.3
16	23	23.9	27.9	28.8
17	23.6	24.3	28.6	29.3
18	24.2	24.6	29.2	29.6

Source: Chinn and Rona, 2002

ergometers or adaptations of the mechanical parts, for example, consideration of different crank lengths for younger children or when measuring $\dot{V}O_2$ appropriate sized mouth-pieces is crucial to account for differences in dead space and ventilation between adults and children. Familiarisation is very important particularly as for many children it might be their first experience of this equipment. As with adult data children's $\dot{V}O_2$ *peak* scores are higher on a treadmill than a cycle ergometer.

AEROBIC PERFORMANCE

Oxygen uptake test

Tests of oxygen uptake are well tolerated by children who have been suitably familiarised and then are capable of reaching limits of voluntary exhaustion. The preferred term when testing children to maximum is $\dot{V}O_2$ *peak* as only a minority of children exhibiting the classic plateau that is used to define $\dot{V}O_{2max}$ in adults. Studies comparing participants who plateau and those who do not plateau have not found a significant difference in final oxygen uptake scores, so it is considered a valid and reliable measure (Armstrong *et al.*, 1996).

Protocol

Both continuous and discontinuous protocols are suitable for children and will depend partially on what other measures are being collected. It is preferable that the younger a child is, that a discontinuous test is performed. This will allow talking and supporting the child during the rest periods and encourage them to maximal effort. This is particularly crucial, as few children will have experienced this level of effort.

For tests on the treadmill children tend to find it harder to run at high speeds therefore it is recommended that increments in intensity is achieved by raising the treadmill gradient. Initial starting speeds can be determined in the warm-up and should seek to elicit <70% heart rate maximum.

For cycle ergometers the principle of increasing intensity is similar to the treadmill, such that increments should not be so large as to induce premature fatigue and consequently these will not be as large as for adults.

Children, on average, reach a steady state in $\dot{V}O_2$ in ~2 min, therefore stage durations of 3 min or longer might only be beneficial if additional measures, for example, blood lactate are being collected.

An example of a discontinuous treadmill protocol is:

1   A warm-up of 3–4 min at 7 km·h^{-1} or 1 km·h^{-1} below the starting speed.
2   Test commences at 8 km·h^{-1} for 2 min duration.
3   A 1 min rest.
4   Repeat with increments of 1 km·h^{-1} every 2 min until the 10 km·h^{-1} stage is completed.
5   Speed remains constant and slope is raised 2.5% every 2 min until voluntary exhaustion.

An example of a continuous cycle protocol is:

1   A warm-up of 3–4 min at 25–50 W.
2   A starting power output of 50 W with increments of 25 W every 2 min until voluntary exhaustion.
3   A pedal cadence of between 60 and 70 rev·min^{-1} is well tolerated by most children.

Finally, in both examples it is important to monitor the child after the test to ensure no ill effects.

CRITERIA FOR $\dot{V}O_2$ PEAK

$\dot{V}O_2$ peak is defined as the highest $\dot{V}O_2$ elicited by the child and usually lacks the demonstration of a plateau. To support the conclusion of maximal effort specific secondary criteria should accompany the $\dot{V}O_2$ peak value. These include an RER >1.0 (treadmill) or 1.06 (cycle), a heart rate which is ≥95% of age predicted maximum and subjective criteria of facial flushing, hypernoea, sweating and unsteady gait. It is not recommended that a blood lactate value be used as an indicator of maximal effort in children, as suggested by some authors (Leger, 1996), as the variability post exercise is too great in children. Perceived exertion is another typical measure taken during a $\dot{V}O_2$ peak test and we recommend using the perceived exertion scale which has been developed for use with children. The Children's Effort rating table (CERT) (Williams *et al.*, 1994) is a numerical scale from 1 to 10 and has verbal exertional expressions that have been developed for children. This scale does not however, appear to be predictive of heart rate.

ANAEROBIC

Anaerobic testing of children is not as well developed as the testing of aerobic performance. The most common test of anaerobic performance is the Wingate test (Bar-Or, 1986) and most protocol guidelines are similar to adults. The most important issue is the load applied most common is 75 g·kg^{-1} body mass (0.74 N·kg^{-1}) although it has been found that loads between 64 and 78 g·kg^{-1} does not significantly alter the peak power obtained. As with adults there is an aerobic contribution to the 30 s Wingate test as high as 36% in some studies, hence shorter tests such as the Force–velocity might be advantageous to assess peak anaerobic power. We would recommend a flying start for the commencement of anaerobic power tests to overcome the inertia of flywheel that is going to be disproportionately higher for children compared to adults. However, software is available to account for these inertial and load corrections. For the youngest children 30 s might be too long as the ensuing fatigue might render the pedal cadence so slow that continuing to turn the pedals becomes extremely difficult. Therefore, a test of 20 s might be more appropriate (Chia *et al.*, 1997).

BLOOD ANALYSIS

The interpretation of children's blood lactate response is not well understood because of the influence of growth and maturation. Differences in methodologies such as venous or capillary samples, whole blood or plasma, and protein-free or lysed blood assays have not helped to clarify these influences (Williams *et al.*, 1992). Although venous sampling has been performed in children for studies investigating cholesterol or free fatty acids, the collection of serum lactates would appear not to be justifiable.

In adults studies the 4 mmol·l^{-1} reference point as an indicator of sub-maximal performance has often been used. However in children, this absolute value is too high and approaches maximal values. More common is a fixed value of 2.5 mmol·l^{-1} (Williams and Armstrong, 1991). For ascertaining peak blood lactate values, a 3-min post-exercise sampling appears to be the most commonly reported. However, it should be noted that this merely reflects the peak value at 3 min and does not necessarily indicate the highest value post exercise.

ISOKINETIC STRENGTH TESTING

Although there is much strength data available for children, much of it is field-based or conducted using purpose-built dynamometers. There is less data on commercially available isokinetic dynamometers, however, children are able to use this equipment if it is adapted. This includes appropriate attention to the back support, length of lever, stabilising straps and the mechanical degree of adjustment for the ergometer. Unlike for adults (Osternig, 1986) there are no set protocols for children when testing for strength (e.g. number of repetitions ranges from 3 to 8) or endurance (number of repetitions ranges from 10 to 50). Suffice that familiarisation and practice will need to be more extensive than for adults. A typical protocol for a maximal isokinetic strength test in children could be:

1   Warm-up 4–5 min including cardiovascular and stretching routine.
2   Practice tests consisting of 5–10 sub-maximal contractions and re-iteration of pre-test instructions.
3   Maximal contractions consisting of 3–6 repetitions.

For a review see de Ste Croix *et al.* (2002).

FIELD MEASURES OF FITNESS/ PERFORMANCE

Coaches, teachers, researchers and allied health professionals commonly use field measures to track the fitness of large populations. During field-testing, fitness is assessed through a battery of tests that are carried out in a predetermined order. The battery usually includes each component of the health (cardiorespiratory, strength, flexibility, body composition, local muscular endurance) and skill (agility, speed, power, balance, reaction time, coordination) related fitness model. These tests have received much criticism over the years because of the limited reliability and validity data and their appropriateness in educational settings. The reliability and validity data that are available have good agreement with criterion and test–retest measures, respectively (see Docherty, 1996 for a review). A better understanding of how to use field tests with youngsters now allows more accurate interpretation and presentation of test results. 'Normative tests' that were popular in the 1970s and 1980s have now been superseded by 'criterion-referenced' tests. Criterion-referenced tests produce performance bands that all children are expected to achieve as opposed to norm tests where percentile scores are attributed to each test. The use of normative referenced tests results in peer comparison that is problematic as results are significantly influenced by physical maturity and genetic endowment whereas criterion-referenced tests use set standards that students can meet and use individually. The other problem with field tests is that there is little empirical evidence to suggest that they are related to any aspect of health or wellness (Riddoch and Boreham, 2000). Given these limitations appropriately designed and delivered field tests of fitness with British

Table 38.2 The EUROFIT fitness test battery

Dimensions	Factor	EUROFIT test	Order of test
Balance	Total body balance	Flamingo balance	1
Speed	Limb speed	Plate tapping	2
Flexibility	Flexibility	Sit and reach	3
Power	Explosive strength	Standing broad jump	4
Strength	Static strength	Hand grip	5
Muscular endurance	Trunk strength	Sit-ups	6
Muscular endurance	Functional strength	Bent arm hang	7
Speed	Running speed agility	Shuttle run 10 m × 5 m	8
Cardiorespiratory fitness	Cardiorespiratory fitness	Endurance shuttle run	9

populations are needed. The largest set of field test fitness data was produced in the Northern Ireland Children's Fitness Survey (Riddoch *et al.*, 1991) and more recently through the Sportlinx project (Taylor *et al.*, 2004). Other than these datasets there is little whole population fitness data available on United Kingdom children. The test most widely used in the United Kingdom is the EUROFIT fitness test battery (Adam *et al.*, 1988). The component of fitness assessed and the order of the tests are outlined in Table 38.2. A detailed description of the tests can be found elsewhere (Adam *et al.*, 1988).

An excellent review of field tests of fitness can be found in Safrit and Wood (1995) and Docherty (1996).

IMPLEMENTATION OF TESTS

When used appropriately fitness testing can be an important aspect of children's education (Cale and Harris, 2005). To achieve a positive testing climate, environments should be inclusive, supportive and conducive to learning where emphasis should be on effort and individual development from test to test. Children should be able to practice the tests before an official measurement starts.

These criteria are important, as field tests have been criticised for de-motivating children who are either unfit or disenfranchised from physical activity. Therefore, care should be taken to ensure that social environments are developed that reward effort as well as performance during fitness testing. For example, setting up an environment where an overweight and physically immature child may be exposed as a failure is clearly bad practice and this should be avoided. When comparisons need to be made these can be done at a group level where boys tend to have better endurance running capacity than girls, heavier children have better grip strength then their lighter peers and girls are more flexible than boys. The uses of field tests by suitably qualified personnel are appropriate at whole population level (for research) or to provide individual feedback about the development of a child's fitness during the growing years.

ADULT CHILD DIFFERENCES

Children will need more time than adults to familiarise themselves with tests. Test results may be more affected by biological age than chronological age making individual comparisons between circumpubescents difficult. Scoring systems in the 20-m multi-stage shuttle run test are different. Instead of using levels (e.g. 7.2) the number of 20-m shuttles are counted, 50 shuttles = 1,000 m.

The use of field tests of fitness to monitor individuals in education, sport and health settings is supported if they are implemented in an appropriate manner by trained individuals who understand their strengths and limitations for use with children.

Key messages for fitness testing in young children:

1   *Individualise* fitness testing.
2   Make fitness testing a *positive* and *fun* experience for *all*.
3   *Teach* concepts during fitness tests. *For example, sit and reach would be linked to flexibility for daily tasks, issues about losing flexibility with age, etc.*
4   Use *developmentally appropriate* tests.
5   Minimise the public nature of testing when you think it may cause embarrassment.
6   Take care to monitor fitness over time and make children aware that sometimes *fitness testing results may be affected by stage of maturation.*
7   Physical activity and fitness test results are *not always* related. A child's fitness may be mainly due to genetic inheritance. This relates to the 'cannot choose my own parents' adage.

PHYSICAL ACTIVITY

Objective and subjective measures

There are over 30 methods of measuring physical activity available but no gold standard is available. Methods can be broadly categorised into objective and subjective areas. Subjective methods include activity diaries, retrospective questionnaires and systematic observation. Objective methods include, pedometers, accelerometers, heart rate monitors and doubly labelled water. The type of monitor chosen will primarily depend on the scientific question being asked but may also be influenced by expediency, accuracy and cost. Paper-based questionnaires whilst being less intrusive are generally thought to be the least robust measure of physical activity particularly in younger children. The most commonly used measures in paediatric populations are accelerometers, pedometers and heart rate telemetry systems. The most important factor to consider when measuring physical activity in children is that 90% of their activity is of high intensity and lasts for 15 s or less (Bailey, 1990). Therefore, scientists need to use sampling rates of 15 s or less if valid results are to be gained. Systems that use sophisticated software also allow detailed analysis of the frequency, duration and intensity of physical activity. The tempo of physical activity is of particular interest for studies that wish to have a detailed measure of behaviour change. For a more detailed description of physical activity measurement see Welk (2002).

REFERENCES

Adam, C., Klissouras, V., Ravazzolo, M., Renson, R. and Tuxworth, W. (1988). *EUROFIT: European Test of Physical Fitness*. Rome: Council of Europe, Committee for the Development of Sport.

Armstrong, N., Welsman, J. and Winsley, R. (1996). Is peak VO_2 a maximal index of children's aerobic fitness? *International Journal of Sports Medicine*, 17: 356–359.

Bailey, R.C., Olson, J., Pepper, S.L., Porszasz, J., Barstow, T.J. and Cooper, D.M. (1995). The level and tempo of children's physical activities: an observational study. *Medicine and Science in Sports and Exercise*, 27: 1033–1041.

Bar-Or, O. (1996). Anaerobic performance. In D. Docherty (ed.), *Measurement in Pediatric Exercise Science*, pp. 161–182. Champaign, IL: Human Kinetics.

Cale, L. and Harris, J. (2005). *Exercise and Young People: Issues, Implications and Initiatives*, pp. 41–80. Basingstoke, UK: Palgrave Macmillan.

Chia, M., Armstrong, N. and Childs, D. (1997). The assessment of children's anaerobic performance using modifications of the Wingate anaerobic test. *Pediatric Exercise Science*, 9: 80–89.

Child Growth Foundation.

Children Act 1999 Craig.

Chinn, S. and Rona, R.J. (2002). International definitions of overweight and obesity for children: a lasting solution? *Annals of Human Biology*, 29: 306–313.

De Ste Croix, M.B.A., Deighan, M.A. and Armstrong, N. (2003). Assessment and interpretation of isokinetic muscle strength during growth and maturation. *Sports Medicine*, 33(10): 727–743.

Department of Health (2000). The Protection of Children Act 1999: a practical guide to the act for all organisations working with children. Department of Health and the NHS Executive, London: Department of Health.

Deurenberg, P., Pieters, J.J. and Hautvast, J.G. (1990). The assessment of the body fat percentage by skinfold thickness measurements in childhood and young adolescence. *British Journal of Nutrition*, 63: 293–303.

Durnin, J.V. and Womersley, J. (1974). Body fat assessed from total body density and its estimation from skinfold thickness: measurements on 481 men and women aged from 16 to 72 years. *British Journal of Nutrition*, 32(1): 77–97.

Docherty, D. (ed.) (1996). *Measurement in Pediatric Exercise Science*. Champaign, IL: Human Kinetics.

Greulich, W.W. and Pyle, S.I. (1959). Radiographic atlas of skeletal development of the hand and wristv (2nd edn) Stanford, CA: Stanford University Press.

Jago, R. and Bailey, R. (2001). Ethics and paediatric exercise science: issues and making a submission to a local ethics and research committee. *Journal of Sports Science*, 19(7): 527–535.

Leger, L. and Gadoury, C. (1989). Validity of the 20 m shuttle run test with 1 min stages to predict VO_{2max} in adults. *Canadian Journal of Sports Science*, 14(1): 21–26.

Lohman, T.G. (1984). Research progress in validation of laboratory methods of assessing body composition. *Medicine and Science in Sports and Exercise*, 16: 596–603.

Matsudo, S.M. and Matsudo, V.R. (1993). Validity of self evaluation on determination of sexual maturation level. In A.C Claessens, J. Lefevre and B. Vanden Eynde (eds), *World Wide Variation In Physical Fitness*, pp. 106–109. Leuven: Institute of Physical Education.

Mirwald, R.L., Baxter-Jones, A.D., Bailey, D.A. and Beunen, G.P. (2002). An assessment of maturity from anthropometric measurements. *Medicine and Science in Sports and Exercise*, 34(4): 689–694.

Morris, N.M. and Udry, J.R. (1980). Validation of a self-administered instrument to assess stage of adolescent development. *Journal of Youth and Adolescence*, 9: 271–280.

Osternig, L.R. (1986). Isokinetic dynamometry; implications for muscle testing and rehabilitation. In K.B. Pandolf (ed.), *Exercise and Sorts Sciences Reviews*, 14: 45–80. New York: Macmillan.

Riddoch, C.J. and Boreham, C. (2000). Physical activity, physical fitness and children's health: current concepts. In A. Armstrong and W. van Mechelen (eds), *Pediatric Exercise Science and Medicine*, pp. 243–252. Oxford, UK: Oxford University Press.

Riddoch, C.J., Mahoney, C., Murphy, N., Boreham, C. and Cran, G. (1991). The physical activity patterns of Northern Irish schoolchildren age's 11–16 years. *Pediatric Exercise Science*, 3: 300–309.

Roche, A.F., Chumlea, W.C. and Thissen, D. (1988). Assessing the skeletal maturity of the hand-wrist: Fels method. Springfield, IL: Charles Thomas.

Safrit, M.J. and Wood, T.M. (1995). *Introduction to Measurement in Physical Education and Exercise Science*, 3rd edn. New York: Mosby.

Tanner, J.M. (1962). *Growth of Adolescents*, 2nd edn. Oxford, UK: Blackwell Scientific.

Taylor, S.R., Hackett, A.F., Stratton, G. and Lamb, L. (2004). SportsLinx: improving the health and fitness of Liverpool's youth. *Education and Health*, 22: 11–15.

Welk, G.J. (2002). *Physical Activity Assessments for Health Related Research*. Champaign, IL: Human Kinetics.

Westrate and Deurenberg (1989).

Williams, C.A. (2003). Ethics in paediatric exercise science. BASES World, March, 10–11.

Williams, J.G., Eston, R. and Furlong, BA.F. (1994). CERT: a perceived exertion scale for young children. *Perceptual and Motor Skills*, 79: 1451–1458.

Williams, J.R. and Armstrong, N. (1991). Relationship of maximal lactate steady state to performance at fixed blood lactate reference values in children. *Pediatric Exercise Science*, 3: 333–341.

Williams, J.R., Armstrong, N. and Kirby, B.J. (1992). The influence and site of sampling and assay medium upon the measurement and interpretation of blood lactate responses to exercise. *Journal of Sports Sciences*, 10: 95–107.

Working Party of Research in Children (1991). *The Ethical Conduct of Research on Children*. London: Medical Research Council.

CHAPTER 39

TESTING OLDER PEOPLE

John M. Saxton

INTRODUCTION

The 2001 census showed that over a fifth of the UK population is now aged over 60. Furthermore, the number of people aged 65 and over is expected to increase at 10 times the overall rate of population growth over the next 40 years and the number of people over the age of 80 is expected to treble in the next quarter of a century (Dean, 2003). The rapid growth of the ageing population, especially amongst the oldest old, means that preventing or delaying the onset of physical frailty and increasing the number of years spent in good health has become an important public health goal.

THE AGE-ASSOCIATED DECLINE IN PHYSIOLOGIC FUNCTION

Ageing is characterised by a decline in cardiorespiratory, muscular, neurological and metabolic capacities (Pendergast *et al.*, 1993). This can severely limit the ability to perform everyday activities, including walking, stair-climbing and even rising from a chair. As many older adults function close to their maximum physical ability level during normal daily activities (Rikli and Jones, 1997), any further decline in physiologic function or small physical set-back could result in the loss of functional independence (Rikli and Jones, 1999a). Shephard (1997) outlined a classification system for the different stages of middle to old age, based on functional status:

- Middle age (40–65 years) – associated with a 10–30% loss of biological function.
- Old age or 'young old age' (65–75 years) – associated with some further loss of biological function, but without any gross impairment of homeostasis.
- Very old age (75–85 years) – characterised by substantial impairment of function in daily activities, but still being capable of functional independence.

- Oldest old age (>85 years) – during which time institutional or nursing care is often required.

The age-associated decline in cardiovascular function is characterised by anatomical and neurological changes affecting the heart and blood vessels, which decrease cardiorespiratory capacity, and hence, aerobic exercise capacity. Aerobic exercise capacity declines at the rate of 7–10% per decade from early adulthood (Fitzgerald et al., 1997; Wilson and Tanaka, 2000), and this can severely reduce sustainable exercise intensity in later years. Changes in arterial structure and vasomotor tone also adversely affect blood pressure, which has a tendency to rise with increasing age in most Western societies and contributes to the age-related increased risk of cardiovascular disorders.

The decline in muscular strength and power with advancing age is judged to have a more profound impact on daily functioning than the decline in cardiorespiratory capacity (Pendergast et al., 1993). The age-related decline in muscular strength occurs sooner and at a faster rate in the lower extremities than in the upper extremities (Frontera et al., 1991), and this can severely affect ambulatory activities. Lower-limb muscle function is considered vital for functional independence and prevention of disability (Pendergast et al., 1993; Guralnik et al., 1995). A direct association between impaired lower-limb physiologic function and everyday activities such as walking and rising from a chair has been demonstrated in the elderly (Judge et al., 1993b; Ferrucci et al., 1997). The decline in leg strength and power with advancing age is also associated with an increased risk of falls and resulting fractures (Whipple et al., 1987; Nevitt et al., 1989; Gehlsen and Whaley, 1990).

FUNCTIONAL FITNESS FOR OLDER ADULTS

Functional fitness for older people has been defined as the physical capacity required to perform normal everyday activities safely and independently without undue fatigue, or with adequate physiologic reserve (Rikli and Jones, 1997). Traditional ergometric tests to volitional exhaustion (developed and validated for younger populations) are generally deemed inappropriate for older adults, as they do not reflect the physical abilities required for common daily activities, including stair climbing, rising from a chair, lifting, reaching and bending. Furthermore, they are likely to be unsafe for the majority of older adults who, on the whole, are likely to be poorly accustomed to exercise ergometers and generally need medical supervision for anything other than light to moderate intensity physical exertion. Traditional ergometer tests are perhaps only suitable for an elite few per cent of the elderly population who are physically fit and/or 'Master' athletes and accustomed to the demands of vigorous exercise. At the other end of the continuum, assessment of functional status in the frail and/or disabled elderly, who constitute ~25% of the elderly population (Rikli and Jones, 1997), requires the use of self-care activity scales, referred to as activities of daily living (Mahoney and Barthel, 1965; Katz et al., 1970; Hedrick, 1995) or instrumental activities of daily living (Lawton and Brody 1969; Lawton et al., 1982).

The physically independent elderly make up the largest sub-group of older people, constituting ~70% of adults over 75 (Spirduso, 1995; Rikli and Jones, 1997). The physically independent elderly exhibit wide variations in physical ability, from those who have enough physical function to participate in voluntary social, occupational and recreational activities, to those who are borderline frail and highly vulnerable to unexpected physical stress or challenge (Spirduso, 1995). Reliable and valid tests that can detect the early stages of functional decline and aid in the prescription of appropriate physical activity interventions in this large heterogeneous sub-group could have the biggest impact on fraily prevention and maintenance of physical independence in older people (Guralnik et al., 1995; Gill et al., 1996; Lawrence and Jette, 1996; Morey et al., 1998).

FUNCTIONAL FITNESS TEST BATTERY ITEMS

A number of functional fitness test batteries have been developed and validated for older adults in the age-range 60 – >90 years, including the American Alliance for Health, Physical Education, Recreation and Dance (AAHPERD) Functional Fitness Assessment Battery (Osness et al., 1990, 1996), the Physical Performance Test (Reuben and Siu, 1990), the MacArthur Physical Performance Scale (Seeman et al., 1994), the Established Populations for Epidemiologic Studies of the Elderly (EPESE) short battery of items to measure strength, balance and gait speed (Guralnik et al., 1994) and the Senior Fitness Test (SFT) (Rikli and Jones, 1999a,b; Rikli and Jones, 2001).

The test items described in this section are typical of those used to assess functional fitness in physically independent older adults of diverse physical ability. Many of the test items have been through extensive validation procedures, although it is recommended that each test centre develop its own test–retest reproducibility data. Normative data for older people on the individual test items can be found in the Allied Dunbar National Fitness Survey (Activity and Health Research, 1992), and in the publications of Osness et al. (1996), Rikli and Jones (1999b) and Holland et al. (2002). It is recommended that a test battery of functional fitness for older adults should include at least one test item from each of the core physiologic function variables that underpin common everyday activities. These were defined by Osness et al. (1990) and Rikli and Jones (1999a) as:

1   Muscle strength/endurance
2   Aerobic endurance
3   Flexibility
4   Balance/agility
5   Body composition.

Muscular strength/endurance

Chair sit-to-stand test

A common method of assessing lower-body muscle function in older adults is the chair sit-to-stand test. Variations of this test exist, but protocols that assess the time it takes to perform a given number of sit-to-stand repetitions (e.g. 5 or 10) have the disadvantage of 'floor' effects because some elderly people might not be able to achieve the number required to complete the test. However, testing the number of repetitions achievable in a set amount of time can overcome this problem. Chair sit-to-stand performance has a good correlation ($r > 0.7$) with one repetition maximum leg-press strength in elderly men and women (Rikli and Jones, 1999a).

The equipment requirements for this test are a stopwatch and a foldable or plastic moulded straight-back chair (without arms or seating cushion) with approximate seating height, width and depth dimensions of 0.45, 0.50 and 0.40 m, respectively (Csuka and McCarty, 1985; Jones *et al.*, 1999; Rikli and Jones, 2001). The chair should have rubber tips underneath each leg to prevent slippage. The chair back is placed against the wall to prevent movement and the participant is seated in the middle of the chair, with back straight and feet approximately shoulder width apart at an angle slightly back from the knees; one foot is placed slightly in front of the other to aid balance and the arms are crossed in front of the chest. This test should be performed either barefooted, or in low-heeled shoes. At the signal to 'go', the participant rises to the full standing position before returning to the seated position as many times as possible in 30 s. Participants are instructed to look straight ahead and to stand up with their weight evenly distributed between both feet. The score is the total number of stands performed correctly (full standing position attained and fully seated between stands) in 30 s. If a participant is more than half way up at the end of the 30 s, this is counted as a full stand (Jones *et al.*, 1999).

Arm-curl test

Adequate upper-body strength and endurance are required for many everyday activities, such as cleaning, carrying food shopping and gardening. A test that reflects the strength requirements of these every activities is the arm-curl test (Osness *et al.*, 1990; Rikli and Jones, 2001). In this test, participants curl a standardised weight using the forearm flexors as many times as possible in a set amount of time. As upper-body strength declines with increasing age and in elderly women is ~50% of that in elderly men (Frontera *et al.*, 1991), these considerations need to be taken into account when deciding on the weight to be used for women and men. The AAHPERD test (Osness *et al.*, 1990) states that weights of 4 lb (1.81 kg) and 8 lb (3.63 kg) should be used for women and men, respectively, whereas the SFT (Rikli and Jones, 2001) uses a weight of 5 lb (2.27 kg) for women.

The equipment requirements for this test are a stopwatch, a foldable or plastic moulded straight-back chair (as for chair sit-to-stand test), and a dumbbell or other suitable weight such as plastic milk cartons filled with sand, water or other material with handles that can be gripped easily. As normative data for the age-ranges 60–94 years are available for the SFT (Rikli and Jones, 2001), weights of 2.27 kg (women) and 3.63 kg (men) are suggested. Velcro wrist straps can be used for individuals with gripping problems resulting from conditions such as arthritis. The participant is seated with back straight and feet flat on the floor, holding the weight in the dominant hand at the side of the body in the fully extended elbow position. The elbow should be braced against the side of the body to stabilise the upper arm. Using good form (the upper arm must remain still throughout the test), the participant curls the weight up and down. The score is the total number of repetitions performed correctly in 30 s. An arm-curl that is more than half way up at the end of 30 s is counted as a full arm-curl (Rikli and Jones, 2001). In the AAHPERD test battery (Osness et al., 1990, 1996), the lower arm must touch the test administrator's hand which is placed on the participant's bicep at termination of the up-phase to be deemed a successful repetition. The number of repetitions achieved in 30 s has a good correlation ($r > 0.77$) with overall upper body strength (as indicated by combined 1TRM biceps, chest press and seated row strength) in elderly men and women (Rikli and Jones, 1999a).

Grip strength

Maximum grip strength can also be included in a functional fitness test battery as an index of upper-limb strength. In this test, a grip-strength dynamometer is gripped between flexed fingers and the base of the thumb with the participant in a seated position, and with the measurement normally being restricted to the dominant hand, unless prevented by injury. The Allied Dunbar National Fitness Survey (Activity and Health Research, 1992) reported a significant decline in hand-grip strength with age, being 30% less in the 65–74 year age group, in comparison to younger adults aged 25–44 years. Handgrip strength of 150 N, or that is equivalent to 20% of body weight, has been suggested as a threshold for performance of everyday tasks requiring a firm grip, as the strength needed to raise body weight onto a raised bus platform is estimated as 17–20% body weight (Activity and Health Research, 1992).

Aerobic endurance

Six-min walk test

Walking is an activity that is fundamental to functional independence. The timed 6-min walk test, which is an adaptation of the 12-min walk-run test originally developed by Cooper (1968) assesses the maximum distance walked in 6 min along a rectangular course (Rikli and Jones, 2001), or up and down a 20–30 m corridor (Simonsick et al., 2001; Steffen et al., 2002). As it is a timed test, walking distance can be obtained for elderly individuals of wide-ranging functional ability. For this test, the 20–30 m corridor or flat 50 m rectangular course (20 m × 5 m) is marked off in 5 m segments with marker cones and foldable or plastic moulded straight-back chairs are positioned at various locations along the course for resting. Participants walk as fast as they can (without running) up and down or around the course, covering as much distance as possible in the 6-min time limit. Standardised encouragement can be given at minutes 1, 3, and 5 to aid pacing (Steffen et al., 2002). At the end of the 6-min time period, participants are told to stop walking and the distance walked, to the nearest meter, is recorded. The timed 6-min walk test has a good correlation ($r > 0.7$) with time to reach 85% predicted maximum heart rate on a progressive treadmill test in elderly men and women (Rikli and Jones, 1998). Alternative tests of lower-limb aerobic endurance (where space is limited) include the 2-min step test (Rikli and Jones, 2001) and the Self-Paced Step Test (Petrella *et al.*, 2001), which were developed for older people.

Flexibility

Lower back and hamstrings flexibility: sit-and-reach test

Impaired flexibility, such as the age-associated decreased range of motion at the hip joint (Roach and Miles, 1991), influences movement dysfunction and disability in the elderly. Variations of the sit-and-reach test have been used for assessing lower back and hamstrings flexibility in older persons, including the conventional 'floor' sit-and-reach test (Osness *et al.*, 1990), a modified sit and reach test (Lemmink *et al.*, 2003) and a seated sit-and-reach test developed by Jones *et al.* (1998).

For this test, the participant sits on the floor in an upright position, with back straight and legs fully extended with the bottom of the bare feet against a sit-and-reach box. The hands are placed one on top of the other and the participant is instructed to slowly reach forward, keeping the hands together and pushing the fingers along the box as far as possible. If participants cannot hold a sitting position on a flat surface with both legs extended, a seated sit-and-reach test can be used (Jones *et al.*, 1998). In this test, the participant sits on the front edge of a chair with one leg extended out in front (knee straight, ankle fully dorsi-flexed, heel resting on the floor) and the other leg bent at the knee with foot flat on the floor. The chair sit-and-reach test has a good correlation ($r > 0.75$) with goniometer-measured hamstring flexibility in elderly men and women (Jones *et al.*, 1998). In both variations of the test, the final position should be held for 2 s and the distance between the finger tips and toes, to the nearest centimetre is recorded. A negative score is assigned to a distance short of reaching the toes and a positive score to a distance reached beyond the toes. Sit-and-reach testing is contra-indicated in participants with extreme kyphosis and in osteoporotic individuals who have previously sustained a vertebral fracture.

Shoulder flexibility: back scratch test

Shoulder flexibility is required for everyday activities such as reaching behind the head and/or lower back to comb hair, to put on, take off or fasten garments, reach into back pockets and wash one's back (Rikli and Jones, 1999a). The back scratch test (Rikli and Jones, 2001) is a convenient way to measure overall shoulder range of motion. This test involves a combination of shoulder abduction, adduction, and internal and external rotation, and measures the distance (or overlap) of the middle fingers behind the back. The only equipment requirement for this test is a 0.5 m ruler or meter stick. The participant places one hand behind the same side shoulder with palm flat on the back and fingers reaching down towards the middle of the back as far as possible. The other hand is placed behind the back (palm facing outwards) and reaches up as far as possible in an attempt to touch or overlap with the fingers of the other hand.

Two test trials are allowed and the score is the distance of overlap (positive score) or distance between the middle fingers (negative score), recorded to the nearest centimetre. Both arm combinations can be measured, but normative data are only generally available for the preferred hand combination (i.e. the hand combination that gives the best score), which can be determined during practice trials.

Balance/agility

Static and dynamic balance

A number of different test protocols have been used to assess static and dynamic balance in the elderly. The Berg Balance Scale assesses the ability to successfully accomplish static, dynamic and weight shifting activities, with each of the 14 items being graded 0–4 by the test administrator (Berg et al., 1989; Berg et al., 1992a,b). A common measure of static balance in the elderly is the ability to maintain balance under conditions of reduced base of support with the eyes open or closed. Different stances are commonly used, including the parallel stance (feet touching side-by-side); semi-tandem stance (from the parallel stance, one foot is moved half a length forward); tandem stance (one foot placed in front of the other, heel to toe) and single-leg stance (Iverson et al., 1990; Verfaillie et al., 1997; Brown et al., 2000). The test score is the maximum length of time that a stance can be held, but usually with the test being terminated after a predetermined length of time (e.g. 10–60 s) for practical reasons. A major problem with static balance tests is that a large proportion of the elderly can achieve perfect scores (ceiling effect). However, caution should be taken with eyes-closed tests to reduce the risk of falls.

Dynamic balance has been measured by counting the number of successful steps or stepping errors while subjects walk toe to heel (tandem walk) over a specified distance or to a maximum number of steps (Nevitt et al., 1989; Topp et al., 1993; Dargent-Molina et al., 1996). Dynamic balance can also be assessed by measuring the time taken to walk along a balance beam placed on the floor (Cress et al., 1999; Brown et al., 2000) or by using the functional reach test, which is a test of the maximum forward displacement of the centre of mass and thus, the 'margin of stability' within the base of support (Duncan et al., 1990). In the latter test, the difference between arm's length and maximum forward reach is measured using a meter stick attached to the wall.

Surrogate measures of dynamic balance include the preferred and maximal walking velocities. Preferred walking velocity decreases linearly with advancing age (Cunningham et al., 1982) and slower preferred and maximal walking velocities are characteristic of older adult fallers (Wolfson et al., 1990; Lipsitz et al., 1991; Wolfson et al., 1995). Preferred and maximal walking velocity is usually measured over a set distance of 6–10 m (Reuben and Siu, 1990; Judge et al., 1993a; Buchner et al., 1996). However, preferred gait velocity has also been measured over 100 m on an indoor oval course (Bassey et al., 1976). This approach may be superior to shorter distance tests, as participants have more time to attain and maintain preferred gait velocity (Table 39.1).

Table 39.1 Common tests of balance and agility in the elderly

Functional fitness dimension	Protocol	References
Static balance	Ability to balance for 10–60 s in various stances: for example parallel stance; tandem stance; single-leg stance	Brown *et al.* (2000) Iverson *et al.* (1990) Verfaillie *et al.* (1997)
Dynamic balance	Tandem walk	Nevitt *et al.* (1989) Topp *et al.* (1993) Dargent-Molina *et al.* (1996)
	Balance beam walk	Cress *et al.* (1999) Brown *et al.* (2000)
	Functional reach test Preferred and maximum walking velocity	Duncan *et al.* (1990) Buchner *et al.* (1996) Judge *et al.* (1993a) Reuben and Siu (1990)
Combined static and dynamic balance	Berg Balance Scale	Berg *et al.* (1989) Berg *et al.* (1992a) Berg *et al.* (1992b)
Agility	Timed up-and-go tests	Mathias *et al.* (1986) Osness *et al.* (1990) Rikli and Jones (2001)

Agility

Timed up-and-go tests measure agility and reflect everyday activities such as disembarking from a bus or car in an efficient and safe manner, quickly getting up to answer the door or telephone, or to tend to something in the kitchen (Rikli and Jones, 1999a). A number of such tests have been described in the literature (Mathias *et al.*, 1986; Osness *et al.*, 1990; Rikli and Jones, 2001) and consist of a timed course, which can be set-up using minimal equipment and space. Variations of timed up-and-go tests involve participants walking as quickly as possible around cones located 2–3 m in front of, or diagonally to the rear of a seated starting position, before returning quickly to the seated position.

Body composition: body mass index

There is evidence of a link between body composition and ability to perform common everyday activities in community-dwelling elderly people and elderly individuals with a high or low body mass index (BMI) are more likely to be disabled in later years than people with normal BMI scores (Galanos *et al.*, 1994). Low BMI is also associated with increased risk of mortality in the elderly, particularly in individuals who lose 10% or more of body weight between the age of 50 and old age (Losonczy *et al.*, 1995). Thus, the inclusion of BMI as a simple index of body composition in a functional fitness test battery for the elderly seems appropriate.

PRE-TEST CONSIDERATIONS FOR ELDERLY PERSONS

The functional fitness test items described should be safe for most community-residing physically independent older adults to perform without medical screening or supervision, as they bear no more risk than common everyday activities. Nevertheless, test administrators (and testing centres, if appropriate) should have a well-defined emergency plan to deal with unexpected medical emergencies and accidents, and have 'first aiders' on hand to manage such situations as they arise. Test administrators should also be vigilant of warning signs that are indicative of undue physiological stress (e.g. excessively high heart rate, nausea, dyspneoa, pallor and pain).

Pre-test procedures should include adequate training and practice sessions for the test administrator(s) and an appropriate gentle warm-up for the participants. The warm-up should provide a period of cardiovascular and metabolic adjustment, followed by mobility exercises in which relevant joints are moved through their comfortable ranges of motion. All tests should be preceded by a full demonstration of the procedure and 1–2 practice trials. Pre-test documentation to be administered should include:

- A 'user-friendly' information sheet, including details of the pre-test instructions (e.g. the need to avoid strenuous exercise, alcohol, heavy meals in the hours preceding the test and the importance of suitable attire).
- A Physical Activity Readiness Questionnaires (PAR-Q) to identify individuals who require a General Practitioner's examination and approval before performing the test battery. Older individuals who have previously experienced chest pain, irregular, rapid or fluttery heart beats or severe shortness of breath should seek the advice of their General Practitioner before undergoing an assessment of functional fitness.
- An informed consent form explaining the risks and responsibilities associated with the testing procedures and informing prospective participants of their right to discontinue testing at any time.

REFERENCES

Activity and Health Research. (1992). *Allied Dunbar National Fitness Survey: Main Findings*. Sports Council and Health Education Authority, London.

Bassey, E.J., Fentem, P.H., MacDonald, I.C. and Scriven, P.M. (1976). Self-paced walking as a method for exercise testing in elderly and young men. *Clinical Science and Molecular Medicine*, 51(Suppl.): 609–612.

Berg, K., Wood-Dauphinee, S., Williams, J.I. and Gayton, D. (1989). Measuring balance in the elderly: preliminary development of an instrument. *Physiotherapy Canada*, 41: 304–311.

Berg, K.O., Maki, B.E., Williams, J.I., Holliday, P.J. and Wood-Dauphinee, S.L. (1992a). Clinical and laboratory measures of postural balance in an elderly population. *Archives of Physical Medicine and Rehabilitation*, 73: 1073–1080.

Berg, K.O., Wood-Dauphinee, S.L., Williams, J.I. and Maki, B. (1992b). Measuring balance in the elderly: validation of an instrument. *Canadian Journal of Public Health*, 83 (Suppl. 2): S7–S11.

Brown, M., Sinacore, D.R., Ehsani, A.A., Binder, E.F., Holloszy, J.O. and Kohrt, W.M. (2000). Low-intensity exercise as a modifier of physical frailty in older adults. *Archives of Physical Medicine and Rehabilitation*, 81: 960–965.

Buchner, D.M., Guralnik, J.M. and Cress, M.E. (1996). The clinical assessment of gait, balance, and mobility in older adults. In L.Z. Rubenstein, D. Wieland and R. Bernabei (eds). *Geriatric Assessment Technology: The State of Art*, pp. 75–89. Milan: Kurtis Editrice.

Cooper, K.H. (1968). A means of assessing maximal oxygen intake. Correlation between field and treadmill testing. *Journal of the American Medical Association*, 203: 201–204.

Cress, M.E., Buchner, D.M., Questad, K.A., Esselman, P.C., deLateur, B.J. and Schwartz, R.S. (1999). Exercise: effects on physical functional performance in independent older adults. *Journals of Gerontology. Series A, Biological Sciences and Medical Sciences*, 54: M242–M248.

Csuka, M. and McCarty, D.J. (1985). Simple method for measurement of lower extremity muscle strength. *American Journal of Medicine*, 78: 77–81.

Cunningham, D.A., Rechnitzer, P.A., Pearce, M.E. and Donner, A.P. (1982). Determinants of self-selected walking pace across ages 19 to 66. *Journal of Gerontology*, 37: 560–564.

Dargent-Molina, P., Favier, F., Grandjean, H., Baudoin, C., Schott, A.M., Hausherr, E., Meunier, P.J. and Breart, G. (1996). Fall-related factors and risk of hip fracture: the EPIDOS prospective study. *Lancet*, 348: 145–149.

Dean, M. (2003). *Growing Older in the 21st Century*. Swindon: Economic and Social Research Council.

Duncan, P.W., Weiner, D.K., Chandler, J. and Studenski, S. (1990). Functional reach: a new clinical measure of balance. *Journal of Gerontology*, 45: M192–M197.

Ferrucci, L., Guralnik, J.M., Buchner, D., Kasper, J., Lamb, S.E., Simonsick, E.M., Corti, M.C., Bandeen-Roche, K. and Fried, L.P. (1997). Departures from linearity in the relationship between measures of muscular strength and physical performance of the lower extremities: the Women's Health and Aging Study. *Journals of Gerontology. Series A, Biological Sciences and Medical Sciences*, 52: M275–M285.

Fitzgerald, M.D., Tanaka, H., Tran, Z.V. and Seals, D.R. (1997). Age-related declines in maximal aerobic capacity in regularly exercising vs. sedentary women: a meta-analysis. *Journal of Applied Physiology*, 83: 160–165.

Frontera, W.R., Hughes, V.A., Lutz, K.J. and Evans, W.J. (1991). A cross-sectional study of muscle strength and mass in 45- to 78-yr-old men and women. *Journal of Applied physiology*, 71: 644–650.

Galanos, A.N., Pieper, C.F., Cornoni-Huntley, J.C., Bales, C.W. and Fillenbaum, G.G. (1994). Nutrition and function: is there a relationship between body mass index and the functional capabilities of community-dwelling elderly? *Journal of the American Geriatrics Society*, 42: 368–373.

Gehlsen, G.M. and Whaley, M.H. (1990). Falls in the elderly: Part II, Balance, strength, and flexibility. *Archives of Physical Medicine and Rehabilitation*, 71: 739–741.

Gill, T.M., Williams, C.S., Richardson, E.D. and Tinetti, M.E. (1996). Impairments in physical performance and cognitive status as predisposing factors for functional dependence among nondisabled older persons. *Journals of Gerontology. Series A, Biological Sciences and Medical Sciences*, 51: M283–M288.

Guralnik, J.M., Simonsick, E.M., Ferrucci, L., Glynn, R.J., Berkman, L.F., Blazer, D.G., Scherr, P.A. and Wallace, R.B. (1994). A short physical performance battery assessing lower extremity function: association with self-reported disability and prediction of mortality and nursing home admission. *Journal of Gerontology*, 49: M85–M94.

Guralnik, J.M., Ferrucci, L., Simonsick, E.M., Salive, M.E. and Wallace, R.B. (1995). Lower-extremity function in persons over the age of 70 years as a predictor of subsequent disability. *New England Journal of Medicine*, 332: 556–561.

Hedrick, S.C. (1995). Assessment of functional status: activities of daily living. In L.Z. Rubenstein, D. Wieland and R. Bernabei (eds), *Geriatric Assessment Technology: The State of the Art*, pp. 51–58. Milan: Editrice Kurtis.

Holland, G.J., Tanaka, K., Shigematsu, R. and Nakagaichi, M. (2002). Flexibility and physical functions of older adults: a review. *Journal of Aging and Physical Activity*, 10: 169–206.

Iverson, B.D., Gossman, M.R., Shaddeau, S.A. and Turner, M.E., Jr. (1990). Balance performance, force production, and activity levels in noninstitutionalized men 60 to 90 years of age. *Physical Therapy*, 70: 348–355.

Jones, C.J., Rikli, R.E., Max, J. and Noffal, G. (1998). The reliability and validity of a chair sit-and-reach test as a measure of hamstring flexibility in older adults. *Research Quarterly for Exercise and Sport*, 69: 338–343.

Jones, C.J., Rikli, R.E. and Beam, W.C. (1999). A 30-s chair-stand test as a measure of lower body strength in community-residing older adults. *Research Quarterly for Exercise and Sport*, 70: 113–119.

Judge, J.O., Lindsey, C., Underwood, M. and Winsemius, D. (1993a). Balance improvements in older women: effects of exercise training. *Physical Therapy*, 73: 254–262.

Judge, J.O., Underwood, M. and Gennosa, T. (1993b). Exercise to improve gait velocity in older persons. *Archives of Physical Medicine and Rehabilitation*, 74: 400–406.

Katz, S., Downs, T.D., Cash, H.R. and Grotz, R.C. (1970). Progress in development of the index of ADL. *Gerontologist*, 10: 20–30.

Lawrence, R.H. and Jette, A.M. (1996). Disentangling the disablement process. *Journals of Gerontology. Series B, Psychological Sciences and Social Sciences*, 4: S173–S182.

Lawton, M.P. and Brody, E.M. (1969). Assessment of older people: self-maintaining and instrumental activities of daily living. *Gerontologist*, 9: 179–186.

Lawton, M.P., Moss, M., Fulcomer, M. and Kleban, M.H. (1982). A research and service oriented multilevel assessment instrument. *Journal of Gerontology*, 37: 91–99.

Lemmink, K.A., Kemper, H.C., de Greef, M.H., Rispens, P. and Stevens, M. (2003). The validity of the sit-and-reach test and the modified sit-and-reach test in middle-aged to older men and women. *Research Quarterly for Exercise and Sport*, 74: 331–336.

Lipsitz, L.A., Jonsson, P.V., Kelley, M.M. and Koestner, J.S. (1991). Causes and correlates of recurrent falls in ambulatory frail elderly. *Journal of Gerontology*, 46: M114–M122.

Losonczy, K.G., Harris, T.B., Cornoni-Huntley, J., Simonsick, E.M., Wallace, R.B., Cook, N.R., Ostfeld, A.M. and Blazer, D.G. (1995). Does weight loss from middle age to old age explain the inverse weight mortality relation in old age? *American Journal of Epidemiology*, 141: 312–321.

Mahoney, F.I. and Barthel, D.W. (1965). Functional evaluation: the barthel index. *Maryland State Medical Journal*, 14: 61–65.

Mathias, S., Nayak, U.S. and Isaacs, B. (1986). Balance in elderly patients: the 'get-up and go' test. *Archives of Physical Medicine and Rehabilitation*, 67: 387–389.

Morey, M.C., Pieper, C.F. and Cornoni-Huntley, J. (1998). Physical fitness and functional limitations in community-dwelling older adults. *Medicine and Science in Sports and Exercise*, 30: 715–723.

Nevitt, M.C., Cummings, S.R., Kidd, S. and Black, D. (1989). Risk factors for recurrent nonsyncopal falls. A prospective study. *Journal of the American Medical Association*, 261: 2663–2668.

Osness, W.H., Adrian, M., Clark, B., Hoeger, W., Raab, D. and Wiswell, R. (1990). *Functional Fitness Assessment for Adults Over 60 Years (A Field Based Assessment)*. Virginia: The American Alliance for Health, Physical Education, Recreation and Dance.

Osness, W.H., Adrian, M., Clark, B., Hoeger, W., Raab, D. and Wiswell, R. (1996). *Functional Fitness Assessment for Adults Over 60 Years (A Field Based Assessment)*. Dubuque, IA: Kendall/Hunt.

Pendergast, D.R., Fisher, N.M. and Calkins, E. (1993). Cardiovascular, neuromuscular, and metabolic alterations with age leading to frailty. *Journal of Gerontology*, 48: Spec No. 61–67.

Petrella, R.J., Koval, J.J., Cunningham, D.A. and Paterson, D.H. (2001). A self-paced step test to predict aerobic fitness in older adults in the primary care clinic. *Journal of the American Geriatrics Society*, 49: 632–638.

Reuben, D.B. and Siu, A.L. (1990). An objective measure of physical function of elderly outpatients. The Physical Performance Test. *Journal of the American Geriatrics Society*, 38: 1105–1112.

Rikli, R.E. and Jones, C.J. (1997). Assessing physical performance in independent older adults: issues and guidelines. *Journal of Aging and Physical Activity*, 5: 244–261.

Rikli, R.E. and Jones, C.J. (1998). The reliability and validity of a 6-minute walk test as a measure of physical endurance in older adults. *Journal of Aging and Physical Activity*, 6: 363–375.

Rikli, R.E. and Jones, C.J. (1999a). Development and validation of a functional fitness test for community-residing older adults. *Journal of Aging and Physical Activity*, 7: 129–161.

Rikli, R.E. and Jones, C.J. (1999b). Functional fitness normative scores for community-residing older adults. *Journal of Aging and Physical Activity*, 7: 162–181.

Rikli, R.E. and Jones, C.J. (2001). *Senior Fitness Test Manual*. Champaign, IL: Human Kinetics.

Roach, K.E. and Miles, T.P. (1991). Normal hip and knee active range of motion: the relationship to age. *Physical Therapy*, 71: 656–665.

Seeman, T.E., Charpentier, P.A., Berkman, L.F., Tinetti, M.E., Guralnik, J.M., Albert, M., Blazer, D. and Rowe, J.W. (1994). Predicting changes in physical performance in a high-functioning elderly cohort: MacArthur studies of successful aging. *Journal of Gerontology*, 49: M97–108.

Shephard, R.J. (1997). *Aging, Physical Activity and Health*, p. 4. Champaign, IL: Human Kinetics.

Simonsick, E.M., Montgomery, P.S., Newman, A.B., Bauer, D.C. and Harris, T. (2001). Measuring fitness in healthy older adults: the Health ABC Long Distance Corridor Walk. *Journal of the American Geriatrics Society*, 49: 1544–1548.

Spirduso, W.W. (ed.) (1995). Physical functioning and the old and oldest-old. In *Physical Dimensions of Aging*, pp. 329–365. Champaign, IL: Human Kinetics.

Steffen, T.M., Hacker, T.A. and Mollinger, L. (2002). Age- and gender-related test performance in community-dwelling elderly people: six-minute walk test, berg balance scale, timed up & go test, and gait speeds. *Physical Therapy*, 82: 128–137.

Topp, R., Mikesky, A., Wigglesworth, J., Holt, W., Jr and Edwards, J.E. (1993). The effect of a 12-week dynamic resistance strength training program on gait velocity and balance of older adults. *Gerontologist*, 33: 501–506.

Verfaillie, D.F., Nichols, J.F., Turkel, E. and Hovell, M.F. (1997). Effects of resistance, balance, and gait training on reduction of risk factors leading to falls. *Journal of Aging and Physical Activity*, 5: 213–228.

Whipple, R.H., Wolfson, L.I. and Amerman, P.M. (1987). The relationship of knee and ankle weakness to falls in nursing home residents: an isokinetic study. *Journal of the American Geriatrics Society*, 35: 13–20.

Wilson, T.M. and Tanaka, H. (2000). Meta-analysis of the age-associated decline in maximal aerobic capacity in men: relation to training status. *American Journal of Physiology*, 278: H829–H834.

Wolfson, L., Whipple, R., Amerman, P. and Tobin, J.N. (1990). Gait assessment in the elderly: a gait abnormality rating scale and its relation to falls. *Journal of Gerontology*, 45: M12–M19.

Wolfson, L., Judge, J., Whipple, R. and King, M. (1995). Strength is a major factor in balance, gait, and the occurrence of falls. *Journals of Gerontology. Series A, Biological Sciences and Medical Sciences*, 50: Spec No. 64–67.

CHAPTER 40

TESTING THE FEMALE ATHLETE

Melonie Burrows

INTRODUCTION

Over the last 30 years, physiological testing of the female athlete has grown dramatically, particularly in assessing the physiological predictors of performance (Lynch and Nimmo, 1999). However, such testing has brought with it many controversies due to the variety of methodological flaws and inconsistencies in the preparation and testing of the female athlete (Burrows and Bird, 2000). Therefore, this chapter aims to cover pertinent issues regarding assessment of the female athlete.

THE MENSTRUAL CYCLE

Unique to the female athlete is the exposure to rhythmic variations in endogenous hormones during the menstrual cycle (Figure 40.1). The textbook length of the menstrual cycle is 28 days. However, the 'normal' menstrual cycle length varies greatly between women from 22 to 36 days between the ages of 20 and 40 years (Vollman, 1977). The cycle is typically divided into three phases, the menses phase, the follicular phase and the luteal phase, during which the levels of gonadotrophins vary considerably. This variation in menstrual cycle hormones and affect on performance has been extensively studied (Burrows *et al.*, 2002; Sunderland and Nevill, 2003; Bambaeichi *et al.*, 2004; Burrows and Bird, 2004), however there is no universal agreement as to the hormonal effects, so precaution needs to be exercised; with hormonal levels and menstrual phases classified and identified prior to, and during, testing of the female athlete. Such classification is essential to ensure that any significant differences found in testing are down to the intervention, and not the variations in hormonal levels at the time of testing. Suitable methods to classify the menstrual cycle phases have been discussed in the literature, of which the main ones are outlined below.

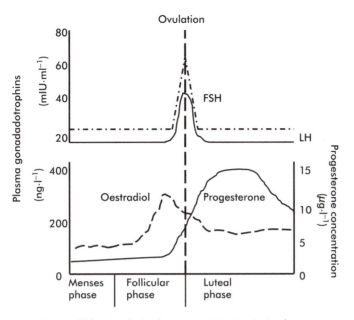

FSH = Follicle stimulating hormone; LH = Luteinsing hormone

Figure 40.1 Diagrammatic representation of the menstrual cycle

MENSTRUAL CYCLE DIARIES AND QUESTIONNAIRES

Menstrual cycle diaries have been used to identify females' menstrual phases by means of a menstrual calendar, in which the female details the dates of menses commencement and cessation. From such information one can estimate the day of ovulation as day 14 from the start of menses (assuming a 28 day text book cycle), and thus the end of the follicular phase and the beginning of the luteal phase (Figure 40.2). Although such a method is non-invasive and easy to administer over consecutive menstrual cycles, the variability of the 'normal' menstrual cycle (22–36 days), makes such a method highly inaccurate (Bauman, 1981). Thus, although the menses phase can be determined very accurately, the end of the follicular phase and the beginning of the luteal phase would lack precision. In addition, no assessment of ovulation can directly take place.

Menstrual cycle questionnaires have also been utilised to report the onset and cessation of menses and thus the occurrence of the follicular and luteal phases, by asking the female to remember past menses. Such questionnaires can assess the female for menstrual history since menarche, age at menarche, menstrual flow in days and cycles experienced per year, so providing useful information on past regularity and irregularity of the females cycle. This may aid in identifying the females menstrual irregularity in the coming months during the testing period as well as highlighting a history of irregularity which may affect current testing results. However, one must keep in mind that such questionnaires provide retrospective data and so the inherent flaws need to be taken into account. Further, no identification of ovulation can take place.

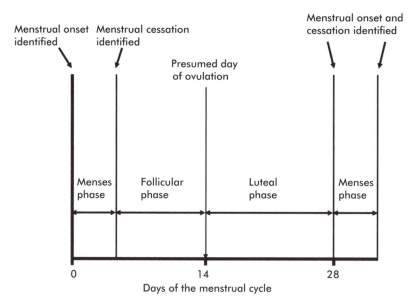

Figure 40.2   The use of menstrual cycle diaries to identify the menstrual phases

Notes
The estimation of ovulation on day 14 is based on assuming a 28-day cycle, so the phases of the menstrual cycle can be identifies as seen above. The start of the next menstrual cycle can be assessed by menstruation onset and cessation identified through a menstrual diary

BODY TEMPERATURE

The cyclical core body temperature (CBT) changes across the menstrual cycle are characterised by a pre-ovulatory rise in temperature and a post-menstrual fall followed by a return to baseline (Birch, 2000). Previous research has indicated that CBT is not an accurate predictor of luteal phase onset, even though there is little doubt that rises in progesterone cause a rise in CBT in the luteal phase of the cycle (Prior *et al.*, 1990). The lack of accuracy could be due to the fact that in many studies the timing and collection protocols for CBT have not been tightly controlled (Eston, 1984). When the collection procedures are strictly controlled, a relationship may be found between body temperature and ovulation (Guida *et al.*, 1999). If females follow a strict protocol for CBT collection it increases the chance of gaining a clear CBT profile. Such a controlled method would be:

- Using a valid and reliable digital thermometer.
- Taking CBT every morning upon awakening (at the same time of day if possible), prior to getting out of bed.
- Placing the thermometer under the tongue for the time taken for the digital thermometer to register the temperature, after which immediately recording the value in a menstrual diary.

- Recording any missed temperature readings, late nights, alcohol usage, cold-symptoms and any factors that they feel might have affect the temperature readings in the menstrual diary.
- Analysing the CBT readings using the Cumulative Sum (CUSUM) method of Lebenstedt *et al.* (1999).

However, it has been reported that 12–22% of apparently 'normal' women do not exhibit a luteal rise of body temperature (Bauman, 1981), and as such these women would not gain a true identification of luteal phases. Additionally, CBT does not provide direct evidence of ovulation.

DIRECT PROGESTERONE ASSESSMENT

Well-controlled research papers have focused on identifying the menstrual phases by classifying the surge of endogenous progesterone or LH around the middle of a 'normal' ovulatory cycle via urine (Ecochard *et al.*, 2001), plasma, serum (Serviddio *et al.*, 2002), or saliva samples (Stikkelbroeck *et al.*, 2003). The medium which has received most attention in the exercise literature to date is the assessment of salivary progesterone concentrations, due to the collection procedures being non-invasive, stress-free, and requiring minimal supervision, allowing multiple participant collections and storage at home (Tremblay *et al.*, 1996l; Chatterton *et al.*, 2005).

When using saliva a strict collection protocol must be followed to ensure clear samples, a low risk of contamination and adequate sample size for analysis. The salivette method of collection has been reported to decrease the possibility of gingival bleeding that is associated with the chewing and spitting methods of saliva collection, and thus lends itself to providing clear samples (Kruger *et al.*, 1996). However, it has received some criticism due to the cotton and polyether rolls affecting saliva composition (Leander-Lumikari *et al.*, 1995; Strazdins *et al.*, 2005). Thus, currently the dribbling method of saliva collection is most popular (Nieman *et al.*, 2005). A strict collection protocol would be as follows:

- Saliva samples should be taken prior to the brushing of teeth and application of make-up (to avoid any contamination in the sample).
- Hands should be washed to ensure no contamination upon handling.
- The mouth should be rinsed with water a few times to remove any debris and then the female should swallow 2–3 times to remove old saliva.
- The females should rest for 2 min prior to taking the sample.
- Unstimulated collection should take place for 4 min by expectoration into a plastic, sterilised vial.
- The female should be urged to pass as much saliva as possible.
- Samples should be immediately frozen and stored in a freezer at −80°C for later analysis.
- There are various assays for the measurement of salivary progesterone and a full review of these is unwarranted here. Please refer to reviews by O'Rorke *et al.* (1994) and Moghissi (1992).

- Total protein should be quantified.
- All samples should be taken at the same time of day.

Saliva samples need to be taken regularly over the menstrual cycle to optimise the identification of the progesterone peak. Although there has been shown to be a strong correlation between progesterone values and ovulation if the plasma concentrations of the hormone exceed 25 nmol·l^{-1} (Abdulla *et al.*, 1983), or 200 pmol·l^{-1} (Simpson *et al.*, 1998), one can never be sure whether the actual peak values are being measured if daily samples are not taken. Daily saliva sampling removes the concern of 'timing' samples to measure the true progesterone peak that exists if venous blood sampling is used. Indeed, the invasiveness of venipuncture means that the amount of samples taken over one month is limited and as a result, many studies have relied on one sample collected on a day estimated to correspond to the middle of the luteal phase, or 2–3 samples around the presumed day of ovulation, to classify menstrual regularity (Sunderland and Nevill, 2003). However, by utilising a low number of blood samples the chance of missing the progesterone peak increases and the likelihood of accurately identifying the luteal phase and ovulation decreases. Thus, although blood samples for the analysis of progesterone concentrations have great validity they are limited by the inconvenience of frequent venipuncture and cost of analysis. Whereas, daily samples of saliva provide a comprehensive profile of the cycle and thus any menstrual irregularities that may be present. Females have a large variation in salivary progesterone measures, thus, the CUSUM method of Lebenstedt *et al.* (1999) should be used for the objective determination of luteal phase onset as it allows for large variations in salivary progesterone concentrations by analysing significant rises from individual's follicular baseline measures. However, the rise in endogenous progesterone is not a direct measure of ovulation with increases in progesterone being reported in anovulatory cycles (Soules *et al.*, 1989). Therefore, the only direct method to document ovulation is the detection of an ovum from ultrasonography or the occurrence of pregnancy in the participant (Israel *et al.*, 1972).

ULTRASONOGRAPHY

Ultrasonography is a direct method of assessing ovulation and thus whether any female is anovulatory. The method uses transvaginal ultrasound, sending out high frequency sound waves into the pelvic region, which bounce of the structures, enabling identification of an ovum if present. The image is of high quality and is a direct measure, and as such is one of the gold standards for assessment of ovulation. However, the method is unavailable in most physiology laboratories due to the expertise and expense required to utilise it. As such, the indirect methods of hormonal classification and diary assessment are often utilised.

Menstrual cycle regularity

What is becoming increasingly clear is that when assessing the female athlete, menstrual cycle status can change from month to month along with the cycle

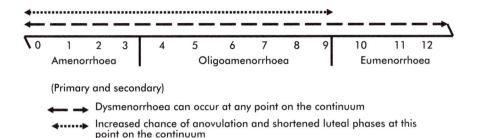

Figure 40.3   The continuum of menstrual cycle irregularities

Table 40.1   Definitions for menstrual terms in female athletes

Term	Definition
Eumenorrhoea	'Normal' cycle: 10–13 menstrual bleeds per year inclusive
Oligoamenorrhoea	4–9 menstrual bleeds per year inclusive, or menstrual cycles longer than 35 days
Amenorrhoea	0–3 menstrual bleeds per year inclusive[a]
Primary amenorrhoea	A female who has never had menstrual bleeding
Secondary amenorrhoea	Females who have had at least 1 episode of menstrual bleeding before loss of the cycle
Delayed menarche	The onset of menses after 16 years old
Dysmenorrhoea	Lower abdominal pain radiating to the lower back or legs, headache, nausea and vomiting across the cycle
Shortened luteal phases	A luteal phase shorter than 10 days
Anovulation	No ovum released at ovulation

Note
a Definitions of amenorrhoea vary greatly so to standaridsed future reports the IOC defined amenorrhoea as 'one period or less in a year'. However, this definition is more rigid than that used in other gynaecological or endocrine literature, and such a definition excludes many athletes with altered endocrinology (Carbon, 2002). Thus, 0–3 cycles per year inclusive is used instead to provide a more encompassing definition

hormone levels. Indeed females may not stay within one category of menstrual regularity, but move between them as the month's progress (Figure 40.3). Due to this variation, when assessing the menstrual cycle phases and ovulation, monitoring of one menstrual cycle may not be adequate to gain an accurate picture. A number of menstrual cycles may be required and a few studies to date have started to assess three menstrual cycles prior to testing the female, continuing the assessment throughout the testing period; gaining a more accurate picture of menstrual regularity (Burrows *et al.*, 2002). Table 40.1 provides some clear definitions of menstrual irregularities to aid identification and standardised interpretation.

SUMMARY

In summary, it would seem that the most accurate, reliable and practical methods to assess the menstrual cycle are a combination of menstrual cycle diaries to assess menses onset and cessation, menstrual cycle questionnaires to assess menstrual history, and salivary progesterone measurement to directly assess hormonal status. Although ultrasonography is a direct measure of ovulation, it is often inapplicable and unavailable in many contexts due to the expertise and expense required. Therefore, saliva progesterone values could offer a more readily available and cheaper method of detecting luteal phase onset and ovulation if the limitations are kept in mind. Longitudinal assessment of menstrual functioning is required over 3+ cycles prior to testing and during the testing period to ensure accurate luteal phase onset and prediction of ovulation across menstrual cycles and thus the occurrence of any menstrual irregularity (Loucks *et al.*, 1992). Such methodology and protocols can be expensive and very time consuming, but it is the only way to ensure the validity and reliability of the testing results from the female athlete.

OTHER IMPORTANT FACTORS TO CONSIDER

Once the menstrual cycle has been accurately identified and classified, other important issues need to be addressed with specific reference to the female athlete. These are circadian rhythms, pregnancy, age, contraceptive use and mood states.

Circadian rhythms

Once the menstrual cycle has been classified and the phases identified, the timing of testing needs to be arranged. As with many physiological variables, the menstrual cycle hormones follow circadian rhythms and thus testing should take place at the same time of day across the menstrual phases. However, in addition to looking across the menstrual cycle phases, the variation of hormones within each phase needs to be taken into account. As such, testing should not only take place at a specified time of day, but on a specified day or days within each of the menstrual phases.

Pregnancy

Pregnancy, in females of reproductive age, must be the first issue discounted prior to testing and/or when assessing menstrual status. Pregnancy can be assessed through a simple urine dipstick test with results available in 5 min. Specific confidentiality issues need to be considered when one is conducting a pregnancy test. If you are testing a female athlete above 18 years of age the results should be given to the adult, along with suggestions to gain advice from the GP. However, if one is working with female athletes below the age of

18 years the confidentiality issues become clouded. Under such conditions it is advisable to solicit advice from the ethical committee or lead physiologist under whom such testing is taking place.

Age

When working with female athletes age needs to be considered. The age of the female will directly affect the regularity of the menstrual cycle. The menstrual cycle should be more regular between the ages of 20–45 years, and thus a large amount of testing on females has been conducted between this age range. Once one tests outside this age range, menstrual cycle irregularity may increase, and present problems with accurately identifying hormonal levels and menstrual phases (Astrup et al., 2004). Obviously, once you go over 45 years, the chance of females progressing towards, or going through, the menopause increases and this needs to be taken into account in the test design or screened for. If working with a minor, the ethical issues that come about with working with children should be followed (see the appropriate section in these guidelines).

Contraceptive use

When testing the female athlete the issue of contraceptive use needs to be addressed. Currently, there are numerous forms of contraceptive agents available to the female, that is, the contraceptive pill, the contraceptive injection and hormonal implants. Due to this array of contraceptives, research is limited into the effects of such agents on the hormones of the menstrual cycle and the concomitant affects on physiological variables. Therefore, when testing the female athlete, one should screen out for any use of contraceptive agents. However, if this is not possible the following issues need to be considered:

1   The type of contraceptive agent the females are taking may affect the test results. If in a research study, all females should be on the same agent, such as the contraceptive injection, implants or the pill. If in a physiological testing service, then over time any changes in contraception should be noted and the test results interpreted accordingly.
2   Whichever agent the females are using should be of the same type. For example, if in a research study, all the females on the contraceptive pill should be using either the Monophasic or Triphasic pills. In addition, all females should be on a similar dose pill, that is, high or low doses of progestrone and/or oestrogen.
3   The duration the female has been on the contraceptive agent should be taken into account prior to testing and controlled for (Lynch and Nimmo, 1998).
4   Any missed pills should be recorded in a diary noting how many pills have been missed, on what day(s) of the packet, and the reason for not taking the pill.

Such information can be gained from well-designed questionnaires, with contraceptive use monitored through a diary. Testing would then need to take place on the same pill day for all females to minimise any hormonal variation. Please refer to a review by Burrows and Peters (2006) for a more detailed discussion on oral contraceptives and performance in female athletes.

Mood states

Mood states should be assessed using a previously validated, prospective mood state questionnaire, prior to and during testing, to ensure any significant changes in variables being assessed are not down to alterations in mood states but the intervention (Choi and Salmon, 1995; Terry, 1995). Participants should complete the questionnaire at an appropriate time, such as prior to the testing session, and follow the appropriate questionnaire instructions.

BONE HEALTH

Bone health is an area of interest that is rapidly growing, particularly with reference to the female athlete. When assessing bone health in the female certain issues need to be addressed, such as the equipment to use, the variables to measure, timing of the measurement and data analysis.

Bone densitometry has been shown to provide a reliable and valid measure of bone mineral content and future fracture risk at all body sites that is far superior to other available methods that have error rates of 30–50% (Cummings et al., 1993). Bone mineral content should be measured over a range of body sites taking into account both cortical and trabecular bone remodelling. However, if time is a limiting factor enabling only one or two body sites to be assessed, the major sites to be measured are the femoral neck and lumbar spine for they are highly correlated with future fracture risk (National Osteoporosis Society, 2002). Prior to any scanning, females should be screened for osteoporosis risk factors, such as long-term corticosteroid use, smoking and alcoholism to aid interpretation of the scan results. Pregnancy should also be assessed using the 28-day rule (i.e. a menses period within the last 28 days), or a negative pregnancy urine test.

When designing a scanning protocol, it should take into account the fact that bone-remodelling cycles (activation–resorption–formation) take about 3–4 months to complete (Eastell et al., 2001). As such, any repeat assessment within this time-span is highly questionable. Indeed, the determination of when to repeat a scan should take into account the sample size and the precision error of the DEXA scanner (Precision error = CV of machine $\times$ 2.8), as well as the intervention. As a result, other short-term measures to assess bone remodelling may be required, and bone biochemical markers may prove useful. A full discussion of these markers is beyond the scope of this chapter so the reader is referred to Eastell et al. (2001) for further information. Suffice to say though, that such markers could provide a short-term indication of the global bone response to an intervention, and be utilised in conjunction with DEXA on a long-term scale.

The interpretation of DEXA scans needs to be accurate, and as bone is a 3D parameter, bone area and volume should be taken into account

(Nevill *et al.*, 2002). Such scaling is particularly important in young females who are still growing and various adjustments for bone size and area are required. Refer to the National Osteoporosis Society guidelines on bone densitometry in children for advice on such scaling issues (National Osteoporosis Society, 2004). All scanning should adhere to Ionising Radiation in Medical Exposure Regulations (2000), and only one trained operator should perform all scans for each distinct study as well as across testing sessions. The analysis procedures should conform to the DEXA manufacturer guidelines.

CONCLUSIONS

This chapter has provided an insight into the many issues that need to be addressed when testing the female athlete and given across some information on specific protocol points. It is up to the physiologist working with the female athlete to decide on the most appropriate course of action for the female at that particular point in time. Although working and researching with the female athlete requires a lot of planning and preparation, it is an area where the complicated physiological mechanisms behind health and sports performance are unclear and as such requires further attention. It is imperative that, as exercise and sports physiologists, we have the evidence-based knowledge to understand the impact of exercise on the females' health and performance. The only way to gaining such evidence is through implementing well-designed and controlled research studies and testing sessions. The extra time and planning needed to achieve this should not be a factor in determining the research conducted in this important area.

REFERENCES

Abdulla, U., Diver, M.J., Hipkin, L.J. and Davis, J.L. (1983). Plasma progesterone levels as an index of ovulation. *British Journal of Obstetrics and Gynaecology*, 90, 543–548.

Astrup, K., Olivarius, N.F., Moller, S., Gottschau, A. and Karlslund, W. (2004). Menstrual bleeding patterns in pre- and perimenopausal women: a population-based prospective diary study. *Acta Obstetrics and Gynaecology in Scandinavia*, 83: 197–202.

Bambaeichi, E., Reilly, T., Cable, N.T. and Giacomoni, M. (2004). The isolated and combined effects of menstrual cycle phase and time-of-day on muscle strength of eumenorrhoeic females. *Chronobiology International*, 21: 645–660.

Bauman, J.E. (1981). Basal body temperature: unreliable method of ovulation detection. *Fertility and Sterility*, 36: 729–733.

Birch, K.M. (2000). Circamensal rhythms in physical performance. *Biological Rhythm Research*, 31: 1–14.

Burrows, M. and Bird, S.R. (2000). The physiology of the highly trained female endurance runner. *Sports Medicine*, 30: 281–300.

Burrows, M. and Bird, S.R. (2005). Velocity at $VO_{2 \, max}$ and peak treadmill velocity are not influenced within or across the phases of the menstrual cycle. *European Journal of Applied Physiology*, Dec. 3 (Eprint ahead of publication).

Burrows, M. and Bird, S.R. (2005). Velocity at $\dot{V}O_{2max}$ and peak treadmill velocity are not influenced within or across the phases of the menstrual cycle. *European Journal of Applied Physiology*. 93(5–6): 575–80.

Burrows, M. and Peters, C.E. (2006). The influence of oral contraceptives on athletic performance in female athletes. *Sports Medicine* (in press).

Burrows, M. Bird, S.R. and Bishop, N. (2002). The menstrual cycle and its effect on the immune status of female endurance runners. *Journal of Sports Sciences*, 20: 339–344.

Chatterton, R.T., Jr, Mateo, E.T., Hou, N., Rademaker, A.W., Acharya, S., Jordan, V.C. and Morrow, M. (2005). Characteristics of salivary profiles of oestradiol and progesterone in premenopausal women. *Journal of Endocrinol*, 186(1):77–84.

Choi, P.Y. L. and Salmon, P. (1995). Symptom changes across the menstrual cycle in competitive sportswomen, exercisers and sedentary women. *British Journal of Clinical Psychology*, 34: 447–460.

Cummings, S.R., Black, D.M., Nevitt, M.C., Browner, W., Cauley, J., Ensrud, K., Genant, H.K., Gluer, C.C., Hulley, S.B., Palmero, L., Scott, J. and Vogt, T. (1993). Bone density at various sites for prediction of hip fractures. *Lancet*, 341: 72–75.

Eastell, R., Baumann, M., Hoyle, N.R. and Wieczorek, L. (2001). *Bone Markers: Biochemical and Clinical Perspectives*. London: Dunitz.

Ecochard, R., Boehringer, H., Rabilloud, M. and Marret, H. (2001). Chronological aspects of ultrasonic, hormonal, and other indirect indices of ovulation. *British Journal of Obstetrics and Gynaecology*, 108: 822–829.

Eston, R.G. (1984). The regular menstrual cycle and athletic performance. *Sports Medicine*, 1: 431–445.

Guida, M., Tommaselli, G.A., Palomba, S., Pellicano, M., Moccia, G., Carlo, C. and Nappi, C. (1999). Efficacy of methods for determining ovulation in a natural planning program. *Fertility and Sterility*, 72: 900–904.

Ionising Radiation (Medical Exposure) Regulations. (2000). *Documents of the National Radiological Protection Board*. Her Majesty's Stationary Office. Didcot, England.

Israel, R., Mishell, D.R., Stone, S.C., Thorneycroft, I.H. and Moyer, D.C. (1972). Single luteal phase serum progesterone assay as an indicator of ovulation. *American Journal of Obstetrics and Gynaecology*, 15: 1043–1046.

Kruger, C., Ulrike, B., Jutta, B.S. and Helmuth, G.D. (1996). Problems with salivary 17-Hydroxyprogesterone determinants using the salivette device. *European Journal of Clinical and Chemical Biochemistry*, 34: 927–929.

Lebenstedt, M., Platte, P. and Pirke, K.M. (1999). Reduced resting metabolic rate in athletes with menstrual disorders. *Medicine and Science in Sports and Exercise*, 31: 1250–1256.

Lenander-Lumikari, M., Johansson, I., Vilja, P. and Samaranayake, L.P. (1995). Newer saliva collection methods and saliva composition: a study of two salivette kits. *Oral Diseases*, 1: 86–91.

Loucks, A.B. and Horvath, S.M. (1992). Exercise-induced stress responses of amenorrhoeic and eumenorrhoeic runners. *Journal of Clinical Endocrinology and Metabolism*, 59: 1109–1120.

Lynch, N.J. and Nimmo, M.A. (1999). Effects of menstrual cycle phase and oral contraceptive use on intermittent exercise. *European Journal of Physiology*, 78: 565–572.

Matthews, K.A., Santoro, N., Lasley, B., Chang, Y., Crawford, S., Pasternak, R.C., Sutton-Tyrrell, K. and Sowers, M. (2006). Relation of cardiovascular risk factors in women approaching menopause to menstrual cycle characteristics and reproductive hormones in the follicular and luteal phases. *Journal of Clinical Endocrinology and metabolism*, 91(5):1789–95.

Moghissi, K.S. (1992) Ovulation detection. *Endocrinology and Metabolism Clinics of North America*, 21(1): 39–55.

National Osteoporosis Society. (2002). *Position Statement on the Reporting of DEXA Bone Mineral Density Scans*. NOS, August.

National Osteoporosis Society. (2004). *Position statement: A practical guide to bone densitometry in children*. NOS, November 2004.

Nevill, A.M., Holder, R.L., Maffulli, N., Cheng, J.C., Leung, S.S., Lee, W.T. and Lau, J.T. (2002). Adjusting bone mass for differences in projected bone area and other confounding variables: an allometric perspective. *Journal of Bone Mineral Research*, 17: 703–708.

Nieman, D.C., Henson, D.A., Austin, M.D. and Brown, V.A. (2005). Immune response to a 30-minute walk. *Medicine and Science in Sports and Exercise*, 37: 57–62.

O'Rorke, A., Kane, M.M., Gosling, J.P., Tallon, D.F. and Fotrell, P.F. (1994). Development and validation of a monoclonal antibody enzyme immunoassay for measuring progesterone in saliva. *Clinical Chemistry*, 40: 454–458.

Serviddio, G., Loverro, G., Vicino, M., Prigigasllo, F., Grattagliano, I., Altomare, E. and Vendemiale, G. (2002). Modulation of endometrial redox balance during the menstrual cycle: relation with sex hormones. *Journal of Clinical Endocrinology and Metabolism*, 87: 2843–2848.

Simpson, H.W., McArdle, C.S., Griffiths, K.M., Turkes, A. and Beastall, G.H. (1998). Progesterone resistance in women who have breast cancer. *British Journal of Obstetrics and Gynaecology*, 105: 345–351.

Soules, M.R., Mclachlan, R.I. and Ek, M. (1989). Luteal phase deficiency: characterisation of reproductive hormones over the menstrual cycle. *Clinical Endocrinology and Metabolism*, 69: 804–812.

Stikkelbroeck, N.M., Sweep, C.G., Braat, D.D., Hermus, A.R. and Otten, B.J. (2003). Monitoring of menstrual cycles, ovulation, and adrenal suppression by saliva sampling in female patients with 21-hydroxylase deficiency. *Fertility and Sterility*, 80: 1030–1036.

Strazdins, L., Meyerkort, S., Brent, V., D'Souza, R.M., Broom, D.H., Kyd, J.M. (2005). Impact of saliva collection methods on sIgA and cortisol assays and acceptability to participants. *Journal of Immunological Methods*, 307(1–2): 167–171.

Sunderland, C. and Nevill, M. (2003). Effect of the menstrual cycle on performance of intermittent, high-intensity shuttle running in a hot environment. *European Journal of Applied Physiology*, 88: 345–352.

Terry, P. (1995). The efficacy of mood state profiling with elite performers: a review and synthesis. *Sport Psychology*, 9: 309–324.

Tremblay, M.S., Chu, S.Y. and Mureika, R. (1996). Methodological and statistical considerations for exercise-related hormones evaluations. *Sports Medicine*, 20: 90–108.

Vollman, N. (1977). *The Menstrual Cycle: Major Problems in Obstetrics and Gynaecology*, 7. Philadelphia, PA: WB Saunders.

INDEX

Note: Page numbers in italics refer to figures and tables.